# THE EQUATORIE OF THE PLANETIS

His tables Tolletanes forth he brought,
Ful wel corrected, ne ther lakked nought,
Neither his collect ne his expans yeeris,
Ne his rootes, ne his othere geeris,
As been his centris and his argumentz
And his proporcioneles convenientz
For his equacions in every thyng.

CHAUCER, *The Franklin's Tale.*
[Ed. F. N. Robinson, V (F) 1273]

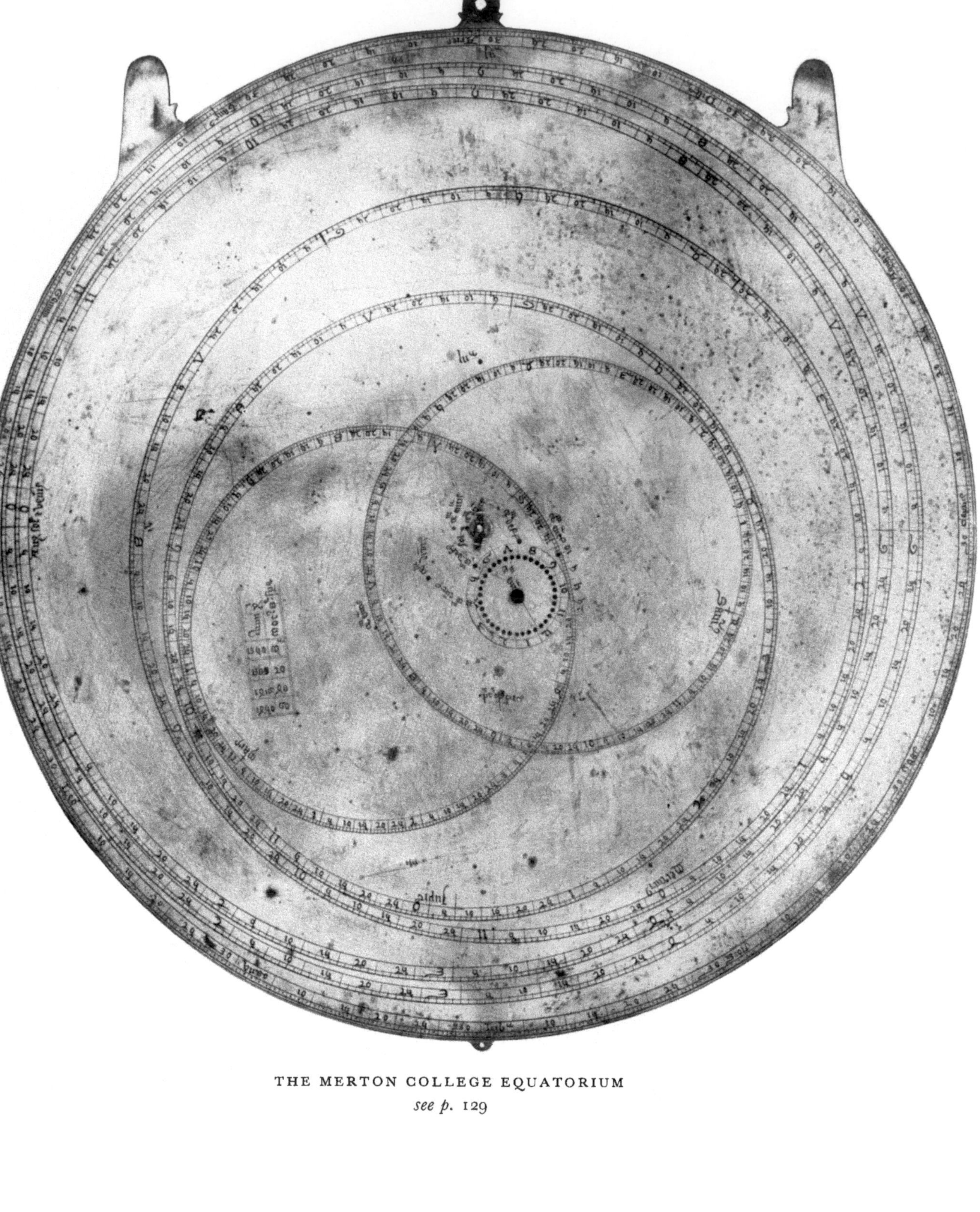

THE MERTON COLLEGE EQUATORIUM

*see p.* 129

# THE EQUATORIE OF THE PLANETIS

EDITED FROM

PETERHOUSE MS. 75. I

BY


DEREK J. PRICE

PH.D. (LOND.), PH.D. (CANTAB.)

*I.C.I. Fellow in the History of Science*

*Christ's College, Cambridge*


WITH A LINGUISTIC ANALYSIS BY


R. M. WILSON, M.A.

*Head of the Department of English Language*

*Sheffield University*



CAMBRIDGE

AT THE UNIVERSITY PRESS

1955

CAMBRIDGE UNIVERSITY PRESS
Cambridge, New York, Melbourne, Madrid, Cape Town,
Singapore, São Paulo, Delhi, Tokyo, Mexico City

Cambridge University Press
The Edinburgh Building, Cambridge CB2 8RU, UK

Published in the United States of America by Cambridge University Press, New York

www.cambridge.org
Information on this title: www.cambridge.org/9781107404274



First published 1955
First paperback edition 2012

*A catalogue record for this publication is available from the British Library*

ISBN 978-0-521-05994-7 Hardback
ISBN 978-1-107-40427-4 Paperback

*TO ELLEN*

# CONTENTS

## LIST OF PLATES

# LIST OF FIGURES

# PREFACE

THE text which is here edited is one of considerable interest to all students of medieval literature and of medieval astronomy. In addition, as pointed out later, it may also be of importance to Chaucerian scholars.

This wide range of interest has necessarily led to the transgression of many academic boundaries which the modern tendency towards specialization has tended to keep almost inviolate. Consequently it early became evident that there were many instances where it was desirable to obtain expert opinion on various specialized points, such as palaeography, medieval astronomy, the history of cipher writings, etc. It is therefore a great pleasure to record that my many requests for assistance from experts in Cambridge and elsewhere were always met with great kindness and consideration.

It was also fortunate that Mr R. M. Wilson, M.A., consented to act as adviser on the linguistic side. In addition to writing Chapter X and compiling the Glossary, he has also provided the section on Punctuation in Chapter IX, and we have collaborated in the translation of the text. Throughout the preparation of this edition we have had frequent consultations, always on the understanding that although advice was freely given on either side, its acceptance or rejection was a matter for the person responsible for that chapter. It need hardly be added that similar considerations apply to advice which we have received from others. Their help has been acknowledged whenever possible, but this must not be taken as committing them to any errors of statement or judgement for which we alone must accept responsibility.

Since this is the first detailed account of a complicated manuscript, it appeared necessary to provide a full description of its varied contents, together with the customary critical apparatus and the special technical explanations which are made necessary by the subject-matter. I have thus attempted to provide the raw materials for the discussions and the assessment of evidence demanded by the manuscript.

It would be out of place to insert these discussions and assessments in the present volume before scholars have had an opportunity of examining the material for themselves, and for this reason the summing-up has been restricted to a tentative account which is designed to assist the reader rather than urge the acceptance of this or that opinion. Particularly in the chapter dealing with the problem of ascription, especial care has been taken to separate the different issues involved and to distinguish those things which appear reasonably certain from those which bear only a greater or a lesser probability of being true.

It is difficult if not impossible to perform this task without betraying the influence of preconceived notions, and it would be dishonest to pretend that a neutral course has been steered throughout the investigation. On the contrary, it would have been impossible, perhaps even undesirable, to embark on this edition without having arrived at some conclusion one way or another.

To explain this it is necessary to outline the early stages of this research. The Peterhouse manuscript was first seen by me at the beginning of December 1951 while collecting data for a general history of scientific instruments. Since the text was in English, and the date 1392 frequently occurred, it seemed at first possible that this might be a missing part of Chaucer's apparently incomplete *Treatise on the Astrolabe* written in 1391—the only instrument tract written in Middle English known to me at that time.

It was soon clear, however, that the instrument described was certainly no type of astrolabe, but a planetary calculator of unfamiliar design, and my interest was further aroused by finding that some of the leaves of the manuscript contained short notes written in cipher; a more detailed study of the manuscript was obviously necessary. In the meantime I was able to decode the notes and found that they were technical rubrics written in Middle English. A full reading of the instrument text revealed that the scientific content was of considerable interest in its own right, though it provided at first sight no evidence for or against my suspicion that it was connected in some way with the *Treatise on the Astrolabe*.

Leaving the text, I turned my attention to the astronomical tables which occupy the greater part of this manuscript volume, and in the course of this study I found, near the inner margin of one of the folios, a note which seemed to contain the word 'chaucer' or possibly 'chancer'. But since nearly half the crucial part of this note was hidden by the very tight binding of the volume it was impossible to read the whole of the short sentence. Application was therefore made to the authorities of Peterhouse, pointing out the importance of the manuscript and its possible connection with Chaucer, and they very kindly consented to have it unbound so as to reveal the hidden part of this note, as well as certain fragments of the text which were similarly obscured.

The full sight of the note indicated that 'chaucer' was the correct reading, and further work with the manuscript led me to think that the text might well be an author's holograph. The obvious thing was to compare the hand with those appearing in certain Chaucer documents in the Public Record Office, and it was clear that there was no agreement at all. My attention was then drawn to another document which had been suggested as a Chaucer holograph by Professor Manly. The relevant file of documents was brought to me, and a casual leafing through showed that only one of them was in a hand similar to that of the Peterhouse manuscript—but this one document displayed a striking similarity. The document was in fact Professor Manly's suggested holograph, and although the manner of finding it does not objectively increase the weight of the evidence, the subjective effect was considerable.

By the end of January 1952 it was possible to bring together photographs of the manuscript and the Public Record Office document, and the detailed comparison proved sufficiently satisfactory for the publishing of a tentative account of my findings.

The research had thus been carried out from the beginning in the hope of the result which had in fact emerged—a most dangerous procedure and one liable to lead to an unconscious weighting of the evidence. I can only hope that I have not fallen into serious error on this account, and there has been an attempt at each stage to lay the evidence before sceptical critics. For their sake as well as for mine I must issue a final caveat: there is, in my opinion, nothing in this book which can by itself be accepted as definite proof of authorship; there is, however, a mass of lesser evidence which has the cumulative effect of suggesting that this is a Chaucer holograph and making it difficult to advance any other reasonable hypothesis to explain all the features of the manuscript. Any single piece of evidence is vulnerable by itself, and the final verdict must therefore depend on the coming to light of fresh evidence, or on the expert assessment of the fabric as a whole.

I must acknowledge first the invaluable advice of Mr R. M. Wilson and Dr A. R. Hall, who have been continuously consulted throughout the progress of this research. On special topics I have had great benefit from the frequent expert advice of Professor Bruce Dickins, Professor R. A. B. Mynors, Mr H. L. Pink, and the staffs of the Cambridge University Library, the University Press, and the Cavendish Laboratory. For other requests which have met with a never-failing courteous response I must acknowledge my thanks to the Keeper of Western Manuscripts of the Bodleian Library, the Public Record Office, the Librarian and the Society of Merton College, the Keeper of Western Manuscripts at the British Museum, and also to Dr B. F. C. Atkinson, Mr B. Colgrave, Dr A. C. Crombie, Dr D. Dewhirst, Dr D. R. Dicks, Dr D. M. Dunlop, Mr I. J. Good, Her Majesty's Nautical Almanac Office, Professor E. S. Kennedy,

Dr H. Lowery, Mr H. A. Mason, Monsieur H. Michel, Professor V. Minorsky, Dr J. Needham, Professor O. Neugebauer, Dr C. T. Onions, Professor Johnstone Parr, Mr B. Penrose, Dr E. Rosenthal, The Royal Museum of Brussels, Dr D. S. Sadler, Mrs D. Waley Singer, Dr W. D. Stahlmann, Professor E. G. R. Taylor, Professor J. R. R. Tolkien, Mr J. E. Tolson, The University of Nebraska, Mr W. Urry, Miss H. Wallis, Professor R. Weiss, Mr D. W. Whitfield, Miss E. Williamson, Dr H. J. J. Winter, Professor E. Zinner, and Mr F. Zloof.

No work on the history of medieval science would be complete without an appreciation of the magnificent tools for research which are provided by the encyclopaedic publications of Professors George Sarton and Lynn Thorndike; in the study of the history of scientific instruments a similar tribute must be paid to the pioneer work of R. T. Gunther.

This research would not have been possible without the co-operation of Rev. J. N. Sanders, then Perne Librarian of Peterhouse, or without the privileges and facilities which have been extended to me by the Master and Fellows of that College. The detailed study of the manuscript has been made much easier for me by the kindness of the Syndics of the Cambridge University Press, who provided a set of photographs taken while the quires of the volume lay detached from their former tight binding. Finally, I should like to thank Professor Sir Lawrence Bragg for his magnificent and much appreciated personal support, and the Managers of the Imperial Chemical Industries Fellowships for permitting me the tenure of a Fellowship for three years for research on the history of scientific instruments.

DEREK J. PRICE

CHRIST'S COLLEGE
CAMBRIDGE

*July 1953*

# THE EQUATORIE OF THE PLANETIS

# I

# INTRODUCTION

THE unique manuscript, Peterhouse (Cambridge) MS. 75. I, from which this text is edited, has considerable claim to attention by virtue of its astronomical subject-matter and also because this unusually technical material is presented in Middle English instead of the medieval Latin which was in use at this date for scholarly writings. These facts alone would be enough to make it desirable that an edited text should be available. The importance of the manuscript is, however, much increased by the possibility that a portion of the volume may be attributed to Geoffrey Chaucer, and was perhaps written in his own hand.

If this claim can be firmly established we should have a complete holograph text which, in addition to its sentimental value, might have great use in textual criticism of the Chaucer canon. All of Chaucer's known works have come down to us only through scribal copies, and of these very few are believed to have had a simple textual history. There are, it is true, certain official documents which have been proposed as Chaucer holographs, but until now it has not been possible to substantiate any of these claims. The hand of this Peterhouse manuscript is sufficiently distinctive to make it possible that other documents and manuscripts might be recognized by this means.

To some it may seem surprising that the subject of this manuscript is neither literary nor an elementary presentation of science or philosophy. It is well known that the poet was intensely interested in the science of his time; he had translated the *De Consolatione Philosophiae* of Boethius into the vernacular, and his poetry contains many allusions and direct references which indicate a detailed understanding of science, particularly of astronomy. But besides these incidental manifestations of 'the learned Chaucer' there has come down to us one completely scientific work, the *Treatise on the Astrolabe* which is said to have been written in 1391 for the poet's ten-year-old son, 'litel Lowis'.

The Peterhouse MS., whoever may have been the author, must be regarded as a companion work to the *Treatise on the Astrolabe*. The astrolabe is an instrument designed to assist all calculations of the apparent positions in the heavens of the Sun and the stars. The new text is concerned with the complementary problem of calculating the positions of the planets, and it describes the construction and use of a special instrument for this purpose; this type of instrument being well known to medieval writers as the *equatorium planetarum*. The astrolabe has long been recognized as one of the most important instruments in the early history of astronomy, and about 1000 European and Oriental examples are preserved in museums and private collections. Fate has been less kind to the equatorium [or equatorie, as this manuscript anglicizes the term], for very little has been published about this equally important instrument, and only two incomplete examples of the actual instrument have been traced. Undoubtedly this is due to the fact that an astrolabe could readily be embellished with all the art of the metal-worker, whereas the equatorium was nearly always so complicated that it could not readily escape a purely utilitarian appearance and the consequent much smaller chance of being preserved.

The connection between the *Treatise on the Astrolabe* and that contained in the present manuscript goes further than the obvious connection between the two instruments described. In its own day the equatorium must have been regarded as a useful, or perhaps necessary, complement to the astrolabe. For the casting of horoscopes, for example, the positions of the stars and the Sun could be found with an astrolabe, but the Moon and the five planets could have their places determined only by a calculation made directly from tables or with the assistance of some type of planetary instrument. As will be shown, the direct calculations on the basis of Ptolemaic theory were too cumbersome and unsuitable for routine use. If Chaucer had laid aside his first treatise and cast around for some similar subject for further writing, it would have been very natural for him to choose the complementary instrument as a topic. Indeed, if we accept a hint contained in the *Astrolabe* that Chaucer had previously written a treatise on the 'Solid Sphere', the three works together would make a very suitable and complete treatise on the basic theory and the instrumentation of medieval astronomy.

Just as the *Astrolabe* is largely borrowed from the *Compositio et operatio astrolabii*, the Latin version of a work by Messahala, the *Equatorie* is clearly derived from a Latin version of some Arabic treatise. Unfortunately, we are unable at present to trace either the Arabic or the Latin texts, but there is a possibility that the Latin treatise was written by Simon Bredon, an astronomer of Merton College who died in 1372. Our manuscript was written about twenty years after Bredon's death, but the Peterhouse library catalogues, stretching back to the beginning of the sixteenth century, agree in ascribing the *Equatorie* to Bredon, in spite of the fact that there is nothing in the text itself to suggest any connection with him. The only evidence indeed is a librarian's note on the end leaf which transcribes biographical details of Bredon from the (incorrect) information supplied in the works of John Bale. If not entirely false, this ascription to Bredon can only be based on a Latin text about the equatorium which was written by that author but is now lost.

Whatever the origin of the treatise contained in the Peterhouse manuscript we still have to determine the author of the Middle English text, whether it be adaptation, translation, or partly original work. Only a small portion of the manuscript volume consists of the text which we have named *The Equatorie of the Planetis* (adapted from *equatorium planetarum*). In addition, there is a set of astronomical tables written in the same hand as the *Equatorie*, and sandwiched between these two companion sections is a long set of astronomical tables written in a contemporary but more formal hand. It seems likely that the text and tables were originally written on loose quires of vellum and possibly remained in this state until they were bound after acquisition by Peterhouse. If this is so it must be supposed that they are not now in their correct order; the logical sequence would be to have the text followed by the tables which are in the same hand and are referred to in that text. The second set of tables must be looked upon as a professionally made copy, probably ordered by the writer to supplement his own short set, and these should have been bound at the end of the volume.

There are many indications that the writer of the text and short set of tables was no scribe or secretary, but rather the original author. He was certainly not the inventor of the instrument in question, or the compiler of the tables—for these are drawn from the famous Alfonsine Tables—but it seems that the author is working out his English version of the text as he proceeds, and modifying it to suit his purposes. But this point, as also the question of the identity of the writer, are dealt with in later chapters.

## II

# PROVENANCE AND PHYSICAL DESCRIPTION OF THE MANUSCRIPT

THE Peterhouse collection of some 276 manuscripts is preserved in the Perne Library of that College, and constitutes an excellent example of a medieval collection, all the more important because of the unusually good records which enable its development to be traced back to the first full register of books, dated Christmas Eve 1418.

MS. 75 of this collection, the subject of our present study, is described in the modern catalogue[1] as consisting of two separate parts; these are related by nothing more than the chance which led to the second, thin manuscript being selected for binding with the larger first part, the folios of which are of approximately the same size. M. R. James gives the following titles of the parts:

I. *Symon Bredon de Equationibus Planetarum*, followed by directions for making an astrolabe (?),[2] in English.

II. 1. *Expositio fr. Nich. Tryvet super Aug. de ciuitate dei.*
  2. *Vegetius de re militari.*

The presence of this composite volume in the Library can be traced back through all the catalogues, printed and manuscript, as far as 1589. In that year, before the death of Dr Perne and the subsequent accession of the large collection bequeathed by him, there was compiled a list of all the books in the College Library, which contains an entry:[3]

No. 29. Classis 10ª in parte Australj.
Simon Bredon equat. Planet: Nicholaus Trivet in August: de civitate Dej.

This description is sufficient to identify the item with the present MS. 75 which must, even then, have contained the two separate parts noted above. The absence of any mention of the Vegetius is not surprising, since this is written in the same hand as the work by Trivet and follows it, without any break, in the middle of a page.

[1] M. R. James, *A Descriptive Catalogue of the Manuscripts in the Library of Peterhouse* (Cambridge, 1899), p. 93.

[2] (?) *Sic* M. R. James.

[3] The list is on f. 12 of what is now MS. 400: *Nomina librorum qui erant in Bibliotheca Coll(eg)ij ante Doctoris Perne mortuum.*

Another point of interest in the 1589 entry is that the title of Simon Bredon's text is given as *equat. Planet.* which might be expanded as either *Equationum Planetarum* or *Equatorium Planetarum.* Although both of these titles are, at first sight, equally likely, it must appear on detailed examination that the latter form was probably the original intention, since similar forms are to be found referring to other treatises on the equatorium, notably that of Campanus of Novara. It seems likely then that the present volume contained the title, only in the abbreviated form, and this has been expanded erroneously in all catalogues from that of 1589 until the present day. A similar error seems to have been committed by John Bale in his *Scriptorum Illustrium maioris Brytannię* (Basel, 1557) where we find on p. 489, 'Simon Bredon,... *Claruit anno domini 1380 sub Ricardo Secundo*', and in the list of his works '*Aequationes planetarum, lib. 1*' (no incipit given), which seems to refer to the same work, since we know of no other planetary writings by this author. As it is also likely that no other copies of the Peterhouse manuscript have ever existed, it would seem that Bale's entry is founded on an examination of this very manuscript. This is interesting for, as will be shown later, the only reason for ascribing the work to Bredon is the existence of a sixteenth-century note at the end of the volume—and this note is copied verbatim from Bale's entry on Bredon. It is known that a copy of Bale was bequeathed to the library by Perne in 1589 and remained there until 1760 at least (former press mark, 07 05 08).

It is possible to take the record back a little further than the catalogue of 1589, for a mention of this manuscript occurs in the *Collectanea* of John Leland (James, *op. cit.* p. 362), where he lists the Peterhouse manuscripts which interested him during a tour of exploration, probably about 1542. No. 43 on this list reads: 'Tabulae aequationum planetarum, autore Simone Bredon', and this almost certainly refers to our MS. 75. I.

Having established that the manuscript has been continuously in Peterhouse from about 1542 until the present day, we must consider the preceding period. It will be shown that the text was written in 1392 and used by the author until 1395, possibly remaining in his possession until about 1400; in the intervening years, from 1400 until 1542, we have little trace of the manuscript. The interval may be narrowed slightly if one can assume that the omission of this manuscript from the earliest Peterhouse catalogue of 1418 implies that the work had not come into the possession of the College by then. This catalogue is unusually precise and yet contains no title which, by any stretch of imagination, might be identified with the present manuscript, and it would seem a safe assumption that the manuscript must have come to Peterhouse some time between 1418 and 1542.

The only clue to its history during this period is an inscription on the manuscript at the foot of the otherwise blank folio 74 verso (see Plate XIII *d*). The first part of this inscription is difficult to decipher, but for what it is worth we may render it as:

Gm...¢ S 1461 .19. meridie of august.

From its placing on the end folio of what is now the first of the two quires containing our text, and from the fact that the same hand has written a minor addition to a table on folio 9 verso, we must infer that this is the mark of some owner who made at least slight use of the manuscript at this date when it was still unbound, or at least bound in a different order which placed folio 74 verso at the end. In the absence of any other evidence, one may suggest that this owner of 1461 represents the last user of the manuscript before it came into the possession of Peterhouse; he might well have been the donor. Significantly enough the period indicated is one during which many astronomical manuscripts of similar type came to the College.

Before considering possible donors from the middle of the fifteenth century we must note a possibility that the manuscript made an earlier appearance in Cambridge. It may well be a significant coincidence that the first flourishing of science in Cambridge was due to the influence and example of John Holbrook, Master of Peterhouse from *c.* 1421 to 1436. He was admitted as a Fellow of the College in 1393, and already by 1406 his work on astronomical tables seems to have been considerable. His famous *Tabulae Cantabrigienses* are said to have been compiled in 1430,[1] but evidence of earlier work in Cambridge, almost certainly attributable to Holbrook, is to be seen in British Museum MS. Royal 12. D. vi at folio 43 verso, where we find:

Notandum q*uod* anno *Christ*i 1406 in mense Julij examinate fueru*n*t cu*m* maxima diligencia iste tabule p*re*cedentes et p*er* 5 ho*mine*s valde morose atq*ue* delib*er*ate op*er*antes in vniuersitate Cantabr*igie* correcte videli*cet*, tabula eq*ua*cionum om*nium* planeta*rum* *secundum* 2 exemplaria Alfonsi, tabule mot*us* sol*is* et lune in una hora *secundum* 2 exemplaria mag. I. de Lineriis, Tabule u*ero* ascens*us* in circulo dir*ecto* et circulo obliquo *secundum* unu*m* exemplar Mag. Ioha*nnis* Mauduth. Et om*nes* iste tabule p*redicte* *secundum* 4 exemplaria mag. Willi*elmi* Reede, vnde p*er* easde*m* restat *cum* audacia calculare.

Since no such collections of astronomical tables appear in the Peterhouse catalogue of 1418, Holbrook and his four companions must have had the

[1] R. T. Gunther, *Early Science in Cambridge* (Oxford, 1937), p. 138. He gives a facsimile of part of the tables from MS. Cam. U. Lib. Ee. 3. 61 and notes that the Preface is also to be found in MSS. Ashmole 340 and 346. A fine copy of Holbrook's tables is in Peterhouse MS. 267.

volumes in their own possession or been able to borrow them from elsewhere for their grand collation. It is possible that the manuscript we are concerned with was first sought out by the Holbrook school in their search for full and accurate tables. Certainly the tables in our volume are remarkably free from the usual scribal errors and therefore of considerable value to a man like Holbrook.

If, however, the manuscript did not come to the College until a later date, we have the choice of at least three men, all of whom are known as donors of similar texts. In the Register of the College which contains the catalogue of 1418 there are a number of entries which record subsequent additions to the Library up to about 1500. Towards the end of this section occur the names of two donors responsible for a good collection of astronomical books, Roger Marshall and John Warkeworth. In this list the following items call for special note:

201 includes a *Theorica Planetarum*, but this seems to be only one short tract in a large collection bound in one volume.

207 *Theorica Planetarum cum diuersis tabulis astronomie*. First words on second leaf 'inferior orbis'. Last leaf but one, blank. These details do not agree with our MS. 75. 1.

210 contains, amongst other things, a *Theorica Planetarum*, but the volume is probably to be identified with the copy of Holbrook's tables, now MS. 267.

211 Another collection containing a *Theorica Planetarum*; again it is probable that this is but one short tract in a large collection.

Following these entries, on p. 22 of the Old Register, a piece of paper has been stitched to the vellum; it is headed in pale ink:

Billa magistri Rogeri Marchall de dono suo scripta propria manu sua a° dni Millesimo 1472.

| | | |
|---|---|---|
| i | Perspectiua Witelonis | ponantur in libraria |
| ii | Arithmetica Jordani cum commento | vestra secretiori in |
| iii | Exposicio super theoricam planetarum | vinculis si vobis |
| iv | Liber animalium Alberti | videatur. |
| v | Tabule equacionum planetarum magne | ponantur in vincula in |
| vi | Logica ffratris Rogeri bacon | apertiori libraria |
| vii | ...etc. | vestra. |

Items iii and v cannot be certainly identified with any of the extant manuscripts, and either is a possible title for MS. 75. 1.

Roger Marshall gave many books to the libraries of King's College and Caius College, as well as to Peterhouse, but all that is known of the provenance of his manuscripts except that many of them came from Bury. The donor,

John Warkeworth, must also be mentioned, for although there are no astronomical manuscripts which bear his name, many of his volumes are marked as having been bought in 1462 and 1463—which is close to our owner's mark of 1461. Warkeworth was a Fellow of Merton College, Oxford, in 1446, and subsequently Principal of Bull Hall and Nevill's Inn; he was Master of Peterhouse from 1473 to 1500.

Another remote possibility is provided by the fact that William Gage, S.T.B., clerk, of Woolpit (west Suffolk), mentions in his will of 1500[1] an astrolabe and various books which he left to Peterhouse. Unfortunately, the books are not described in sufficient detail to make identification possible, but the astrolabe is recorded in the Old Register. A similar instrument is said to have been given by Warkeworth. All these entries tell us nothing more definite than the popularity of such astronomical texts and instruments at Peterhouse towards the close of the fifteenth century. It seems likely that if our manuscript was not acquired by Holbrook *c.* 1400, it must have come from Roger Marshall or one of his contemporaries after 1460.

The absence of markings in other hands on such a useful working set of tables may be taken as evidence (albeit negative) that the manuscript was not in use by any other person for some decades after leaving the author's hands. If it had been used it is almost certain that the user would have added up-to-date radices to facilitate his calculations. Furthermore, after an interval of about forty or fifty years, such tables might have become so erroneous as to require revision of their constants; such a period may be taken as the natural 'working life' of such a table without revision, and after this time a set of tables of this nature would tend to be disregarded as a scientific aid and be preserved (if at all) only as a curiosity or through inertia.

The state of the volume therefore would suggest that it lay unused for a few decades from the time when it left its writer's hands in 1400 or just before. By the middle of the fifteenth century part of the tables would have become obsolete, and the owner of 1461 would probably have regarded it as a manuscript of no great importance. Possibly this owner, or a person to whom the manuscript passed shortly afterwards, presented it to Peterhouse where it is still preserved. It is probable that the manuscript was bound, or even rebound, after it came to the College, and the part which is now MS. 75. II was almost certainly added at this time.

Part of the original binding, or perhaps an early wrapper, has been preserved as a vellum fly-leaf used in the nineteenth-century rebinding of the volume. It seems that a large sheet of thick vellum was folded down the

[1] Canterbury Cathedral Chapter Archives. Register F, f. 131 verso.

middle, and one side was stuck to the inner surface of the front board; the other half of the sheet then acted as a protective fly-leaf. Unfortunately, the volume was a victim of that vandalism so notorious in libraries, and at some time after this rebinding the free fly-leaf was crudely slashed out by some person seeking a fine sheet of vellum for some other purpose. The cut is about one inch from the fold, and it has been made so violently that the three folios following this fly-leaf have been almost severed. Fortunately it is clear that no folios have been lost, since there is no break or companionless folio at the end of the first quire in the volume.

The remaining thick vellum sheet still bears traces of geometrical drawing which indicates that the original sheet was marked out as some sort of astronomical instrument, and following these traces it is possible to reconstruct part of the markings and determine the probable size of the original sheet (Fig. 1). So far as can be seen the markings would accord well with those required in laying out an astrolabe projection, for we have circles corresponding to the tropics of Cancer and Capricorn and to the Equator (between them) and the Ecliptic (running from one to the other). The only part of the inscription which can now be read is the word '*ver*nale' which occurs near a position which would correspond to the vernal equinox at the first point of Aries.

The verso of this vellum sheet is much worn and discoloured, but seems free of any trace of marking or inscription. Two small holes at the edge of the sheet may have held a tape so that the sheet might be used as a wrapper for the manuscript. If this is the case it would seem that the sheet was first used for the astrolabe drawing and later pressed into service as a wrapper; the drawing would hardly have been made on a sheet which had been spoiled by bending and exposure. The astrolabe drawing may possibly indicate that the wrapper was supplied by the person who wrote the manuscript, and in accordance with our previous suggestion it was probably used to protect the loose quires of the manuscript.

In the early nineteenth century all the Peterhouse manuscripts were subjected to a uniform rebinding in vellum over card boards, the vellum being readily available in the form of obsolete College leases.[1] When MS. 75 was first examined by the editor in 1951, it had the form of a large vellum-bound folio volume, about 14 in. by 11 in. and $1\frac{1}{4}$ in. thick, and on the spine were written the old press-mark '0.7.4' and the names of the authors 'Bredon,

[1] It is perhaps worth noting that the lease used for MS. 75 appears to be a twenty-year lease of farming land to one Simon Perdue; the boards have first been covered with waste sheets from the printing of an early nineteenth-century book on the history of Transylvania, of which p. 631 can be seen clearly through the spine.

Trivet, Vegetius'. The modern press-mark '75' was on a label at the lower end of the spine.

The volume has suffered greatly at the hands of the nineteenth-century binder. Four saw-cuts, each about ½ in. deep, have been made in the back of

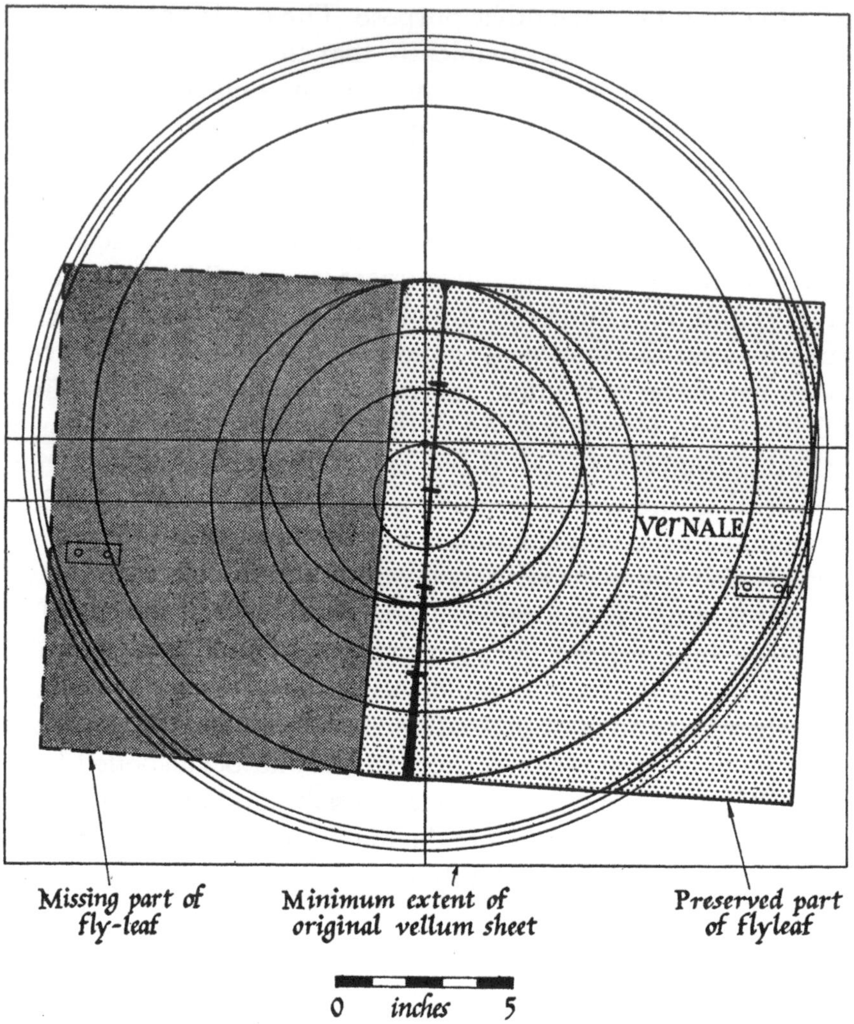


Fig. 1. Reconstruction showing surviving part of fly-leaf, original extent of double fly-leaf, and original extent (minimum) of sheet containing astrolabe construction. The full sheet must have been almost two feet square.

the quires, and the cover has been fastened on and held tight by strings inserted at the bottom of these cuts. This resulted in much damage, for many of the folios are written right up to the fold of the quire and therefore letters have been lost when the slots were cut. Another result was that the tightness of the binding rendered almost an inch of each folio invisible; indeed, without using considerable force it was difficult to turn the pages so as to leave less than one or even two inches obscured.

To investigate the manuscript fully, and in particular to read the whole of the important note mentioning Chaucer by name, it was necessary to cut these binding strings and separate the folios of the volume. Taking advantage of this, the manuscript has now been completely repaired and rebound, and the unrelated MS. 75. II containing the works of Trivet and Vegetius has been separated from the astronomical portion. The manuscript with which we are concerned has now been put into a fine binding of natural leather and is protected by a cloth case; MS. 75. II has been rebound in the earlier binding of the composite manuscript.

The last leaf of the original volume (now the last leaf in 75. II) had been stuck to the back board of the old binding; on its free surface a bibliographical note has been added, apparently by a late sixteenth (?)-century librarian:

In isto vol { Simonis Bredon æquat plan
Nic. Trivett in Aug$^{m}$ de civ. Dej
Vegeti*us* de re militari }

Bredon è coll. Marton. Oxon Math. Medic. Dr
floruit a$^{o}$ 1380 Bal Cent. 6. f. 489
Nicolas Triveth Norfolkei' studio Oxon. Thos...[1]
obijt 7ginari*us* 1328...[1]
expos Aug de civ dej nihil...[1]

This is of great importance to us, since it gives the only explicit mention of Simon Bredon as the author of our manuscript, and it will receive further attention in our discussion of the ascription. Here it is only necessary to point out that the title appears contracted as 'aequat plan' which has already been suggested as the most likely form of the original designation, and that the facts about Bredon (including the erroneous date) are taken almost verbatim from the cited place in the works of John Bale. Since Leland quotes this title and author before Bale's book was written, the note just transcribed must have been preceded by an earlier inscription giving a similar form of title and author. Most likely this inscription was on the part of the fly-leaf which was slashed away, or on some portion of the old binding.

The cutting of the binding cords also permitted a more complete collation to be made than that which was possible for M. R. James in 1899. The arrangement of the quires is as set out on p. 14.

Continuity of subject-matter indicates the existence of three groups of quires which belong together and make up the two sets of tables and the textual

[1] Illegible in manuscript, probably a reference from Bale.

portion. This possibility is confirmed (and supplemented) by the presence of leaves which are stained and discoloured, as if they had been outer leaves at some stage when the groups had an independent existence. Such outer leaves occur at the front of quires a and i, and at the back of g and h. This indicates that at some time there must have been two separate portions, one consisting of quires a to g which contain the two sets of tables, and a second containing quires i, k, h (in that order) which describe the equatorie instrument, outline its practice, and add a table of the ascensions of signs.

| | Quire | Folios |
|---|---|---|
| Tables I | $a^4$ | 1–4 |
| | $b^{14}$ wants 8–11, 13 (see diagram) | 5–13 |
| Tables II | $c^{12}$ | 14–25 |
| | $d^{12}$ wants 6 (see diagram) | 26–36 |
| | $e^{12}$ | 37–48 |
| | $f^{12}$ | 49–60 |
| Addition | $g^4$ | 61–64 |
| | $h^8$ wants 1, 2 (see diagram) | 65–70 |
| Text | $i^4$ | 71–74 |
| | $k^4$ [smaller folios] | 75–78 |

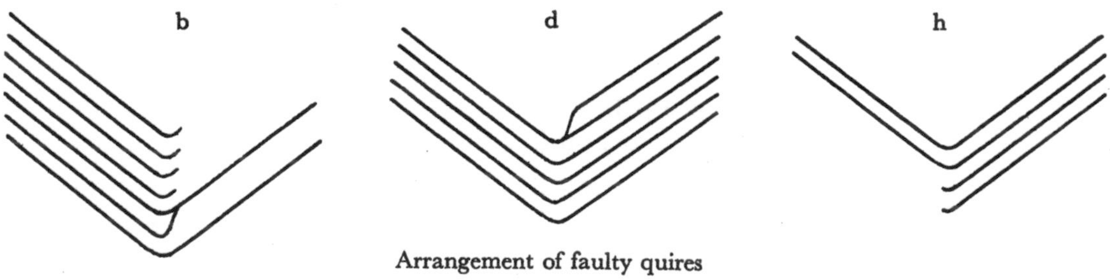

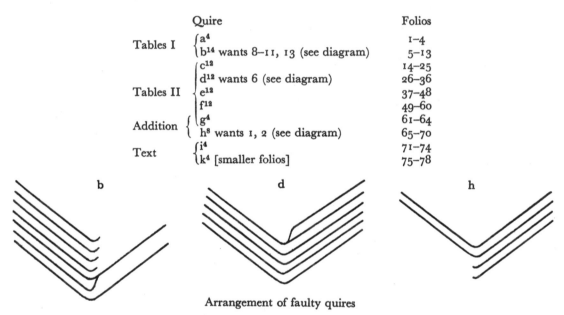

Arrangement of faulty quires

Reference is made in the text to a 'table following' which might indicate that the original order should be i, k, h followed by a–g, but there is no indication that the quires were ever bound thus. The collation marks on the folios seem to date from the sixteenth century, and they indicate the same order as is found today.

It is difficult to say whether the missing folios in quires b, d and h represent any loss of material or whether they are due to the writer having removed spoilt pages or added extra leaves. Certainly there is no obvious discontinuity or omission at any point, though possibly the first two folios of h might have contained a heading or even a canon (explanation) for the table of ascensions of signs. Some support for this possibility may be found in the fact that only in this quire have the missing leaves been removed exactly at the middle fold—a procedure which endangers the strength of fastening of the halves which are left unattached.

Since it will later be suggested that this manuscript is the first draft, by the author, of a self-contained work which was left unfinished, it will be of interest to describe the procedure and the stages of writing by which it was probably put together.

The author wrote quire i containing the description of the instrument and left a blank leaf at the end. At some later date he added quire k (of quite different vellum) describing the practice of the instrument, possibly abandoning the attempt at the end of the quire because of a heavy increase in the complexity of the remaining subject-matter which had to be treated. At some other time (it is not clear whether before or after writing k) he compiled or copied the Tables I found in quires a and b which must have been written together. Later he obtained the longer Tables II to supplement his work, probably having them professionally written specially for this purpose, and proceeded to annotate them in cipher. The last two folios of this set had been left blank by the scribe, and our author may have started by using the end folio for a rough draft, in Latin, of an explanation of a certain type of judicial horoscope. After that, being careful to leave a blank at the end of Tables II, the author filled two sides with additional tables and then continued on quire h for a complete table of ascensions of signs. At some other date the author used the front folio of quire i for a draft (unfinished) of a table of fixed stars embodying his own observations made in London.

Although it is thus evident that the different texts were originally written on loose quires, there is nothing to suggest that these were to be bound in any definite order. On the contrary, it seems there are separate sections destined for recopying in some quite different scheme. For all we know there may have been other quires of text or tables which have been lost.

The sixteenth-century librarian of Peterhouse has done his best with this disordered collection, making only the error of binding the tables before, instead of after, the text. One can well imagine that a similarly confused state of the author's draft could give rise to the variations in content and order which are found in even the earliest copies of the *Canterbury Tales*, or perhaps more strikingly the whole sets of alternative sections which are found in the manuscripts of the *Treatise on the Astrolabe*.

## PLATE I

FOLIO 71 VERSO (A)
Actual size

In the name of god pitos & merciable seide the largere þt thow makest this
instrument the largere ben thi chef divisiouns the largere þt ben tho divisiouns
in hem may ben mo smale fracciouns & evere the mo of smale fracciouns
the ner the trewthe of thy conclusiouns / tak therfore a plate of metal or elles
a bord þt be smothe shave by level & evene polised / of which whan it is
rownde by compas the hole diametre shal contene 72 large enches or elles 6 fote of
mesure / the whiche rownde bord for it shal nat werpe ne krooke, the egge of
the circumference shal be bownde wt a plate of yren in maner of a carte whel.
this bord yif the likith may be vernissed or elles glewed wt perchemyn for
honestyte / tak thanne a cercle of metal þt be 2 enche of brede & þt the hole
dyametre wt in this cercle shal contene ~~the forseide~~ 68 enches or 5 fote & 8 enches & subtili lat this cercle
be nayled up on the circumference of this bord or ellis mak this cercle of
glewed perchemyn. this cercle wole I clepe the lymbe of myn equatorie
þt was compowned the yer of crist 1392 complet the laste meridie of decembre
this lymbe shaltow devyde in 4 quarters by 2 diametral lynes in maner of
the lymbe of a commune astrelabye & lok thy croys be trewe proued by

geometrien [illegible]
point on the myddel of the bord, which myddel shal be nayled a plate of
metal rownd / the hole diametre of this plate shal conteue 18 enches large
for in this plate shollen ben perced alle the centris of this equatorie / & ek in
proces of tyme may this plate be turned a bowte after þe auges of planetes
ben moeued in the 9 spere / thus may thin instrument laste perpetuel. tak thanne
as I haue seid by forn the fix fot of thy compas & set it in the myddel of
this plate & with the moeuable point of thi compas descriue a cercle in the
ferthest circumference of thy lymbe & nota þt the myddel poynt of this plate
wher as the fix fot of thy compas stondith wole I calle centre aryn. mak
thanne a nother cercle þt be descriued vp on the same centre aryn but
litel quantite fro tho ferthest forseid cercle in the lymbe in which space
shollen ben deuysed mynutes of the lymbe / mak thanne a nother cercle som
what ferther distant fro the laste seid cercle in which shal be deuyded the
degres of the same lymbe mak yit a nother cercle som what ferther
distant fro this laste seid cercle in which shal ben writen the nombres of

III

# TRANSCRIPT AND FACSIMILES

### *Conventions of Transcription*

EDITORIAL interference with the transcription has been kept to the minimum permitted by normal typography. Even obvious errors in the manuscript have been retained, and the original punctuation has been reproduced as closely as possible.

In handwriting of this period there is often no distinction between the initial and capital forms of a letter (especially with *m*, *r* and *s*); in all cases of doubt this transcription avoids the capital form. The two ampersands, & and $\overline{\&}$, have been distinguished, for although their use in this text appears to be without difference of meaning, there are some Latin manuscripts where $\overline{\&}$ corresponds to *etiam* rather than to *et*.

Contractions have been expanded and set in *italic* in the usual fashion. The numerous corrections and glosses in the text are difficult to reproduce; supralinear insertions and glosses are inserted in the appropriate place in (round brackets), and a caret, ‸, is only given when this occurs in the manuscript. Marginal additions and glosses are inserted in the same manner, but attention is drawn to the fact that the point of insertion is merely editorial conjecture.

It is impossible to specify accurately all the places where the present text is written over an erased passage, but there are many instances in which it is possible to read (sometimes with the aid of ultra-violet light) such words removed by the writer. In these cases the reading has been printed in [square brackets] at the appropriate place, and the following conventions have been used to indicate the means by which the passage was removed: [cancelled]$^c$, [erased]$^e$, [expunged], [obliterated]$^o$. Such passages have not been included in the translation of the text. In a few places in the manuscript initial letters have been omitted, presumably for the later addition of embellished capitals; in some cases a guide letter is given in the margin. Such initial letters have been printed in {curled brackets}.

A bar | in the printed transcript marks the end of lines in the manuscript; the beginnings of lines are numbered on the left of the page.

An asterisk * after the line number draws attention to the existence of one or more textual notes in Chapter V (pp. 62 ff.).

A

1* In the name of god pitos & m*er*ciable seide [leyk]$^{e}$ the largere þ$^{t}$ thow makest
2 this | instrume*n*t / the largere ben thi chef deuisiou*ns* / the largere þ$^{t}$ ben tho
3 deuisiou*ns* / | in hem may ben mo smale fracciou*ns* / $\overline{\&}$ euere the mo of smale
4* fracciou*ns* | the ner the trowthe of thy *con*clusiou*ns* / tak ther fore a plate of
5 metal or elles | a bord þ$^{t}$ be smothe shaue / by leuel / & euene polised ¶ of which
6* whan it is | rownd (‸ by compas) / the hole diametre shal *con*tene .72. large
7 enches or elles .6. fote of | mesure / the whiche rownde bord for it shal nat
8 werpe ne krooke / the egge of | (‸ the) circu*m*fere*n*ce shal be bownde w$^{t}$ a plate
9* of yren in maner of a karte whel. / | ¶ this bord yif the likith may be vernissed
10 or elles glewed w$^{t}$ *per*chemyn for | honestyte ¶ tak thanne a cercle of metal
11 þ$^{t}$ be .2. enche of brede / & þ$^{t}$ the hole | dyametre (‸ w$^{t}$ in this cercle shal)
*con*tene [the forseide]$^{c}$ .68. enches / or .5 fote (& .8. enches) / & subtili lat
12* this cercle | be nayled vp on the circu*m*fere*n*ce of this bord or ellis mak this
13* cercle of | glewed *per*chemyn. / this cercle wole I clepe the lymbe of myn
14* equatorie | þ$^{t}$ was *com*powned the yer of crist .1392. *com*plet the laste m*er*idie
15 of dece*m*bre | ¶ this lymbe shaltow deuyde in 4 quarters by .2. diametral
16* lynes in man*er* of | the lymbe of a comune astrelabye $\overline{\&}$ lok thy croys be trewe
17* proued by | geometrical *con*clusiou*n* / tak thanne a large *com*pas þ$^{t}$ be trewe
18 & set the ffyx | point ou*er* the middel of the bord (‸ on) which middel shal be
19 nayled a plate of | metal rownd ¶ the hole diametre of this plate shal *con*tiene
20 .16. enches large | for in this plate shollen ben perced alle the centris of this
21 equatorie / & ek in | *pro*ces of tyme may this plate be turned a bowte after þ$^{t}$
22* auges of planetes | ben moeued in the .9. spere thus may thin instrume*n*t laste
23 *per*petuel. ¶ tak thanne | as I haue seid by forn the fix fot of thy *com*pas $\overline{\&}$ set
24 it in the middel of | this plate & w$^{t}$ the moeuable point of thi *com*pas descriue
25 a cercle in the | ferthest circu*m*ference of thy lymbe ¶ $\overline{\&}$ no*ta* þ$^{t}$ the middel
26* poynt of this plate | wher as the fix fot of thy *com*pas stondith / wole I calle
27* centre .aryn. / mak | thanne a narwer cercle þ$^{t}$ be descriued vp on the same
28 centre aryn but | litel q*ua*ntite fro the forthest forseid cercle in the lymbe in
29* whiche space | shollen ben deuyde mynutes of the lymbe / mak thanne a narwere
30 cercle som | what ferther distau*n*t fro the laste seid cercle / in which shal be
31 deuyded the | degres of the same lymbe mak yit a narwere cercle som what
32 ferthere | distau*n*t fro this laste seid cercle in which shal ben writen the
nombres of

**line 29** deuyde *sic for* deuyded.

In the name of god pitos & merciable seide the lasse þt thow makest this
instrument the lasse ben thi chef deuisiouns the lasse þt ben tho deuisiouns
in hem may ben mo smale fracciouns & euere the mo of smale fracciouns
the ner the trewthe of thy conclusiouns / tak ther fore a plate of metal or elles
a bord þt be smothe shaue by leuel & euene polised / of which whan it is
rownd [by compas] the hole diametre shal contene 72 large enches or elles 6 fote of
mesure / the whiche rownde bord for it shal nat werpe ne krooke / the egge of
the circumference shal be bownde wt a plate of yren in maner of a karte whel
this bord yif the likith may be vernissed or elles glewed wt perchemyn for
honestyte. tak thanne a cercle of metal þt be 2 enche of brede & þt the hole
dyametre [wt in this cercle shal] contene ~~the forseide~~ 68 enches or 5 fote [& 8 enches] & subtili lat this cercle
be nayled vp on the circumference of this bord or ellis mak this cercle of
glewed perchemyn. this cercle wole I clepe the lymbe of myn equatorie
þt was compowned the yer of crist 1392 complet the laste meridie of decembre
this lymbe shaltow deuyde in 4 quarters by 2 diametral lynes in maner of
the lymbe of a commune astrelabie & lok thy croys be trewe proued by
geometrical conclusioun tak thanne a large compas þt be trewe & set the ffyx
point [on] the middel of the bord / which middel shal be nayled a plate of
metal rownd / the hole diametre of this plate shal contene 18 enches large
for in this plate shollen ben perced alle the centris of this equatorie & ek in
proces of tyme may this plate be turned a bowte after þt auges of planetes
ben moeued in the 9 spere thus may thin instrument laste perpetuel. tak thanne
as I haue seid by forn the fix fot of thy compas & set it in the middel of
this plate & wt the moeuable point of thi compas descriue a cercle in the
ferthest circumference of thy lymbe & now þt the middel poynt of this plate
wher as the fix fot of thy compas stondith wole I calle centre aryn. mak
thanne a narwer cercle þt be descriued vp on the same centre aryn but
litel quantite fro the forthest forseid cercle in the lymbe in which space
shollen ben deuyded mynutes of the lymbe mak thanne a narwere cercle som
what ferther distaunt fro the laste seid cercle in which shal be deuyded the
degrees of the same lymbe mak yit a narwere cercle somwhat ferthere
distaunt fro this laste seid cercle in which shal ben writen the nombres of

B

1 degres / mak yit anarwere cercle som what ferther distau*n*t fro this laste
2* seid | cercle in which shollen ben writen the names of 12 signes / & no*ta* þ$^{t}$
3* this laste | seid cercle wole I calle the closere of the signes / now hastow .5.
4 cercles | in thy lymbe / & alle ben descriued vp on centre aryn / and eu*er*ich
5 of the | .4. quarters in thi lymbe shal ben deuided in 90 degres þ$^{t}$ is to sein
6* .3. signes | & eu*er*i degre shal be deuided in 60 mi$^{a}$ // & shortly thi lymbe is
7* deuided | in man*er* of the lymbe in the bakside of an astrelabie / deuyde
8* thanne thilke | lyne þ$^{t}$ goth fro centre aryn vn to the cercle closere of the
9* sygnes (vers*us* fine*m* ge*minorum*) in 32 | parties eq*ua*les. whiche p*ar*ties ben
10* (‸ cleped) degres of the semydiametre / marke thise | p*ar*ties dymli (ut
postea deleantur) // and no*ta* þ$^{t}$ this diametral lyne deuided in 32 p*ar*ties
11* shal be | cleped lyne alhudda / set thanne the fix point of thy *com*pas vp on
12 the ende | of the firste deuysiou*n* fro centre aryn in lyne alhudda / & the
13 moeuable | point vp on the ende of the 30 deuisiou*n* fro the fix poynt of thi
14 *com*pas in | the same lyne / so dwelleth ther but .1. deuisiou*n* by twixe thy
15 moeuable | point & the closere of the signes / &̅ .1. deuysiou*n* bi twixe thy
16 fix poynt | & the centre aryn / & descryue thus a cercle / & tak ther the
17 ecce*n*trik | cercle of the sonne / scrape thanne awey the deuysiou*n*s of lyne
18 alhudda // | // deuyde yit (‸ dymly oculte) the same lyne alhudda fro centre
19 aryn / vn to the closere | of the signes in 60 parties eq*ua*les / set thanne the
20* fix poynt of thy | *com*pas in centre aryn / &̅ the moeuable point in .12. degres
21 & 28 mi$^{a}$ of lyne | alhudda & descriue a cercle / & þ$^{t}$ is the centre deffere*n*t
22* of the mone /| p*er*ce thanne (‸ al) the circu*m*fere*n*ce of this deffere*n*t in 360
23 subtil holes equales | of space & thise spaces by twixe the holes / ben deuyded
24* owt of the | degres of the lymbe // &̅ no*ta* þ$^{t}$ the yer of crist 1392 (‸ *com*plet
25 [– vlt*i*mo .10.b*re* in m*er*idie london]$^{e}$) the aux of saturn*us* | was the (‸ last)
26* m*er*idie of decembre at londone I seye the aux of saturne in the | .9. spere
27* was .4. dowble signes 12 g$^{a}$ 7 mi$^{a}$ .3. 2$^{a}$ &*cetera* / the remenau*n*t of | auges sek
28 hem in the table of auges folwynge / tak thanne a Rewle | & ley þ$^{t}$ on ende
29* in centre aryn / & þ$^{t}$ other ende in the lymbe / in the ende | of the minut wher
30 as endith the aux of the planete / & draw ther a lyne | w$^{t}$ a sharp instrume*n*t
31 fro centre aryn vn to the closere of the signes / &̅ | no ferthere for empeiryng
32 of the lymbe &̅ fasteby this lyne writ the | name of the planete (cui*us* e*st* aux)
33* this rewle is gen*er*al for alle planetis / sek thanne | in thi table of centris / the
34 distau*n*ce of the centre equant of saturne | fro centre aryn which is .6. g$^{a}$
35 50 mi$^{a}$ / set thanne the fix point of thy | *com*pas in centre arin / &̅ the moeuable
36 poynt in 6 g$^{a}$ &̅ 50 mi$^{a}$ in lyne | alhudda fro centre aryn // turne than softely
37 thy *com*pas abowte. til þ$^{t}$ the | moeuable poynt [til it]$^{c}$ towche the lyne of the
aux of saturne &

**line 24** *before* vltimo *there is one letter which is illegible even under ultra-violet light.*

degres. mak yit another cercle somwhat ferther distant fro this laste seid
cercle in which shollen ben writen the names of 12 signes & no{ta} that this laste
seid cercle whole I calle the closere of the signes. now hastow 4 cercles
in thy lymbe & alle ben descriued vp on centre aryn. And euch of the
4 quarters in thi lymbe shal ben deuided in 90 degres that is to seyn 3 signes
& eueri degre shal be deuided in 60 minutes & shortly this lymbe is deuided
in maner of the lymbe in the bakside of an Astrelabie. deuide thanne thilke
lyne that goth fro centre aryn vn to the cercle closere of the signes [vers fine deleantur] in 32
parties equales. which parties ben [cleped] degres of the semidiametre. mak thise
parties dymli [vt postea deleantur] And no{ta} that this diametral lyne deuided in 32 parties shal be
cleped lyne alhudda. set thanne the fix point of thy compas vp on the ende
of the fifte deuision fro centre aryn in lyne alhudda & the moeuable
point vp on the ende of the 30 deuision fro the fix poynt of thi compas in
the same lyne so dwelleth ther but 1 deuision bytwixe thy moeuable
point & the closere of the signes & 1 deuision bitwixe thi fix poynt
& the centre aryn & descriue thus a cercle & tak ther the eccentrik
cercle of the sonne. scrape thanne awey the deuisiones of lyne alhudda.
deuide yit [by myn centre] the same lyne alhudda fro centre aryn vn to the closere
of the signes in 60 parties equales. set thanne the fix poynt of thy
compas in centre aryn & the moeuable point in 12 degres & 28 minutes of lyne
alhudda & descriue a cercle & that is the centre deferent of the mone.
perce thanne [al] the circumference of this deferent in 360 smal holes equales
of space & thise spaces bytwixe the holes ben deuided [aplace] after the
degres of the lymbe. & no{ta} that the yer of crist 1392 the aux of saturne
was the [laste] midie of decembre at londone I seye the aux of saturne in the
9 spere was 4 dobble signes 12 dg' 7 m' 3 2 & c the remenaunt of
auges sek hem in the table of auges folwynge. tak thanne a sekle
& ley that on ende in centre aryn & that other ende in the lymbe in the ende
of the minut wher as endith the aux of the planete & draw ther a lyne
with a sharp instrument fro centre aryn vn to the closere of the signes &
no ferthere for empeiryng of the lymbe & fastby this lyne writ the
name of the planete. this [aux & aux] sekle is general for alle planetis. sek thanne
in thi table of centris the distaunce of the centre equant of saturne
fro centre aryn which is 6 dg' 40 m'. set thanne the fix point of thy
compas in centre aryn & the moeuable poynt in 6 dg' & 40 m' in lyne
alhudda fro centre aryn. turne than softely thy compas aboute til that the
moeuable poynt til it touche the lyne of the aux of saturne &

C

1 stondinge alwey (^ stille) the fix poynt of thy *com*pas in centre aryn / marke
2 w$^{t}$ thy moeuable poynt | in the lyne of the aux of saturn*us* a dep prikke / for
3 in þ$^{t}$ prikke shal be p*er*ced a smal hole | for the centre equant of sa*turnus* / &
4* faste by this hole mak an E in signefyeng of eq*u*ant // | ⁋ thanne tak awey thy
5 *com*pas / & loke in thi table of centris / the distau*n*ce of the centre | differe*n*t
of saturn*us* / & þ$^{t}$ is .3. g$^{a}$ & 25 mi$^{a}$ / set (a centre aryn) thanne the fix point
6 of thy *com*pas in | centre aryn & thy moeuable point in .3. g$^{a}$ & 25 mi$^{a}$ in
7 lyne alhudda / & torne softely | thi *com*pas til þ$^{t}$ the moeuable point towche
8 the (^ forseide) lyne of the aux of saturne / & stonding | stille thy fix poynt
9 of thi *com*pas in centre aryn / marke w$^{t}$ the moeuable poynt in | the lyne of
10* the aux of saturne a dep prikke for ther in shal be p*er*ced a smal hole | for
11 the centre differe*n*t of saturn*us* / & fasteby this hole mak an .D. for differe*n*t | ⁋ &
12* no*ta* þ$^{t}$ by this ensample of saturn*us* shaltow make the centres differe*n*tes / & | ek
the equantes of alle the planetis / after hir distau*n*ces (in tabul*is*) fro centre
13 aryn | & prikke hem in the lynes of hir auges ⁋ thanne shaltow sette the fix
14 point | of thy *com*pas in the lyne of the aux of m*er*curie euene by twixe the
15* centre .E. | & centre .D. of mercuri*us* / & strid the moeuable poynt til it wole
16 towche bothe | centre E & ek centre .D. of m*er*curi*us* / & descryue ther a litel
17 cercle & thanne | shaltow se þ$^{t}$ the lyne of the aux of m*er*curie departith this
18 litel cercle in | .2. arkes equals / this is to seye / þ$^{t}$ the lyne kerueth this litel
19 cercle euene | amidde / this litel cercle shal be perced ful of smale holes (in
20 circu*m*fere*n*cia circuli) by euene p*ro*porciou*n* | as is the centre differe*n*t of the
21* mone in 360 holes yif it be possible or in | .180. or in 90 atte leste / but sothly
22 the spaces by twixe the holes (ne) shal nat | be deuided owt of the grete lymbe
23 of the instrume*n*t / as is the ce*n*tre differe*n*t | of the mone / but owt of the
24* circu*m*fere*n*ce of the same litel cercle it shal be | deuided by thy *com*pas ⁋ scrape
25 thanne awey thilke 60 deuysiou*n*s in lyne alhudda | ⁋ & yit deuyde the same
26 lyne alhudda in .5. p*ar*ties eq*u*ales (by *com*pas) fro centre aryn vn | to the
27* cercle þ$^{t}$ is closere of the signes / & eu*er*ych of thilke .5. parties shal be | deuided
28 in 60 p*ar*ties / thise diuisiou*n*s ne shal nat ben scraped awey / deuyde | thanne
29* the line þ$^{t}$ goth fro centre aryn to the hed of capricone which lyne | is cleped
30 in the tretis of the astrelabie the midnyht line / I seye deuyde this | midnyht
31 lyne in .9. parties equals fro centre aryn (^ vn) to the closere of the | signes / &
32* eu*er*ich of thise deuysiou*n*s shal be deuided by thy *com*pas in 60 p*ar*ties | equales ⁋
33 thise deuysiou*n*s ne shal nat be scraped awey ⁋ laus deo vero | now hastow the
34 visage of this p*re*cios eq*ua*torie / no*ta* þ$^{t}$ thise last seid .9. diuisiou*n*s | in the
midnyht lyne shollen seruen for Eq*u*aciou*n* of the 8$^{e}$ sp*er*e

**line 28** capricone *sic for* capricorne.

stondinge alwey the fix point of thy compas in centre aryn, make with thy moevable poynt in the lyne of the aux of Saturnus a dep prikke, for in that prikke shal be perced a smal hole for the centre equant of Saturnus, and faste by this hole mak an E in signefyenge of equant. //

Thanne tak awey thy compas and loke in thi table of centris the distance of the centre defferent of Saturnus, and that is 15-3-8 and 24 min; set thanne the fix point of thy compas in centre aryn, and thy moevable point in 3-8 and 24 min in lyne alhudda, and torne softely thi compas til that the moevable point touche the forseide lyne of the aux of Saturne, and stonding stille thy fix poynt of thi compas in centre aryn make with the moevable poynt in the lyne of the aux of Saturne a dep prikke, for ther in shal be perced a smal hole for the centre defferent of Saturnus, and faste by this hole mak a D for defferent.

And note that by this ensample of Saturnus shaltow make the centres defferentes and ek the equantes of alle the planetes after hir distances in tabul fro centre aryn, and prikke hem in the lynes of hir auges. Thanne shaltow sette the fix point of thy compas in the lyne of the aux of Mercurie evene bytwixe the centre E and centre D of Mercurius, and styd the moevable poynt til it wole touche bothe centre E and ek centre D of Mercurius, and descryve ther a litel cercle; and thanne shaltow se that the lyne of the aux of Mercurie departith this litel cercle in 2 partes equals, this is to seye that the lyne kerveth this litel cercle evene amidde. This litel cercle shal be perced ful of smale holes by evene proporcioun in circumferencia circuli as is the centre defferent of the mone in 360 holes yif it be possible, or in 180 or in 90 atte leste; but sothly the spaces bytwixe the holes ne shal nat be devided with the grete lymbe of the instrument as is the centre defferent of the mone, but with the circumference of the same litel cercle it shal be devided by thy compas. Scrape thanne awey thilke 60 devisiouns in lyne alhudda, and yit devyde the same lyne alhudda in 9 parties equales by compas fro centre aryn un to the cercle that is clepid the cercle of the signes, and everich of thilke 9 parties shal be devided in 60 parties; thise divisiouns ne shal nat ben scraped awey. Devyde thanne the line that goth fro centre aryn to the hed of capricorne, which lyne is cleped in the tretis of the astrelabie the midnyght line, I seye devyde this midnyght lyne in 9 parties equals fro centre aryn un to the cercle of the signes, and everich of thise devisiouns shal be devided by thy compas in 60 parties equales; thise devisiouns ne shal nat be scraped awey. Laus deo vero.

Now hastow the makinge of thin equatorie. Nota that thise last seid 9 divisiouns in the midnyght lyne shollen serven for equacioun of the 8 spere.

See below after l. 5 { a b c

D

{n}ow for the *com*posiciou*n* of the Epicicle for the visage of thyn eq*ua*torie / thow shalt | make a cercle of metal of the same brede &̄ of the same widnesse in circu*m*|ference in diametre / &̄ in alle thinges lik to the lymbe of thin instrume*n*t | &̄ in the same maner*e* shal it be deuyded in mynut*is*. in degres in nombres | in names of signes / &̄ in 5 cercles *com*pased / as is the firste seid lymbe / saue þ$^{t}$ .a. // | // .b. the eccentrik of the sonne / ne shal nat be in the epicicle &̄ also | þ$^{t}$ it be nat filed to ney to the closere of his signes list thow p*er*ce the hole (forame*n*) | of thi *com*mune centre defferent amys / or elles list the hole breke | [þ$^{t}$]$^{c}$ / this epicicle mot haue suffisau*n*t thikkenesse to sustene hym self | ⁋ Tak thanne this epicicle & ley it (‸ sadly &) euene vp on the visage of thin eq*ua*torie | so þ$^{t}$ (capud) aries of thin epicle lie euene vp on the hed of aries in the lymbe | of thin eq*ua*torie / &̄ libra vp on libra & canc*er* vp on cancer & cap*ri*corn*e* vp | on cap*ri*corn*e* & (‸ euery) signe vp on signe this is to seyn the hed of eu*er*y signe | vp on hed of eu*er*y signe / tak thanne a Renspyndle or a boydekyn / & in | direct of the hed of canc*er*. (‸ thow shalt) in the cercle þ$^{t}$ is closere of the signes (perce) make a | litel hole thorw the epicicle / & thanne shaltow se (þ$^{t}$) yif thow haue trewely *com*pased | thy cercles / þ$^{t}$ the poynt of thy renspindle shal haue towched the closere | of the signes in direct of the hed of cancer in thyn equatorie / this | litel hole þ$^{t}$ is no grettere than a smal nedle shal be cleped the comune | centre defferent of planetes / tak thanne a barre of metal of the brede of | a large enche & of suffisau*n*t thyknesse / (‸ of) the whiche barre þ$^{t}$ on ende | shal be sowded to the closere of (‸ the) signes in direct of aries in this epicicle | & þ$^{t}$ other ende shal be sowded to the closere of the signes in direct of libra | in the same epicicle / draw thanne by thi rewle a lyne fro the hed of aries | to the hed of libra endelong the barre / & draw swich another lyne (‸ ouerthwart) [in]$^{c}$ | the barre fro the hed of canc*er* to the hed of cap*ri*corn*e* / & in the secciou*n* | of this crois is the centre of the epicicle / tak thanne a rewle of latou*n* | that (ne) be nat ful thykke / &̄ lat it be the brede of an enche & the lengthe shal | be as long as al hol the diametre of the Epicicle this rewle mot be | shape in man*er* of a label on an astrelabie ⁋ the centre of this rewle | shal be nayled to the centre of the forseide barre in swich a man*ere* | þ$^{t}$ this label may torne abowte as doth the label of an astrelabie / in | middes of this nayl þ$^{t}$ fastnyth the barre & the label to gidere ther mot | be a smal prikke þ$^{t}$ be dep / which prikke (i*d est* punct*us*) is the centre of thin epicicle | ⁋ tak thanne by thy large *com*pas the distau*n*ce by twixe centre aryn & the | closere of the signes / which distau*n*ce is the lengthe of lyne alhudda | ⁋ &̄ be it on a long rewle or elles be it on a long p*er*cemyn / marke w$^{t}$ thy | *com*pas the forseide distau*n*ce / & deuyde it in 60 p*ar*ties equals &̄ than hastow | a newe lyne alhudda / sek thanne in thy table of centres / the semydiamet*re* | of the epicicle of saturn*us* / & þ$^{t}$ is .6. g$^{a}$ & 30 mi$^{a}$ of swiche degres as ben | 60 in line alhudda / tak thanne w$^{t}$ thy *com*pas the space of .6. g$^{a}$ &̄ 30 mi$^{a}$. | of lyne alhudda / & set the fix point of thy *com*pas in the centre of thin

**line 1** Space left for initial; guide-letter *n* in inner margin.

**line 8** epicle *sic for* epicicle. **line 34** percemyn *sic for* perchemyn.

.b. the centre of the sonne / ne shal nat be in the epicicle & also [for same]
þt it be nat filed to ney to the closere of his signes lest thow perce the hole
of thi comune centre defferent amys. or elles lest the hole breke

As for the composicioun of the Epicicle for the visage of thin equatorie thow shalt
make a cercle of metal of the same brede & of the same thiknesse in circu-
ferente in diametre / & in alle thinges lik to the lymbe of thin instrument
& in the same manere shal it be deuyded in mynutis in degres in nombres
in names of signes & in 4 cercles compased as is the forseid lymbe. saue þt
þt this epicicle mot haue suffisant thikkenesse to sustene hym self
Tak thanne this epicicle & ley it sadly & euene vp on the visage of thin equatorie
so þt caput aries of thin epicicle lie euene vp on the hed of aries in the lymbe
of thin equatorie & libra vp on libra & cancer vp on cancer & capricorn vp
on capricorn & euery signe vp on signe this is to seyn the hed of euery signe
vp on hed of euery signe. tak thanne a fersyndle or a bodekyn & in
direct of the hed of cancer thow shalt in the cercle þt is closere of the signes perce make a
litel hole thurgh the epicicle & thanne shaltow se þt yif thow haue trewely compased
thy cercles þt the poynt of thy fersyndle shal haue towched the closere
of the signes in direct of the hed of cancer in thyn equatorie / this
litel hole þt is no grettere than a smal nedle shal be cleped the comune
centre defferent of planetes / tak thanne a barre of metal of the brede of
a large enche & of suffisant thiknesse / the whiche barre þt on ende
shal be sowded to the closere of the signes in direct of aries in this epicicle
& þt other ende shal be sowded to the closere of the signes in direct of libra
in the same epicicle / Draw thanne by thi rewle a lyne fro the hed of aries
to the hed of libra endelong the barre & draw swich another lyne ouerthwart
the barre fro the hed of cancer to the hed of capricorn / & in the section
of this croys is the centre of the epicicle / tak thanne a rewle of laton
that ne be nat ful thikke & lat it be the brede of an enche & the lengthe shal
be as long as al hol the diametre of the Epicicle this rewle mot be
shape in manere of a label on an astrelabie / the centre of this rewle
shal be nayled to the centre of the forseide barre in swich a manere
þt this label may torne abowte as doth the label of an astrelabie / in
myddes of this nayl þt fastnyth the barre & the label to gidere ther mot
be a smal prikke þt be depe. which prikke or poynt is the centre of thin epicicle
/ tak thanne by thy large compas the distaunce by twixe centre aryn & the
closere of the signes / which distaunce is the lengthe of lyne alhudda
& be it on a long rewle or elles be it on a long percemyn make wt thy
compas the forseide distaunce / & deuyde it in 60 parties equals & than hastow
a newe lyne alhudda / sek thanne in thy table of centres / the semydiametre
of the epicicle of saturnus & þt is 6 gr & 30 mi of whiche degres as ben
60 in line alhudda / tak thanne wt thy compas the space of 6 gr & 30 min
of lyne alhudda. & set the fix point of thy compas in the centre of thin

E

1 Epicicle þ$^{t}$ is the poynt (pu*nctus*) in the hed of the nail / & endelong the label
2 set the | moeuable poynt of thi *com*pas & w$^{t}$ þ$^{t}$ moeuable poynt mak a marke
3 astrik in the | label & fasteby the strik writ .sa. for saturne / this ensample of
4 saturne | techith how to maken in the label alle the semydiametres of Epicicles |
5* of alle the planetis / no*ta* þ$^{t}$ the sonne (ne) hath non epicicle ¶ & no*ta* þ$^{t}$
6*, 7 alwey | as the label turnyth / so shewith it the epicicle of eu*ery* planete / laus | deo
8* vero now hastow *com*plet thyn equatorie w$^{t}$ alle hise membris | ¶ and no*ta* þ$^{t}$
eccen*tri*k of the sonne shal nat be *com*passed in this epicicle / Explicit |
9 the face of the eq*ua*torie

10 ☞ no*ta* þ$^{t}$ eu*er*y centre mot ben also smal
11 as a nedle / & in eu*er*y equant mot
12 be a silk thred
13* ☞ no*ta* þ$^{t}$ the eccen*tri*k of the sonne is
14 *com*paced on the bord of the instrume*nt*
15 & nat on the lymbe for sparing
16 of metal / ☞ no*ta* shortly þ$^{t}$ but so be
17 þ$^{t}$ bothe the closeres of the signes
18 ben p*re*cisly ilike of widnesse / &
19 but so be þ$^{t}$ centre aryn stonde p*re*cise
20 as fer fro his closere of the signes as
21 the centre of thin epicicle stondith fro
22 the comune centre defferen*t* p*re*cise / thyn
23 epicicle is fals / but natheles
24 yif thow myshappe in this
25 cas i shal teche the aremedie
26* ¶ knokke thi centre defferen*t*
27 innere or owtre til it stonde
28 p*re*cise vp on the closere of the
29 signes in the lymbe of thin
30 eq*ua*torie / so wole thanne the
31 centre of thin epicicle p*re*cise
32 stonde vp on centr*e* aryn

33* the sixte
34 cercle is the
35 eccentric of the sonne
36 & the .5. cercle þ$^{t}$ is red
37 is the closere of the signes
38 & the secciou*n* of the crois
39 is centre aryn ¶ & þ$^{t}$ other
40 centr*e* is the centre of the
41 eccen*tri*k of the sonne
42 ¶ & the lyne deuyded in .9.
43 is the midnyht lyne I
44* wot wel it is figured
45 boistosly / & the cercle
46 abowte centre aryn is
47 the ce*n*tre defferen*t* of the
48 mone / the litel cercle
49 is the defferen*t* of m*er*curie //
50 ¶ the smale lynes ben lynes
51 of auges

52 the prikkes in the lynes
53 ben the ce*n*tris eq*uantis* &
54 defferen*tis* / & alle thise
55 centres saue the eq*ua*nt
56 of mars ben by twixe
57 centre aryn & the centr*e*
58* defferen*t* of the mone
59 the owterest space is mynut*is*
60 & the nexte space is degres
61 & the thridde space is nombres
62 of degres & the ferthe space is
63 for names of signes but
64 natheless the narwere cercle
65 of the signes is cleped the
66 closere of the signes / & it is
67 compased with red

Epicicle that is the poynt in the hed of the nail [provided] & endelong the label set the
moevable poynt of thi mars & wher that moevable poynt mak a marke a strik in the
label & write by the strik wyt. sa. for saturne / this ensample of saturne
techith how to maken in the label alle the semydiametres of epicicles
of alle the planetes / nota that the sonne hath non epicicle / & nota that alwey
as the label turneth so sheweth it the epicicle of every planete / laus
deo vero / now wostow complet thyn equatorie with alle hise membres
/ and nota that the eccentrik of the sonne shal nat be compassed in this epicicle / Explicit

the face of the equatorie

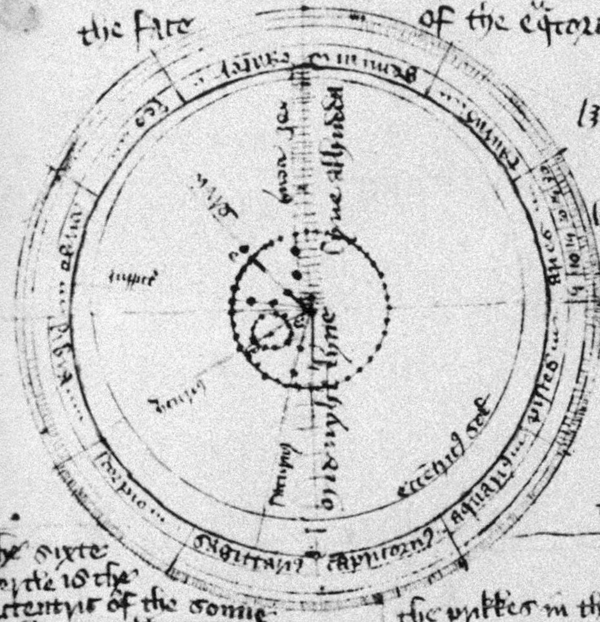

13 nota that every centre mot ben also smal
as a nedle / & in every equant mot
be a silk thred
nota that the eccentrik of the sonne is
compassed on the bord of the instrument
& nat on the lymbe for sparing
of metal / nota shortly that but so be
that bothe the closures of the signes
ben precisly ilike of wydnesse / &
but so be that centre aryn stonde precise
as fer fro bothe closures of the signes as
the centre of thin epicicle stondith fro
the commune centre deferent precise / thyn
epicicle is fals / but natheles
yif thow myshappe in this
cas I shal teche the the remedie
/ knowe thi centre deferent
inner or owtre til it stonde
precise vp on the closure of the
signes in the lymbe of thin
equatorie so wole thanne the
centre of thin epicicle precise
stonde vp on centre aryn

the sixte
cercle is the
eccentrik of the sonne
& the .4. cercle that is red
is the closure of the signes
& the section of the cros
is centre aryn / & that other
centre is the centre of the
eccentrik of the sonne
& the lyne devyded newe
is the midnyght lyne
wel wel it is figured
westowrd & the cercle
abowte centre aryn is
the centre deferent of the
mone / the litel cercle
is the deferent of mercurie //
/ the smale lynes ben lynes
of auges

the prikkes in the lynes
ben the centres equantes &
deferentz / & alle thise
centres saue the equant
of mars ben by twixe
centre aryn & the centre
deferent of the mone
the owterest space is mynut
& the nexte space is degrees
& the thridde space is nombres
of degrees & the ferthe space is
for names of signes but
natheles the narwere cercle
of the signes is cleped the
closure of the signes & it is
compassed with red

F

the epicicle    no*ta* file nat to ney the rede cercle þ$^{t}$ is closere of
the signes / list the *com*mune centre defferen*t* breke
lat stonde a litel lippe as shewith in direct of
the hed of canc*er*
no*ta* I *con*seile the ne write no names of signes (i*d est* in epiciclo)
til þ$^{t}$ thow hast proued (^þ$^{t}$) thi comune
centre defferent is treweli & justli set
in direct of the closere of the signes
of thin eq*ua*torie

¶ this
epicicle
is deuyded
& *com*pased in
alle thinges
lik to the lymbe of the eq*ua*torie
¶ but it hath non
eccentrik of the
sonne / the prikke
þ$^{t}$ stant in the
closere of the
signes i*n* direct
of the ende of
ge*min*is is the
*com*mune centre
defferen*t* ——
¶ but natheless thus lith thin
instrume*n*t whan thow makest
equaciou*n* of thy mone

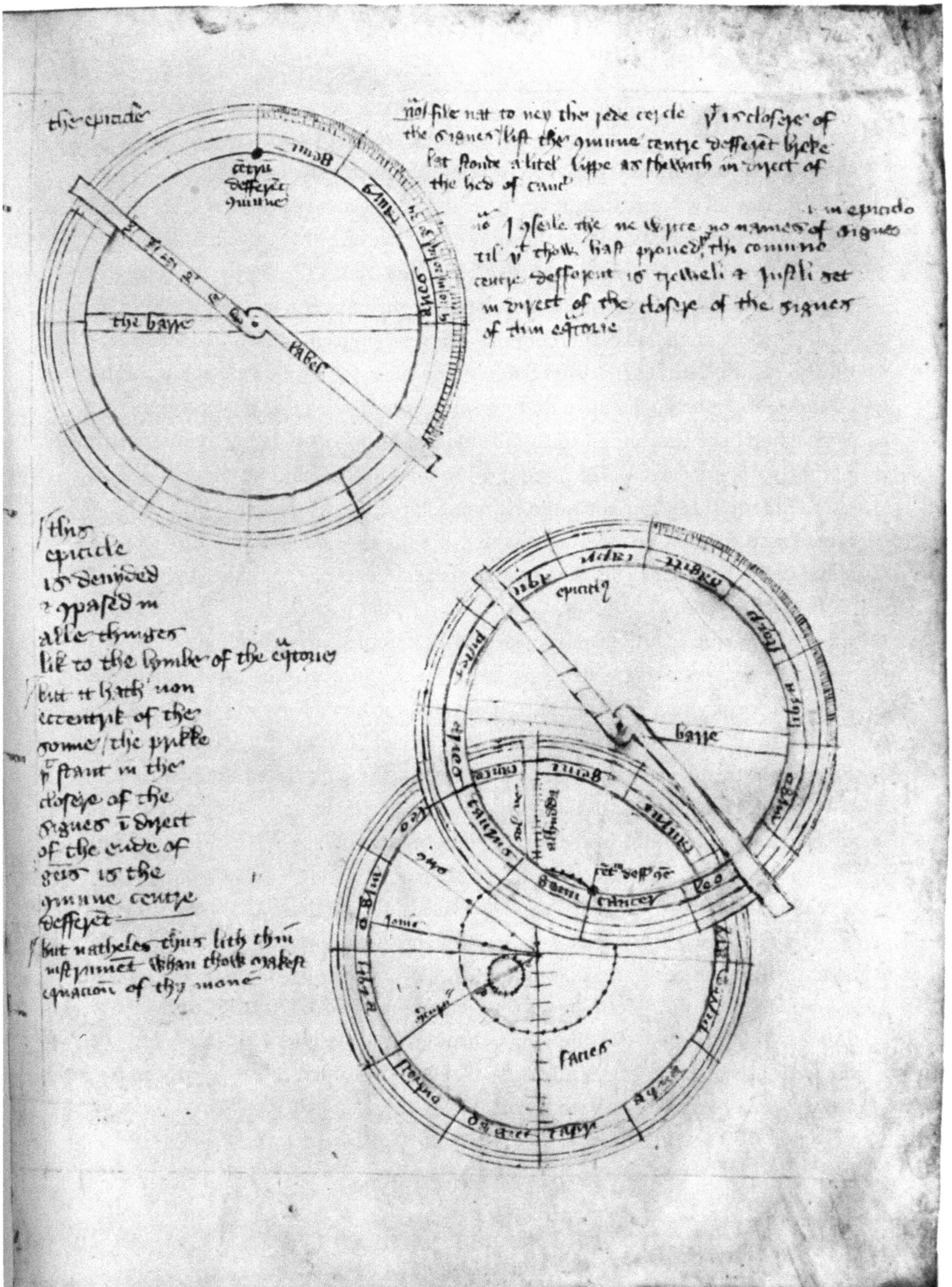


FOLIO 74 RECTO

F

G

p*ro* argume*n*tis triu*m* sup*er*ior*um* minue eor*um* med' motꝫ de med' motꝫ sol*is* / & remane*t* argu*mentum* | [sat*ur* .io. & mart*is* & ven*er*]$^{e}$ | {s}ek medi*us* mot*us* of sat*urnus* Juppit*er* mars &̅ ven*us* / & hir mene argu|ment*is* in thy tables / & writ hem in thy sklat / put thanne a | blak thred in centre aryn (terre) & a whit thred in centre equant of | any pl*anet*e þ$^{t}$ the list haue of equacion. // and put the comune | centre defferent of thyn Epicicle / vp on the centre different | in thy plate / of thilke pl*anet*e þ$^{t}$ thow desirest to haue equaciou*n* | I sey þ$^{t}$ w$^{t}$ a nedle thow shalt stike the comune centre deffe|rent of thin Epicicle / vp on the centre defferent þ$^{t}$ is p*er*ced | on thy plate for swich a pl*anet*e a the list to haue of equacou*n* | ¶ loke thanne (p*ro* success*ione* signo*rum*) fro the hed of aries wher the mene mot*us* of thy pl*anet*e | endith / in the grete lymbe of thy plate / & ley ther thy blake thred. | ¶ ley thanne thy white thred equedistant by the blake thred in | the same lymbe. &̅ proeue by a *com*pas þ$^{t}$ thy thredes lyen equedista*nt* // | vnder whiche white thred ley the pool of thyn Epicicle ¶ &̅ stond=|inge thyn Epicicle stille in this maner / I seye stondinge | the pool of thin epicicle vndir thy white thred stille / & the | *com*mune centre different fix w$^{t}$ thy nedle to the foreseide ce*n*tre | defferent of the pl*anet*e desired / tak than thy blake thred | & ley it so þ$^{t}$ it kerue the centre of the Epicicle / & streche | forth vp vn to vpperest part of the same epicicle / and | than shal this blake thred shewe bothe the verrey mot*us* (locu*m*) | of the epicicle in the grete lymbe / & ek the verrey aux of | the pl*anet*e in the epicicle / & thanne the ark by twixe medios | mot*us* of the pl*anet*e / & the verrey mot*us* of the epicicle / is | cleped the equacio*n* of his centre in the lymbe (zodiac*us*) // to whom | is lik the equacio*n* of his argume*n*t in his epicicle. // | þ$^{t}$ is to sein the ark by twixe his mene aux & his verrey | aux. ¶ for sothly the mene aux is shewed in the Epi|cle by the white thred / vnder which thow puttest the | pol of the epicicle // & the verrey aux is shewed in the | epicicle by the blake thred ¶ &̅ stondinge stille thin Epi|cicle in this same disposiciou*n* / ley the ende of thy label | þ$^{t}$ is grauen // fro the white thred as many signes | degres & mi$^{a}$. as shewith the mene argume*n*t in thy | [grene]$^{c}$ tables for þ$^{t}$ day of thy pl*anet*e desired ¶ & rekne | this mene argument fro the white thred after suc|cessiou*n* of signes of eu*er*y pl*anet*e saue only of the mone | [sa 1 31 19 22 44]$^{e}$

line 1 Space left for initial; guide-letter *s* in inner margin. **line 9**, second a *sic for* as.
**line** 9 equacou*n* *sic for* equaciou*n*. **line 28/29** Epi|cle *sic for* Epi|cicle.

¶ argumentis thus supposit omne tempus [illegible] not de med mot / set a semilatus argum

[illegible]

¶ Eke mediu(s) mot(us) of saturnus iupiter mars & venus is by mene argu
mentz in thi tables & writ hem in thi sclat / put thanne a
blak thred in centre aryn & a whit thred in centre equant of
any planete that the list haue of equacion and put the commune
centre deferent of thin Epicicle vp on the centre different
in thi plate of thilke planete that thow desirest to haue equacion
/ I sey that with a nedle thow shalt stike the commune centre defe
rent of thin Epicicle vp on the centre defferent that is p(ro)pred
on thi plate for which a planete that the list to haue of equacion
loke thanne fro the hed of aries wher the mene mot of thi planete
endeth in the grete lymbe of thi plate & ley ther thi blake thred
/ ley thanne thi white thred equedistant by the blake thred in
the same lymbe & procede by a manere that thi thredes ben equedistant
vnder which white thred ley the pool of thin Epicicle / & stond
inge thin Epicicle stille in this manere I seye stondinge
the pool of thin epicicle vnder thi white thred stille & the
comune centre different fix with thi nedle to the forseide centre
defferent of the planete desired, tak than thi blake thred
& ley it so that it kerue the centre of the Epicicle & streche
forth vp on to uttereste part of the same epicicle And
than shal this blake thred shewe bothe the verrey mot
of the epicicle in the grete lymbe & ek the verrey aux of
the planete in the epicicle / & thanne the ark by twixe medius
mot of the planete & the verrey mot of the epicicle is
cleped the equacion of his centre in the lymbe / to whom
is lik the equacion of his argument in his epicicle //
is to seyn the ark by twixe his mene aux & his verrey
aux / for sothly the mene aux is shewed in the Epi
cicle by the white thred vnder which thow puttest the
pol of the epicicle & the verrey aux is shewed in the
epicicle by the blake thred / & stondinge stille thin Epi
cicle in this same disposicion, ley the ende of thy label
that is grauen fro the white thred as many signes
degres & m(inute)s as shewith the mene argument in thy
grene tables for that day of thy planete desired / & rekne
this mene argument fro the white thred after suc
cession of signes of euery planete saue only of the mone /

FOLIO 75 RECTO

H

1*, 2 ¶ &̄ ligginge the marked ende of thy label vpon [the] | the ende of this mene
3 argum*en*t in the epicicle / ley | thy blake thred vp on the marke of thy pl*anet*e
4, 5 þ$^t$ is | grauen in thi label. &̄ wher as the same blake | thred keruyth the lymbe
6 of thy plate tak ther | the verrey place (locu*m*) of the pl*anet*e in the .9. spere |
7, 8 ¶ and the ark by twixe the verrey place (locu*m*) of the | pl*anet*e (in limbo)
9 & the verrey place of the epicicle *con*sidered in | the lymbe is cleped eq*ua*ciou*n*
10* of his argum*en*t this man*er* | of equaciou*n* is for saturn*us* Juppit*er* mars &
11, 12* ven*us* / but | in the remenau*n*t of pl*anet*es in some thinges it varieth | sol ¶ the
13 mene mot*us* of the sonne ben rekned fro the hed of aries | after successiou*n* of
14 signes ¶ the sonne hath non Epicicle ne non | Equant &̄ therfor the pol of the
15 epicicle mot ben in stide of the | body of the sonne in the .9. spere ¶ the white
16 thred þ$^t$ thow puttest | in his centre deffere*n*t in the plate (lamina) // mot ben
17 in stide of the white | thred þ$^t$ othre pl*anet*es han in hir centres equant*is* ¶ the
18 blake thred | þ$^t$ eu*er*mo stant in centre aryn / mot be leid at the ende of his |
19, 20 mene mot*us* ¶ tak thanne his white thred & lei it equedistant | in the lymbe
21 by the blake thred ¶ whiche blake thred shewith the | mene mot*us* of the
22 sonne ¶ fixe thanne w$^t$ thy nedle the *com*mune | centre defferent of thyn
23 Epicicle to centre aryn. &̄ remew nat thy (^ nedle) | ¶ &̄ vnder this white thred ley
24 softely the pol of the Epicicle | & wher as the white thred keruyth the grete lymbe
25, 26* tak ther | [ther] the verrey place of the sonne in the .9. spere / the ark | of the
27 lymbe by twixe his aux þ$^t$ is now in canc*er*. & the blake | thred is the argume*n*t
28 of the sonne ¶ the ark by twixe the | blake thred & the white in the lymbe is
29* the equacio*n* of the | sonne / which ark nis but litel. / the mene mot*us* of the
30, 31 sonne | is the ark in the lymbe by twixe the hed of aries &̄ the blake | thred in
32 the same lymbe ¶ the verrey mot*us* of the sonne is | the ark of the lymbe by
33 twixe the hed of aries & the blake | thred whan it (i*d est* filu*m*) is remewed fro
34 the mene mot*us* & crossith the | white thred in the in the pol of the Epicicle ¶ the
35 same | verrey mot*us* was shewed erst by the white thred of (^ the) deffere*n*t |
36, 37 whan it lay equedistant by the blake thred i*n* the limbe ~~~| &̄ no*ta* þ$^t$ the
38 markes i*n* thy label descriuen the Epicicles of pl*anet*es as the | label turneth

**line 12** Sol *in outer margin.* **line 34** in the in the *sic for* in the.

//& liggynge the marked ende of thy label upon the
the ende of this mene argument in the epicicle / ley
thy blake thred up on the marke of thy planete that is
writen in thi label. & wher as the same blake
thred kervyth the lymbe of thy plate tak ther
the verrey place of the planete in the 9 spere
//And the ark by twixe the verrey place of the
planete & the verrey place of the epicicle considered in
the lymbe is cleped equacion of his argument this manere
of equacioun is for saturnus Jupiter mars & venus / but
in the remenant of planetes in some thinges it varieth

//the mene mot of the sonne ben rekned fro the hed of aries
after successioun of signes / the sonne hath non Epicicle ne non
Equant & ther for the pol of the epicicle mot ben in stede of the
body of the sonne in the 9 spere / the white thred that thow puttest
in his centre deferent in the plate / mot ben in stede of the white
thred that othre planetes han in hir centres equantz / the blake thred
that comunly stant in centre aryn / mot be leid at the ende of his
mene mot / tak thanne his white thred & lei it equedistant
in the lymbe by the blake thred / which blake thred sheweth the
mene mot of the sonne / fixe thanne with thy nedle the comune
centre deferent of thyn Epicicle to centre aryn & remeve nat thy
& under this white thred ley softely the pol of the Epicicle
& wher as the white thred kervyth the grete lymbe tak ther
ther the verrey place of the sonne in the 9 spere / the ark
of the lymbe by twixe his aux that is now in cancer & the blake
thred is the argument of the sonne / the ark by twixe the
blake thred & the white in the lymbe is the equacion of the
sonne / which ark nis but litel. the mene mot of the sonne
is the ark in the lymbe by twixe the hed of aries & the blake
thred in the same lymbe / the verrey mot of the sonne is
the ark of the lymbe by twixe the hed of aries & the blake
thred whan it is remewed fro the mene mot & crossith the
white thred in the in the pol of the Epicicle // the same
verrey mot was shewed erst by the white thred of the deferent
whan it lay equedistant by the blake thred in the lymbe
& nota that the markes in thy label bytwixen the Epicicles of planetes as the
label turneth

1*

g* [m*er*curi*us*]°                    this canon is fals

1,2 {R}ekne after succession of signes fro the hed of aries in the | lymbe / the mene

3 mot*us* of m*er*curi*us* // & *con*sidere ek how mochel in | the same lymbe is by twixe

4* the hed of aries. & the lyne of his | aux / þt yit is (‸ in) the lattere ende of

5 libra // & rekne alwey after | successiou*n* of signes // wt draw thanne the

6,7 quantite in | the lymbe. by twix the hed of aries & the forseid aux / owt | of

8 his mene mot*us* // & *con*sidere how moche is the remnau*n*t | of his mene mot*us*

9 whan this aux is thus wt drawe owt | of al the hoole mene mot // & so mochel

10 rekne after successi*on* | of signes in his litel cercle / fro the lyne of his aux þt

11*, 12 keruyth | the same litel cercle // I seye rekne after successiou*n* of | signes. from

13 l*ette*re .D. þt is grauen in his lytel cercle / & p*ro*cede | in the same litel cercle

14 to ward l*ette*re .E. opposit to .D. / | I sey rekne thilke remnau*n*t (‸ of the mene

15 mot*us*) þt dwelde whan the quantite | of his aux was wt draw owt of his hole

16*, 17 mene mot*us* | as I haue seid by forn // & wher as —— thilk remnau*n*t | forseid

18 endith in the litel cercle / tak ther the verrey centre | deffere*n*t of m*er*curie. as

19 it happith diu*er*sely som tyme in on | hole & som tyme in an other // for l*ette*re

20 .D. ne seruyth of nothyng | ellis but for to shewe the wher thow shalt by

21 gynne thy | reknyng in thy litel cercle // ne l*ette*re (‸ E) ne seruyth nat but

22*, 23 for | to shewe the which wey þt thow shalt p*ro*cede fro l*ette*re .D. / now | hastow

24 founde thy deffere*n*t & thin equant // in which eq*ua*nt put | a whit thred // & stike

25 wt anedle the          comune centre deffere*n*t | vp on his centre defferent in the

26 plate (lamina). // & wt thin Epicle wirk | & wt thy thredes as thow workest

27 wt saturn*us* Juppit*er* mars | & ven*us* // no*ta* þt yif the aux of m*er*curie be fro the

28, 29 hed of aries | more than his mene mot*us* fro the same hed. / than shaltow | adde

30 .12 signes to his mene mot*us* / than maistow wtdraw | his aux owt of his mene

31 mot*us* // & no*ta* gen*er*aly þt thy nedle | ne be nat remewed whan it is stikyd

32, 33 thorw the *com*mune | centre deffere*n*t in to any centre different on thy plate | til

34 thin equaci*on* of the pl*anet*e be endid // for yif thy *com*mune | centre different

35 stirte fro the centre ([deffer]ᵉ) on thy plate | al thin equaci*on* of thy pl*anet*e

36* desired is lorn // ☞ hic no*ta* þt the ce*ntre* | defferent of mercurie hath but

37 24. holes as i*n* myn i*n*strme*nt* | whcr for .I. rekne but .2. holes for a signe as

38 in the | gretter cerkle of mercur*ie* fro the lyne of his aux

**line 1** A space has been left for an initial R.
**line 25** Epicle *sic for* Epicicle.
**line 35** The hand is in the outer margin.
**line 36** i*n*strme*nt* *sic for* i*n*strume*n*t.

this canon is fals

rekne after succession of signes fro the hed of aries in the
lymbe the mene mot of mercurie & considere how mochel in
the same lymbe is by twixe the hed of aries & the lyne of his
aux that yit is in the latter ende of libra & rekne alwey after
succession of signes wt which thanne the quantite in
the lymbe by twix the hed of aries & the forseid aux owt
of his mene mot & considere how moche is the remenant
of his mene mot when this aux is thus wt drawe owt
of al the hool mene mot & so mochel rekne after successio
of signes in his litel cercle fro the lyne of his aux that kerveth
the same litel cercle I sey rekne after succession of
signes from the D. that is graven in his lytel cercle & procede
in the same litel cercle toward the E. opposit to D. I sey rekne thilke remenant that I spak of when the quantite of the mene mot
of his aux was wt drawe owt of his hole mene mot
as I have seid by forn & wher as thilke remenant
forseid endeth in the litel cercle tak ther the verrey centre
defferent of mercurie as it happith diversly som tyme in on
hole & som tyme in an other for the D. ne serveth of nothing
ellis but for to shewe the whether thow shalt bygynne thy
rekning in thy litel cercle ne the E. ne serveth nat but for
to shewe the which wey that thow shalt procede fro the D. whan thow
hast fownde thy defferent & thin equant in which centre putte
a whit thred & stike wt a nedle the comune centre defferent
up on his centre defferent in the plate & wt thin cercle
wt thy thredes as thow workest wt saturnus jupiter mars
& venus & note that yif the aux of mercurie be fro the hed of aries
more than his mene mot fro the same hed than shalt thow
adde 12 signes to his mene mot than maistow wt drawe
his aux owt of his mene mot & note generaly that thy nedle
ne be nat remewed whan it is ficched thorgh the comune
centre defferent in to any centre different on thy plate
til thin equacion of the planete be endid for yif thy comune
centre different sterte fro the centre on thy plate
al thin equacion of thy planete desired is lorn & note that the centre
defferent of mercurie hath but 2 holes as I sayde erst & never
wher for I rekne but 2 holes for a signe as in the
gretter cercle of mercurie fro the lyne of his aux

J

h

luna

1, 2 - luna {R}ekne after succession of signes fro the hed of aries in the | lymbe
3 the mene mot*us* of the mone ¶ &̄ rekne in the same | man*ere* the mene mot*us*
4 of the sonne (a capite ariet*is*) as fer as it strechcheth // | w$^{t}$ draw thanne the
5* mene mot*us* of the sonne owt of the mene | mot*us* of the mone ¶ [&̄ as moche
6 as the mene mot of the | mone is more than the mene mot of the sonne]$^{c}$
7 (‸ &̄) *con*sidere þ$^{t}$ | difference / ¶ and the quantite of that difference þ$^{t}$ I clepe |
8*, 9 the remenau*n*t / rekne it fro the (‸ ende of the) mene mot*us* of the sonne in | the
10 lymbe bakward agayn successiou*n* of signes / & wheras | endith this remenau*n*t /
11 mak a mark in the lymbe ¶ draw | thanne thy blake thred to this forseide mark //
12, 13 & wher as | thy blake thred keruyth the cercle defferent of the mone | in that
14* same hole is the centre defferent of the mone | as it happith. ¶ &̄ in the nadyr
15 of this hole / is the ce*ntre* | equant / put thanne in this centre eq*ua*nt a whit
16, 17 thred | ¶ now hastow thy two centres / stike thanne thy *com*mune | centre
18 defferent vp on the centre differe*n*t of the mone / | w$^{t}$ thy nedle ¶ yit rekne
19 agayn the mene mot*us* of the | mone fro the hed of aries aft*er* successiou*n* of
20, 21 signes | & ley ther thy blake thred ¶ & ley thy white thred | equedista*n*t by the
22 blake thred in the lymbe // moeue | thanne softely the pool of thyn Epicicle
23 vnder thy | blake thred (‸ tak thanne thy white thred & ley it ouer the pol of
24 the epicicle) & wheras thy white thred keruyth | the cercle of the Epicicle
25, 26 tak ther the mene aux in | thyn Epicicle & fro this white thred rekne in | thyn
27 Epicicle bakward a gayns successiou*n* of | signes thy mene argume*n*t / I seye
28, 29 rekne it in the | degres of thin Epicicle / &̄ where as endith thy | reknynge in
30 the Epicicle. ley ther the marked | ende of thy label. / &̄ ley thy blake thred
31, 32 vp on the | mark of the mone in thy label. & wher as this same | blake thred
33 keruyth the lymbe tak (‸ ther) the verrey place of | the mone in the .9.
34 spere / no*ta* þ$^{t}$ the pool of the Epicicle | ne shal nat ben leyd vnder the blake
35 thred of non other | .pl*anete*. saue only of the mone

**line 1** luna *in outer margin.*

**line 1** A space has been left for an initial R.

rekne after succession of signes fro the hed of aries in the
lymbe the mene mot of the mone / & rekne in the same
manere the mene mot of the sonne as fer as it streccheth //
wt draw thanne the mene mot of the sonne out of the mene
mot of the mone / & ~~as mochel as the mene mot of the~~
mone ~~is more than the mene mot of the sonne~~ / considere pt
difference / and tho quantite of that difference pt I clepe
the remenant rekne it fro the ende of the mene mot of the sonne in
the lymbe bakward agayn succession of signes / & wher as
endith this remenant / mak a mark in the lymbe / Draw
thanne thy blake thred to this forseide mark // & wher as
thy blake thred kervyth the cercle deferent of the mone
in that same hole is the centre deferent of the mone
as it happith / & in the nadyr of this hole is the centr
equant / put thanne in this centre equant a whit thred
/ now hastow thy two centres / stik thanne thy comune
centre deferent up on the centre deferent of the mone
wt thy nedle / / yit rekne agayn the mene mot of the
mone fro the hed of aries after succession of signes
& ley ther thy blake thred / & ley thy white thred
equedistant by the blake thred in the lymbe // meve
thanne softely the pool of thyn Epicicle under thy
tak thanne thy white thred & ley it ouer the pol of the epicicle
blake thred / & wher as thy white thred kervyth
the cercle of the Equantcle tak ther the mene aux in
thyn Epicicle & fro this white thred rekne in
thyn Epicicle bakward agayns succession of
signes thy mene argument / & so rekne it in the
degrees of thin Epicicle / & where as endith thy
rekenynge in the Epicicle / ley ther the merked
ende of thy label / & ley thy blake thred up on the
mark of the mone in thy label / & wher as this same
blake thred kervyth the lymbe tak ther the verrey place of
the mone in the .9. spere / nota & the pool of the Epicicle
ne shal nat ben leyd under the blake thred of non other
plan(ete) save only of the mone

K

i* þis canon is fals

1*, 2 and no*ta* þ$^{t}$ yif the mene mot*us* of the sonne is more than | the mene mot of

3 the mone than shaltow adde .12. signes | to the mene mot of the mone & thanne

4, 5* maistow w$^{t}$drawe | the mene mot of the sonne owt of the mene mot*us* of | the

6* mone. &̄ shortly for to speken of this theorike I sey | þ$^{t}$ the centre of hir (lu*ne*)

7 epicicle (‸ in voluella) moeuyth eq*ua*ly aboute the ce*ntre* | of the zodiac þ$^{t}$

8* is to sein aboute the pol of the epicicle þ$^{t}$ | is thy riet / & thy blake thred whan

9, 10 it first leid thorw | the pol of thyn Epicicle it shewith the verrey aux of | the

11 pl*anet*e (in epiciclo) riht as the white thred shewith the mene aux i*n* | the same

12 epicicle. Item whan thow (‸ hast) rekned the argume*nt* | of apl*anet*e in thin

13 epicicle. thanne is the body of the pl*anet*e | in thin Epicicle at the ende of thyn

14 argume*nt* &̄ whan (‸ thy) blak thred | is leid thorw the marke of a pl*anet*e in

15, 16 thi label. in man*er* | forseid than shewith thy blake thred the verre place of | the

17 pl*anet*e at regard of the .9. spere as shewith in thy lymbe | &̄ the ark by twixe

18 the verrey mot*us* &̄ the mene mot*us* | of the mone is the equacio*n* of his argume*nt*

19, 20 in the lymbe / | &̄ the ark by twixe his mene aux & his verrey aux is | the

equacio*n* of his argume*nt* in the Epicicle

21 ¶ to knowe the latitude of the mone by thyn

22 instrume*nt* loke in thyn almenak the v*er*rey

23 mot*us* of the mone &̄ the v*er*rey mot*us* of caput .dra.

24 .lune. at same tyme / &̄ yif so be þ$^{t}$ thy

25 verre mot of thy mone be lasse than .6.

26 signes fro caput draco*n*is // w$^{t}$draw the

27 verrey mot*us* of caput owt of the verey

28 mot*us* of the mone / &̄ writ þ$^{t}$ differe*n*ce (i*d est* v*erum* arg*umentum* latitu*dinis* lune)

29 for þ$^{t}$ is hir (lu*ne*) | verrey argume*nt* (i*d est* latitudi*n*is) &̄ so many signes

30 degres & mi$^{a}$ as | thow hast i*n* the (‸ verrey) argume*nt* (‸ of hir latitude) rekne

31, 32 hem fro the | hed of aries after successiou*n* of signes i*n* thy lymbe / | &̄ wher

33* as endith thy reknyng ley þ$^{t}$ on end of thy thred | & þe middel of thy thred

shal kerue the m*e*ridio*nal* lyne / &̄ strechche

**line 28** MS. has *latitu* with no abbreviation for ending. This gloss is written in the right-hand margin.

 

And no yt yif the mene mot of the sonne is more than
the mene mot of the mone than shaltow adde 12 signes
to the mene mot of the mone & thanne maistow wt-drawe
the mene mot of the sonne owt of the mene mot of
the mone. & shortly for to speken of this theorike I sey
yt the centre of his epicicle in his ecuantie moeuyth egaly abowte the centre
of the zodiac yt is to seyn abowte the pol of the epicicle
is thy net & thy blake thred whan it first leid thorw
the pol of thyn Epicicle it shewith the verrey aux of
the planete in epicicle as the white thred shewith the mene aux in
the same epicicle. Item whan thow hast rekned the argument
of a planete in thin epicicle. thanne is the body of the planete
in thin Epicicle at the ende of thin argument & whan thy blak thred
is leid thorw the marke of a planete in thi label in maner
forseid than shewith thy blake thred the verre place of
the planete at regard of the 9 spere as shewith in thy lymbe
& the ark by twixe the verrey mot & the mene mot
of the mone in the equacion of his argument in the lymbe
& the ark by twixe hir mene aux & hir verrey aux is
the equacion of hir argument in the Epicicle

// to knowe the latitude of
instrument loke in thyn
mot of the mone & the verey
lunes at same tyme & yif
verre mot of thy mone be
signes fro caput draconis //
verrey mot of caput owt

the mone by thyn
almenak the verey
mot of caput dra-
cis be yt thy
lasse than 6
wt-drawe the
of the verey

mot of the mone & write yt difference for yt is hir folio
verrey argument of latitude & so many signes degrees & minutes as
thow hast in the verrey argument of hir latitude rekne hem fro the
hed of aries after successiou of signes in thy lymbe
& wher as endith thy rekenyng ley yt on end of thy thred
& ye middel of thy thred shal kerve the midd lyne & precisly

in margin: thy /; in margin: ... latitu line

L

1,2 so forth ou*er*thwart al the dyametre of thy plate vnto | the lymbe / as thus
3 I suppose þ$^{t}$ on ende of thy thred laye | after succession of signes .10. g$^{a}$. fro
4 the hed of aries in | the lymbe / þ$^{t}$ other ende of thy thred shold lye .20 g$^{a}$.
5,6 of | virgo / in the lymbe / *con*sidere thanne how many .g$^{a}$. & mi$^{a}$ | þ$^{t}$ the middel
7 of thy thred lith fro centre aryn. wher as | eu*er*mo by gynnith this reknyng / I seye
8 *con*sidere in the | secci*on*s of the m*er*idio*nal* lyne how many g$^{a}$ & mi$^{a}$ lith the
9,10 middel | of thy thred fro centre aryn / & tak ther the nombre of the | latitude
11 .7. trio*nal* of thy mone fro the Ecliptik / which latitu(de) | ne passith neu*er*
12 .5. g$^{a}$. ⁋ & yif the verrey mot*us* of the mone | be more than .6. signes fro the
13 verrey mot of caput / than | shaltow w$^{t}$draw the verrey mot*us* of cauda owt
14,15 of the | verrey mot*us* of the mone. and by gynne thy reknynge | at the hed of
16 libra / & p*ro*cede bakward agayns successiou*n* | of signes / as thus þ$^{t}$ yif þ$^{t}$ on
17 ende of thy thred laye | agayn successiou*n* of signes .10. g$^{a}$ fro the hed of libra |
18,19 than sholde þ$^{t}$ other ende lye in the .10. g$^{a}$ fro the hed | of aries after successiou*n*
of signes / *con*sidere thanne in

| | | |
|---|---|---|
| 20 | the m*er*idio*nal* lyne | the quantite m*er*idio*nal* of the lati |
| 21 | tude of thy mone | fro the Ecliptik / as I haue |
| 22 | told by forn þ$^{t}$ | the quantite of g$^{a}$ & mi$^{a}$ þ$^{t}$ the |
| 23 | middel of thy | thred in the m*er*idio*nal* lyne lith |
| 24 | fro centre aryn | the same q*ua*ntite of degres & |
| 25 | mi$^{a}$ is the latitude | of the mone fro the eclyptik |
| 26 | be it north be it | sowth ⁋ & no*ta* þ$^{t}$ gen*er*aly eu*er*mo |
| 27 | bothe endes of | of thy thred shollen lyen |
| 28 | equedistant fro | thilke diametre þ$^{t}$ keruyth the |

29,30 heuedes of aries & libra ⁋ yit quykly vnderstond this | canon. I sey whan the
31 forseide verrey argume*n*t of the | mone is p*re*cisly .90. degres fro the hed of
32 aries in the | lymbe after succession of signes / tak ther the grettest

**line 27** of of *sic for* of.

so forth on thilke arc al the dyametre of thy plate vnto
the lymbe/ as thus I suppose that on ende of thy thred laye
after successioun of signes 10 gr fro the hed of aries in
the lymbe/ that other ende of thy thred sholde lye 20 gr of
virgo/ in the lymbe/ considere thanne how many gr & min
that the middel of thy thred lith fro centre aryn whe[r] as
elmuo by azimuth thin rekenyng/ I seye considere in the
seccioun of the midnight lyne how many gr & min lith the middel
of thy thred fro centre aryn/ & tak ther the nombre of the
latitude arid of thy mone fro the ecliptik which latitude
ne passith neuere 6 gr / & yif the verrey mot of the mone
be more than 6 signes fro the verrey mot of capitis/ than
shaltow wit[h] rekne the verrey mot of cauda draconis of the
verrey mot of the mone and by gynne thy rekenynge
at the hed of libra & procede bakward agayns successioun
of signes as thus that yif that on ende of thy thred lay
agayn successioun of signes 10 gr fro the hed of libra
than sholde that other ende lye in the 10 gr fro the hed
of aries after successioun of signes/ considere thanne in
the midnight lyne
the quantite midnight of the latitude of thy mone
fro the ecliptik as I haue
told by forn that
the quantite of gr & min that the
middel of thy
thred in the midnight lyne lith
fro centre aryn
the same quantite of degrees
min is the latitude
of the mone fro the ecliptik
be it north be it
south & nota that generaly elmuo
bothe endes of
of thy thred shollen lyen
equedistant fro
thilke diametre that kervith the
heuedes of aries & libra
that yit quykly vnderstonde this
canoun I sey whan the forseide verrey argument of the
mone is precisly 90 degrees fro the hed of aries in the
lymbe after successioun of signes tak ther the grettest

M

1 latitude of the mone 7*trional* (*id est* ab ecliptica) / & yif so be þ$^{t}$ hir verrey
2 *argument | passe (‸ anything) .90. degres fro the hed of aries / styrt ouer the*
3 m*er*idio*na*l lyne | in to the firste of cancer & ley ther þ$^{t}$ on ende of thy thred
4, 5 & þ$^{t}$ | other ende in to the laste of gem*in*is / & so forth day by day | shaltow
6 descende in the m*er*idio*nal* lyne after þ$^{t}$ the reknynge | of thy verrey argume*n*t
7* requerith. / til thow come agayn to | centre aryn. for than hastow mad equacio*n*
8 of latitudes for .6. | signes as I first seide / & eu*er*e mo lith thy thred equedista*n*t |
9, 10 fro the diametre þ$^{t}$ keruyth the heuedes of aries & libra / & | eu*er* mo as many
11 degres & mi$^{a}$ as the midel of thy thred lith in | the m*er*idio*nal* lyne fro centre
12 aryn. so many g$^{a}$ & mi$^{a}$ is the lati|tude of the mone fro the Ecliptik / & whan
13, 14 thy verrey argu|me*n*t passith .6. signes wyrk w$^{t}$ cauda as I tawhte the | &
15 ascende vpward in the m*er*id*ional* lyne day by day to the laste | of gem*in*is in
16 the lymbe / & fro thennes discende agayn as | I haue seid by forn ¶ & no*ta*
17 þ$^{t}$ whan the mone is direct | w$^{t}$ caput or cauda. she hath no latitude & whan
18 she passith | caput til she be 3 signes (‸ in) dista*n*ce (p*ro* successio*ne*ȝ sig*norum*)
19 fro caput she is 7trio*nal* | ascendinge / & in hir gettest latitude 7trio*nal* / & fro
20, 21 the | ende of thilke 3 signes she is 7trio*nal* descending til she | come to the
22* opposit of caput þ$^{t}$ is to seyn cauda draco*n*is | & fro cauda til she come mid
23 wey (in medio) by twix cap*ud* & cauda | & fro thennes is she m*er*idio*nal*
24* assending til she come agayn | at cap*ud* ¶ (1391 .17. dece*m*bris) Ensample
25 my mone was .12. g$^{a}$ .21. mi$^{a}$ of virgo / & | caput was 4 g$^{a}$ 46 mi$^{a}$ of aries tho
26 drow I the verrey mot*us* | of cap*ut* þ$^{t}$ is to seyn .0. in signes .4. g$^{a}$ 46 mi$^{a}$ owt
27, 28 of the | verrey moeuyng / of the mone þ$^{t}$ is to sein owt of 5 signes | .12 g$^{a}$ 21 mi$^{a}$ /
29 tho fond I þ$^{t}$ the verrey argume*n*t of the mone | dwelde .5. sig$^{a}$ 7 g$^{a}$ 35 mi$^{a}$ tho
30 rekned I after successiou*n* | of signes fro the hed of aries in the lymbe (‸ the
31 same) 5 signes .7 g$^{a}$ | 35 m$^{a}$ / & ther leide I þ$^{t}$ on ende of my thred & þ$^{t}$ other
32 ende | lay in .22 g$^{a}$ 35. mi$^{a}$ of aries / tho karf (‸ the) midel of my thred the |
33, 34 m*er*idio*nal* lyne .1. degre & 54 mi$^{a}$ fro centre aryn by which I | knew þ$^{t}$ the
35 latitude of my mone was .1. degre & 54 mi$^{a}$ 7*trional* | descending fro the
Ecliptik

**line 18** successio*ne*ȝ *for* successio*ne*?
**line 19** gettest *sic for* grettest. **line 24** The date is in the left-hand margin.

latitudes of the mone north / & yif so be that hir verrey argument
passe 90 degres fro the hed of aries / styrt ouer the midel lyne
in to the fyfte of cancer & ley ther that on ende of thy thred & that
other ende in to the laste of gemini / & so forth day by day
shaltow descende in the midel lyne after that the rekenynge
of thy verrey argument requereth / til thow come agayn to
centre aryn · for than hastow mad equatio of latitudes for 6
signes as I first seide / & eue mo lith thy thred equedistant
fro the diametre that kerueth the heuedes of aries & libra / &
al so as many degres & minutes as the midel of thy thred lith in
the midel lyne fro centre aryn · so many degres & minutes is the lati
tude of the mone fro the ecliptik / & whan thy verrey argu
ment passith 6 signes wryt it with cauda as I tawhte the
& ascende upward in the midel lyne day by day to the laste
of geminis in the lymbe & fro thennes descende agayn as
I haue seid byforn / & note that whan the mone is direct
with caput or cauda · she hath no latitude & whan she passith
caput til she be 3 signes in distance fro caput she is north
ascendinge / & in hir grettest latitude north / & fro the
ende of thilke 3 signes she is north descending til she
come to the opposit of caput that is to seyn cauda draconis
& fro cauda til she come amiddes bytwixe caput & cauda
& fro thennes is she midel ascending til she come agayn
at caput / Ensample my mone was 12 grades 21 minutes of virgo / &
caput was 4 grades 26 minutes of aries tho drow I the verrey mot I
of caput that is to seyn 0 signes 4 grades 26 minutes owt of the
verrey moeuyng of the mone that is to seyn owt of 5 signes
12 grades 21 minutes / tho fond I that the verrey argument of the mone
twelue 5 signes 7 grades 35 minutes tho rekned I after succession
of signes fro the hed of aries in the lymbe the same 5 signes 7 grades
35 minutes / & ther leide I that on ende of my thred & that other ende
lay in 22 grades 35 minutes of aries / tho kast the midel of my thred the
midel lyne · 1 · degre & 49 minutes fro centre aryn / by which I
knew that the latitude of my mone was · 1 · degre & 49 minutes north
descending fro the ecliptik

N

1* (1391 .19. februar*ii*) ¶ another ensample // I fond (*scilicet* in almenak) my
2 mone in .8. g$^{a}$d .13 m$^{a}$ of virgo | & caput drac*onis* in 20 g$^{a}$ & 42. mi$^{a}$ of aries
3 (^ .tho.) drow I the verrey | mot*us* of —— caput fro the verrey mot*us* of the
4 mone in this | manere. / I say wel þ$^{t}$ I myht nat drawe .20 degres owt of |
5,6 .8. degres / ne 42 minut*is* owt of 13 mi$^{a}$ tho added I .30. degres | to the forseide
7 .8. degres of virgo & 60 mi$^{a}$ to the .13. mi$^{a}$ of the same | virgo. & tho drow
8 I the verrey mot*us* of caput owt of the verrey | mot*us* of the mone / tho dwelde
9 me the verrey argume*n*t of the latitude | of the mone þ$^{t}$ is to seyn (^ .4. sig$^{a}$)
10 .17. g$^{a}$ 31 mi$^{a}$ tho leide I þ$^{t}$ on ende of my | thred 4 signes 17 degres
11 (^ 31 minuta) fro the hed of aries in the lymbe after | successiou*n* of signes
12 & þ$^{t}$ other ende (^ lay) equedistant fro the diametr*e* | þ$^{t}$ passith by the heuedes
13 of aries & libra / & tho fond I the middel | of my thred karf the m*e*ridi*onal* lyne
14* at .3. degres & 22 mi$^{a}$ fro the | centre of the Erthe þ$^{t}$ is centre aryn / wher for
15 I knew wel þ$^{t}$ | my mone was 3 degres .22. mi$^{a}$ in latitude 7trio*nal* descendinge |
16 fro the Ecliptik ¶ (1391 23 februar*ii*) the thridde ensample is this I fond
17*, 18 in myn | almenak the verrey mot*us* of the mone was 6 degres 24 mi$^{a}$ of | Scorpio
19 & the verrey mot*us* of caput was 20 degres .29. mi$^{a}$ of | aries. tho moste I wirke
20 w$^{t}$ cauda by cause þ$^{t}$ verre mot*us* of | my mone passed mor than .6. signes tho
21 drow I the verrey | mot*us* of cauda owt of the verrey mot*us* of the mone / in
22, 23 this | man*er* / I added 30 g$^{a}$ to .6. g$^{a}$ of scorpio & 60 mi$^{a}$ to 24 mi$^{a}$ of the | same
24 scorpio / tho dwelde me the verrey argume*n*t (of latitude) of the mone | .o. in
25 signes 15 g$^{a}$ & 55 mi$^{a}$ / tho leide I þ$^{t}$ on ende of my thred | o in signes 15 g$^{a}$
55 mi$^{a}$ fro the hed of lib*ra* (^ agains succession of signes) by cause þ$^{t}$ I wirke
26, 27 w$^{t}$ | cauda / & þ$^{t}$ other ende of my thred lay equedistant fro the diametr*e* | þ$^{t}$
28 passith by the heuedes of aries & libra / & tho fond I þ$^{t}$ the middel | of my thred
29* karf the m*e*ridi*onal* lyne .1. degre 22 mi$^{a}$ fro centre aryn | bi which I knew the
30 latitude of my mone was —— .1. g$^{a}$ 22 m$^{a}$ fro | the Ecliptil m*e*ridi*onal*
31 disce*n*dinge ¶ thus shaltow p*ro*cede day by day vpward | fro the hed of libra
32 vn to 90 degres agayns succession of signes | þ$^{t}$ is to seyn vnto the firste of
33 canc*er* & thanne stirt ou*er* the m*e*ridi*onal* lyne | whan thy verrey argume*n*t of
34 thy latitude of the mone passit any|thing 90 g$^{a}$ / & ley þ$^{t}$ on ende of thy thred
35 in gem*in*i & þ$^{t}$ other ende | in cancer / & so com downward day bi day til
36 thow come agayn | at centre aryn / & thanne wirk w$^{t}$ caput as I haue told
37* by fore | ¶ & no*ta* þ$^{t}$ whan any eclips (lu*ne*) fallith in aries taur*us* gem*in*i.
38 ca*n*c*er*. *leo*. v*irgo* / than is | [than is][c] the Eclips in caput / & the remenant of
39* the Eclipses ben | in cauda

**line 1** The date is written in the left-hand margin.
**line 16** The date is written in the left-hand margin. **line 30** Ecliptil *sic for* Ecliptik.

.s. in almenak

// Another ensample. I fond my mone in 8 degrees 13 min of virgo
1391 [illegible] & caput draconis in 20 degrees & 22 min of aries. tho drowe I tho verrey
mot of caput fro the verrey mot of the mone in this
manere. I wiste wel þt caput was 20 degrees abak of
8 degrees and 22 minut abak of 13 min. tho addede I 30 degrees
to the forseide 8 degrees of virgo & 60 min to the 13 min of the same
virgo & tho drowe I the verrey mot of caput out of the verrey
mot of the mone. tho bilefte me the verrey argument of the latitude
of the mone þt is to seyn 5 signes 17 degrees 31 min. tho leide I þt on ende of my
thred 2 signes 17 degrees 31 minut fro the hed of aries in the lymbe after
succession of signes & þt other ende lay equedistant fro the diametre
þt passith by the hevedes of aries & libra & tho fond I the middel
of my thred kerf the midel lyne at 3 degrees & 22 min fro the
centre of the erthe þt is centre aryn. wher for I knew wel þt
my mone was 3 degrees 22 min in latitude north descendinge
fro the ecliptik. // The thridde ensample is this. I fond in myn
1391 23 februarii almenak the verrey mot of the mone was 6 degrees 29 min of
scorpio & the verrey mot of caput was 20 degrees 29 min of
aries. tho wiste I wel þt cauda by cause þt verre mot of
my mone passed mor than 6 signes. tho drowe I the verrey
mot of cauda abak of the verrey mot of the mone in this
manere. I addede 30 degrees to 6 degrees of scorpio & 60 min to 29 min of the
same scorpio. tho bilefte me the verrey argument of latitude of the mone
0 in signes 14 degrees & 49 min. tho leide I þt on ende of my thred
0 in signes 14 degrees 49 min fro the hed of libra by cause þt I wiste wel
cauda, & þt other ende of my thred lay equedistant fro the diametre
þt passith by the hevedes of aries & libra & tho fond I þt the middel
of my thred kerf the midel lyne 1 degree 22 min fro centre aryn.
by which I knew the latitude of my mone was 1 degree 22 min fro
the ecliptik north descendinge. // thus shaltow procede day by day upward
fro the hed of libra unto 90 degrees ayeyns succession of signes
þt is to seyn unto the fyrste of cancer & thanne kerf on the midel lyne
whan thy verrey argument of thy latitude of the mone passith any
thing 90 degrees & ley þt on ende of thy thred in [illegible] & þt other ende
in cancer & so com dounward day by day til thow come agayn
at centre aryn, & thanne wirk wt caput as I have told by fore
// & no þt whan any eclips fallith in aries taurus [illegible], than is
than is the eclips in caput / & the remenant of the eclipses ben
in cauda

## IV

# TRANSLATION

The asterisks indicate that a note on the text occurs in Chapter V

In the name of God, pitiful and merciful.* (Leyk)† said: The bigger you make this instrument the bigger will be your main divisions. The bigger these divisions, the smaller the fractions into which they can be divided, and the smaller these fractions, the more accurate your calculations.* A

Take therefore a metal plate, or else a board planed smooth, tested with a level, and polished evenly; and when it has been made into a perfect circle by your compasses, it shall be 72 large inches* or six feet in diameter. The circumference of this circular board should be bound with an iron rim, just like a cart wheel, so it does not warp or become crooked. If you wish, the board can be varnished, or parchment can be glued over it* to give a good surface.

Next take a circle of metal or of glued parchment, two inches broad and 68 inches or 5 feet 8 inches in internal diameter,* and skilfully nail this circle to the circumference of the board. I will call this circle the 'limb' of my Equatorie,* which was constructed in the year of Our Lord 1392 complete,* on thc last noon of December.

Divide this limb into four quarters by means of two diametral lines, in the way in which the limb of an ordinary astrolabe is divided,* and see that the right angle is true by testing it geometrically.* Then take a large pair of compasses, in good order, and place the fixed point* in the centre of the board. Over this centre shall be fastened a round plate of metal, and since all the centre holes of the Equatorie are to be pierced in this plate, it must be 16 large inches in diameter. Also, in course of time, this plate can be turned around to follow the motion of the *auges* in the ninth sphere, and so your instrument will remain valid perpetually.*

Then, as I have said before, take the fixed point of your compasses and place it on the centre of this plate, and with the movable point of your compasses

† Erased in manuscript but legible in ultra-violet light. It may be a proper name, or the sense might be 'It is said' or 'Dixit'. See p. 165.

describe a circle as near the outer edge of the limb as possible. Note that I shall call the centre of the plate, where the fixed point of your compasses stands, centre *aryn*.*

Next describe a smaller circle about the same centre *aryn*, only a slight distance from the aforesaid outer circle on the limb.* The space between the two circles shall be divided in minutes of the limb.*

Then make a smaller circle, a little further from the last mentioned circle, and the space between them shall be divided in degrees of the limb.

Then make a still smaller circle, a little further from the last mentioned circle, and in this the degrees shall be numbered.

B Then make a still smaller circle, a little further distant from the last mentioned circle, and in this shall be written the names of the twelve signs;* and note that I shall call this last mentioned circle the 'Encloser of the Signs'.*

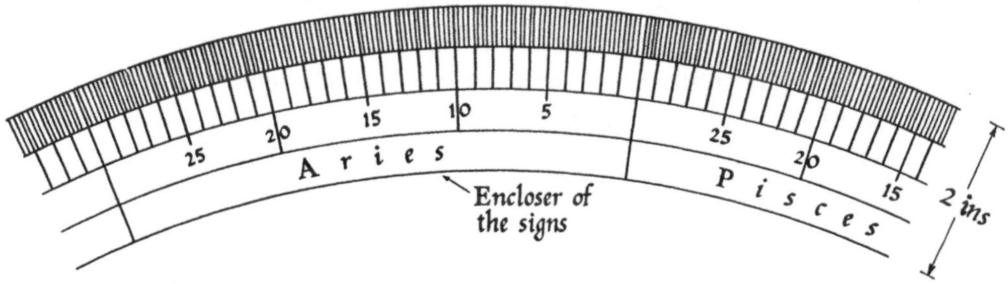


Fig. 2. The five circles on the limb of the equatorie instrument.

Now you have five circles upon your limb, all of them described about centre *aryn*. And each of the four quarters of your limb shall be divided into 90 degrees, that is, into three signs; and each degree shall be divided into 60 minutes.* In brief, your limb shall be divided just like the limb on the dorsum of an astrolabe.*

Next, divide into 32 equal parts that line which goes from centre *aryn* to the circle called Encloser of the Signs (*versus finem Geminorum*); these parts are called degrees of the radius. Mark these divisions faintly (*ut postea deleantur*), and note that this radial line, divided into 32 parts, will be called line *alhudda*.*

Set the fixed point of your compasses at the end of the first division from centre *aryn* on this line *alhudda*; and set the movable point of your compasses at the end of the thirtieth division from the fixed point. There will thus remain just one division between the movable point and the Encloser of Signes, and just one division between the fixed point and centre *aryn*. With your compasses placed thus describe a circle, and take it to be the Eccentric Circle of the Sun. Then erase the divisions of the line *alhudda*.

Next divide (faintly, *occulte*) the same line *alhudda*, from centre *aryn* up to the Encloser of the Signs, into 60 equal parts. Then place the fixed point of your compasses upon centre *aryn*, and the movable point at 12° 28′ of line *alhudda*, and describe a circle; and that is the (circle of the) Centre Deferent of the Moon.*

Then make 360 small holes equally spaced around the entire circumference of this circle,* this spacing of the holes being marked out by lines running from the degrees on the limb.

Note that in the year of Our Lord 1392 complete, the *aux* of Saturn, on the last noon of December at London*—I repeat, the *aux* of Saturn in the ninth sphere—was $4^{ss}$ 12° 7′ 3″ &c.* You must find the rest of the *auges* from the table of *auges* which follows.* Take a ruler and place one end on centre *aryn* and the other end on the limb at the minute where the *aux* of the planet is located;* and then, with a sharp instrument, rule a line from centre *aryn* to the Encloser of the Signs—but no further lest you should spoil the limb—and beside this line write the name of the planet (*cuius est aux*). This procedure holds good for all planets.

Look up your table of centres to find the distance of the centre equant of Saturn from centre *aryn*; it is 6° 50′.* Then place the fixed point of your compasses on centre *aryn*, and the movable point on line *alhudda* at 6° 50′ from centre *aryn*. Then turn your compasses carefully, until the movable point touches the line of the *aux* of Saturn. Keeping the fixed point of your compasses C firmly on centre *aryn*, make a deep prick with the movable point where it falls on the line of the *aux* of Saturn, for at that point a fine hole must be made for the centre equant of Saturn; and near this hole write an E for equant.

Next remove your compasses, and look up in your tables of centres* the distance of the centre deferent of Saturn; it is 3° 25′ (of line *alhudda*). Then place the fixed point of your compasses on centre *aryn* and the movable point at 3° 25′ of line *alhudda*, and turn the compasses carefully until the movable point touches the previously mentioned line of the *aux* of Saturn. Keeping the fixed point of your compasses on centre *aryn*, make a deep prick with the movable point on the line of the *aux* of Saturn, for there you must make a small hole for the centre deferent of Saturn; and near this hole write a D for deferent.*

And note that, following this example of Saturn, you must construct the centre deferent and the centre equant of each planet according to their distances (*in tabulis*)* from centre *aryn*, and make holes for them on the line of their *auges*.

Next place the fixed point of your compasses on the line of the *aux* of Mercury, mid-way between the centre E and the centre D of Mercury,* and extend the movable point until it can touch both the centre E and also the centre D of Mercury, and describe a small circle. You will then notice that the line of the *aux* of Mercury divides this small circle into two equal arcs; that is to say the line cuts the small circle exactly in two.

This small circle must be pierced full of fine holes, evenly spaced (*in circumferentia circuli*), just as in the case of the (circle of the) centre deferent of the Moon. If possible there should be 360 holes, or 180, or 90 at least.* But the spacing of the holes shall not be marked out by lines running from the great limb of the instrument—as was the procedure with the (circle of the) centre deferent of the Moon—but by lines running from the circumference of this same small circle, which circumference shall be divided by your compasses.*

Next erase the 60 divisions on the line *alhudda*, and then, by means of your compasses, divide that same line *alhudda*, from centre *aryn* to the circle that is called the Encloser of the Signs, into five equal parts.* Each of these five parts shall be divided into 60 parts. These divisions shall not be erased.

Next divide the line that goes from centre *aryn* to the head of Capricorn, which line in the Treatise of the Astrolabe is called the Midnight Line.* Divide this midnight line, from centre *aryn* to the circle called the Encloser of the Signs, into nine equal parts; and each of these parts shall be divided by means of your compasses into 60 equal parts.* These divisions shall not be erased.

*Laus Deo vero*. Now you have the Face of this precious Equatorie.

*N.B.* The nine divisions of the midnight line just mentioned will serve for the equation of the eighth sphere.

D Now for the construction of the Epicycle† for the Face of your Equatorie.* You shall make a circular strip of metal of the same breadth, width, circumference, and diameter, as the limb of your instrument. And it shall be divided in the same way as the first limb, with minutes, degrees, numbers, names of signs, and with five circles described with compasses, except that* the eccentric circle of the Sun shall not be figured on the Epicycle. Also the edge of the Epicycle shall not be filed too near to the Encloser of the Signs lest you misplace the hole (*foramen*) of your common centre deferent, or lest it should break. This Epicycle must be thick enough to be rigid.

Next take this Epicycle and place it firmly and evenly upon the Face of your Equatorie, so that the head of Aries on the Epicycle is exactly on the head of Aries on the limb of the Equatorie, and Libra on Libra, Cancer on Cancer, Capricorn on Capricorn, and every sign on every sign; that is to say the head of every sign on the head of every sign.

† A capital letter has been used to distinguish the Epicycle portion of the Equatorie instrument from the epicycle circle of the planetary theory.

Then take a pivot-rod or a large bodkin, and make a small hole in the Epicycle at the head of Cancer in the circle that is called the Encloser of the Signs. If you have drawn your circles accurately, you will see that the point of the pivot-rod touches the Encloser of the Signs at the head of Cancer in the Equatorie.* This small hole, which is no bigger than (one made with) a fine needle, shall be called the Common Centre Deferent of the Planets.

Next take a metal bar, one large inch in breadth, and of sufficient thickness;* one end of this bar must be soldered* to the Encloser of the Signs at Aries on this Epicycle, and the other end must be soldered to the Encloser of the Signs at Libra on the same Epicycle. Then draw a line with your ruler from the head of Aries to the head of Libra along the bar, and draw a similar line from the head of Cancer to the head of Capricorn across the bar; and where the two lines intersect is the centre of the Epicycle.

Then take a strip of brass,* not very thick, an inch in breadth, and as long as the diameter of the Epicycle. This strip must be shaped like the Label of an Astrolabe.* The centre of this strip must be nailed to the centre of the aforesaid bar in such a way that it can turn round like the label of an astrolabe. In the centre of the head of the nail that fastens the strip of brass and the bar together there must be a fine, deep prick; and this prick (*i.e. punctus*) is the centre of your Epicycle.

Next open your large compasses to the distance between centre *aryn* and the Encloser of the Signs: this distance is the length of the line *alhudda*. And then, either on a long ruler, or on a long strip of parchment, mark off this distance with your compasses, and divide it into 60 equal parts. This will give you a new line *alhudda*.*

Then, in your table of centres, look up the radius of the epicycle of Saturn; this is 6° 30′ of those degrees which number 60 in the line *alhudda*.* Measure with your compasses 6° 30′ of the line *alhudda*, and place the fixed point of the compasses on the centre of the Epicycle, that is, on the point (*punctus*) in the head of the nail. Then place the movable point of the compasses along the label, and with this movable point make a mark across the label; and close by this mark write Sa for Saturn. This example with Saturn shows you how to mark the radii of the epicycles of all the planets on the label.

*N.B.* The Sun has no epicycle.*

*N.B.* As the label turns it shows the epicycle of each planet.*

*Laus Deo vero.** Now you have the complete Equatorie with all its parts.

*N.B.* The eccentric circle of the Sun is not to be drawn on this Epicycle.

*EXPLICIT*

## The Face of the Equatorie.

**(Here follows the diagram on f. 73 v.)**

*N.B.* Every centre must be as fine as a needle, and there must be a silk thread in every equant.

*N.B.* The eccentric circle of the Sun is drawn on the board of the instrument, not on the limb, so as to save metal.*

*N.B.* In brief, unless both the Enclosers of the Signs have exactly the same width, and unless centre *aryn* is exactly the same distance from the Encloser of the Signs as the centre of the Epicycle is from the common centre deferent, your Epicycle will be inaccurate. Nevertheless, if you have made this mistake, I can show you a remedy: knock your common centre deferent further in or further out until it is exactly on the Encloser of the Signs on the limb of your Equatorie.* Then the centre of your Epicycle will be precisely over centre *aryn*.

*The sixth circle is the eccentric circle of the Sun; the fifth circle, the one in red, is the Encloser of the Signs; and where the lines cross at right angles is centre *aryn*. The second centre is the centre of the eccentric circle of the Sun; and the line divided into nine parts is the midnight line. I know very well it is only roughly sketched.* The circle round centre *aryn* is the (circle of the) centre deferent of the Moon; the little circle is the (circle of the centre) deferent of Mercury; and the thin lines are the lines of the *auges*. The dots on these lines are the centres, both equant and deferent, and all these centres, except the equant of Mars, are between centre *aryn* and the (circle of the) centre deferent of the Moon.* The outermost circle is minutes, the next degrees; in the third the degrees are numbered, and in the fourth are the names of the signs. It is however the smaller of the circles of the signs which is called the Encloser of the Signs, and I have drawn it in red.

## F The Epicycle.

**(Here follows the upper diagram on f. 74 r.)**

*N.B.* Do not file too near to the red circle which is the Encloser of the Signs; lest the common centre deferent should break, allow a small tongue of metal to remain, as you can see (on the accompanying diagram) at the head of Cancer.*

*N.B.* I advise you not to write in the names of the signs (*i.e. in Epiciclo*) until you have tested whether your common centre deferent is properly and truly placed on the Encloser of Signs on your Equatorie.

This Epicycle is divided and the circles are drawn in every way like the limb of the Equatorie, except that it has no eccentric circle of the Sun. The hole on the Encloser of the Signs at the end of Gemini is the common centre deferent. Furthermore, this is how your instrument should be placed when you want to calculate the position of the Moon.* **(Here follows the lower diagram on f. 74 r.)**

*Pro argumentis trium superiorum minue eorum med' mot' de med' mot' solis et remanet* G
*argumentum.* (For the arguments of the three superior planets, take away their *medius motus* from the *medius motus* of the Sun, and the argument will remain.)*

Look up in your tables the mean *motus* of Saturn, Jupiter, Mars, and Venus, and their mean arguments, and write them down on your slate.* Then place a black thread in centre *aryn* (*terre*),* and a white thread in the centre equant of any planet whose position you wish to calculate; and place the common centre deferent* of your Epicycle upon the centre deferent, on the plate, of the planet whose position is sought.

I say that you should transfix with a needle the common centre deferent of your Epicycle, and the centre deferent which has been pierced on the plate for that planet whose position is sought.

Then proceed (*pro successione signorum*)* from the head of Aries to the place where the mean *motus* of that planet ends, on the great limb of the Face, and lay your black thread there.

Then lay your white thread parallel to the black thread on the same limb, and with your compasses test whether the threads are parallel. Place the pole of your Epicycle under the white thread; and keeping your Epicycle firmly in place—I repeat, keeping the pole of your Epicycle firmly in place under the white thread, and with the common centre deferent fixed by your needle to the aforesaid centre deferent of the planet whose position is sought—take your black thread and lay it so that it passes over the centre of the Epicycle, and extends to the further part of the same Epicycle.

This black thread will show on the great limb the true *motus* (*locus*) of the epicycle, and also it shows on the Epicycle the true *aux* of the planet. Then the arc between the mean *motus* of the planet and the true *motus* of the epicycle is called the Equation* of its Centre, on the limb (zodiac), and the Equation of its Argument on the Epicycle is equal to this; that is to say, it equals the arc between its mean *aux* and its true *aux*. For in fact the mean *aux* is shown on the Epicycle by the white thread under which you placed the pole of the Epicycle, and the true *aux* is shown on the Epicycle by the black thread.

Keeping your Epicycle fixed in this same position, set the marked end of the label away from the white thread, that number of signs, degrees, and minutes shown in your tables for that day's mean argument of the planet sought.* And reckon this mean argument from the white thread anticlockwise for every planet, except only for the Moon.*

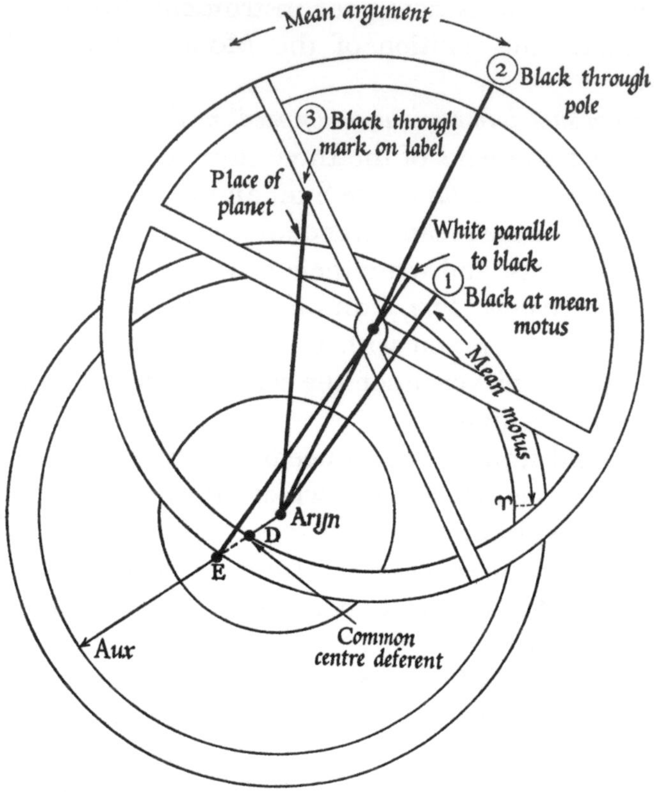


Fig. 3. Procedure in setting the equatorie for computing the longitude of a major planet. The three stages in setting the white and black threads are numbered.

H Placing the marked end of the label upon the end of this mean argument on the Epicycle, lay your black thread upon the mark of the planet which is engraved on the label; and wherever the same black thread crosses the limb of the Face, take there the true place (*locum*) of the planet, in the ninth sphere. And the arc between the true place (*locum*) of the planet (*in limbo*) and the true place of the epicycle, as found on the limb, is called the Equation of its Argument. This manner of determining the position holds good for Saturn, Jupiter, Mars, and Venus, but for the remaining planets it differs to some extent.*

*Sol.*

The mean *motus* of the Sun is to be reckoned anticlockwise from the head of Aries. The Sun has neither an epicycle nor an equant, and therefore the pole of the Epicycle must replace the body of the Sun in the ninth sphere. The white thread that you place in its centre deferent on the plate must replace the white thread in the centre equant which is used for the other planets.

The black thread that remains as usual in centre *aryn* must be placed at the end of its mean *motus*. Then take the white thread and lay it on the limb, parallel to the black thread which shows the mean *motus* of the Sun.

Then transfix with your needle the common centre deferent of the Epicycle and centre *aryn*, and do not remove the needle. Carefully place the pole of the Epicycle under this white thread; and wherever the white thread crosses the great limb, that is the true place of the Sun in the ninth sphere.

The arc of the limb, between its *aux*—which is now in Cancer*—and the black thread, is the argument of the Sun. The arc on the limb, between the black and the white thread, is the equation of the Sun, and this arc is quite small.*

The mean *motus* of the Sun is the arc on the limb between the head of Aries and the black thread on this limb. The true *motus* of the Sun is the arc on the limb between the head of Aries and the black thread (*i.e. filum*) when it is removed from the mean *motus*, and crosses the white thread at the pole of the Epicycle. The same true *motus* was shown previously by the white thread of the deferent, when it lay on the limb parallel to the black thread.

*N.B.* The marks on your label, as it turns, trace out the epicycles of the planets.

This explanation is wrong.*

Reckon the mean *motus* of Mercury anticlockwise from the head of Aries on the limb, and then ascertain the arc on the same limb between the head of Aries and the line of the *aux* of Mercury, which *aux* is currently towards the end of Libra.* And always reckon anticlockwise.

Then subtract from the mean *motus*, that angle between the head of Aries and the aforesaid *aux*; and calculate what remains of the mean *motus* when the *aux* is thus subtracted from the whole of the mean *motus*.

Proceed anticlockwise* round its little circle, reckoning from the line of the *aux* which cuts this same little circle; (and to the extent of this angle,) I repeat, reckon anticlockwise from the letter D that is engraved in the little circle, and continue in the same little circle towards the letter E which is opposite to the D.

I repeat, calculate the remainder of the mean *motus* when the extent of the *aux* is subtracted from the whole mean *motus*, as I have said before. And wherever this same remainder ends in the little circle, take there the true centre deferent of Mercury, which comes sometimes in one hole, sometimes in another. For the letter D only indicates where to begin your reckoning in the little circle, and the letter E only indicates which way to go round from the letter D.*

Now you have found your (centre) deferent and your (centre) equant. Place a white thread in your equant, and transfix with a needle the common centre deferent and the centre deferent on the plate (*lamina*), and continue with your Epicycle and your threads in the same way as you did for Saturn, Jupiter, Mars, and Venus.

*N.B.* If the *aux* of Mercury is further from the head of Aries than is the mean *motus*, you must add 12 signs to the mean *motus*, and then you will be able to subtract the *aux* from the mean *motus*.

*N.B.* In general, when your needle transfixes the common centre deferent and any other centre deferent on the plate, it must not be removed until you have finished finding the position of the planet. For if the common centre deferent shifts from the centre on the plate, then all your calculations of the sought planetary position are wasted.

*N.B.* In my instrument, the (circle of the) centre deferent of Mercury has only 24 holes,* and so I reckon only two holes to a sign in the greater circle of Mercury from the line of its *aux*.

J *Luna.*

Reckon the mean *motus* of the Moon anticlockwise from the head of Aries on the limb, and in the same way reckon the mean *motus* of the Sun (*a capite Arietis*) as far as it extends. Then subtract the mean *motus* of the Sun from the mean *motus* of the Moon, and note the difference. The amount of this difference I call the Remainder.* Reckon it backwards from the end of the mean *motus* of the Sun in a clockwise direction, and where this remainder ends make a mark on the limb.

Then draw the end of the black thread to this mark; and wherever the black thread cuts the (circle of the) centre deferent of the Moon, in that hole is the centre deferent of the Moon for the time being, and at the nadir of this hole is the centre equant.* Then place a white thread in this centre equant. Now you have your two centres.

Next transfix with a needle your common centre deferent and the centre deferent of the Moon. Once more reckon the mean *motus* of the Moon anticlockwise from the head of Aries, and lay there your black thread. And lay your white thread on the limb, parallel with the black thread.

Then carefully move the pole of your Epicycle under the black thread, and take the white thread and lay it over the pole of the Epicycle. Wherever the white thread cuts the circle of the Epicycle, take there the mean *aux* in the Epicycle; and from this white thread reckon backwards in a clockwise direction the mean argument on your Epicycle.

I repeat, reckon it in degrees on your Epicycle, and wherever your reckoning ends on the Epicycle, place there the marked end of your label. And lay the black thread upon the mark of the Moon on your label; and wherever this same black thread crosses the limb, read there the true place of the Moon in the ninth sphere.

*N.B.* The pole of the Epicycle must not be placed beneath the black thread of any planet but the Moon.

This explanation is wrong.* K

*N.B.* If the mean *motus* of the Sun is greater than the mean *motus* of the Moon, you must add 12 signs to the mean *motus* of the Moon, and then you can subtract the mean *motus* of the Sun from the mean *motus* of the Moon.*

And to put the matter briefly, I say that the centre of its (*Lune*) epicycle (*in volvella*)* turns uniformly about the centre of the zodiac, that is to say about the pole of the Epicycle which is your *rete*.*

The black thread, when it is first passed over the pole of the Epicycle, shows the true *aux* of the planet (*in Epiciclo*), just as the white thread shows the mean *aux* in the same Epicycle. Then again, when you have reckoned the argument of a planet on your Epicycle, then the body of the planet is at the end of the argument in your Epicycle. And when the black thread is laid through the mark of a planet on the label, in the way described above, then the black thread shows the true place of the planet with respect to the ninth sphere, as is indicated on your limb.

The arc between the true *motus* and the mean *motus* of the Moon is the Equation of its True Argument on the limb; and the arc between its mean *aux* and its true *aux* is the Equation of its Argument on the Epicycle.

To find the latitude of the Moon by means of your instrument, look up in your almanac the true *motus* of the Moon and the true *motus* of Caput Draconis Lune for the same time. If the true *motus* of the Moon is less than six signs from Caput Draconis, subtract the true *motus* of Caput from the true *motus* of the Moon, and write down that difference (*i.e. verum argumentum latitudinis Lune*), for that is its (*Lune*) true argument (*i.e. latitudinis*).

Reckon the signs, degrees, and minutes of the true argument of its latitude, from the head of Aries anticlockwise on your limb, and place one end of your thread where the reckoning ends, and the thread must cross the meridional line* and stretch right across the diameter of your plate as far as the limb. L

If, for instance, one end of the thread lies on the limb 10° anticlockwise from the head of Aries, then the other end of the thread should lie at 20° of Virgo on the limb.

Next see how many degrees and minutes the middle of your thread is from centre *aryn*—from which distances are always reckoned. I repeat, by means of the divisions of the meridional line, see how many degrees and minutes the middle of your thread is from centre *aryn*, and this will give you the latitude of the Moon, north of the ecliptic; and this latitude never exceeds 5°.*

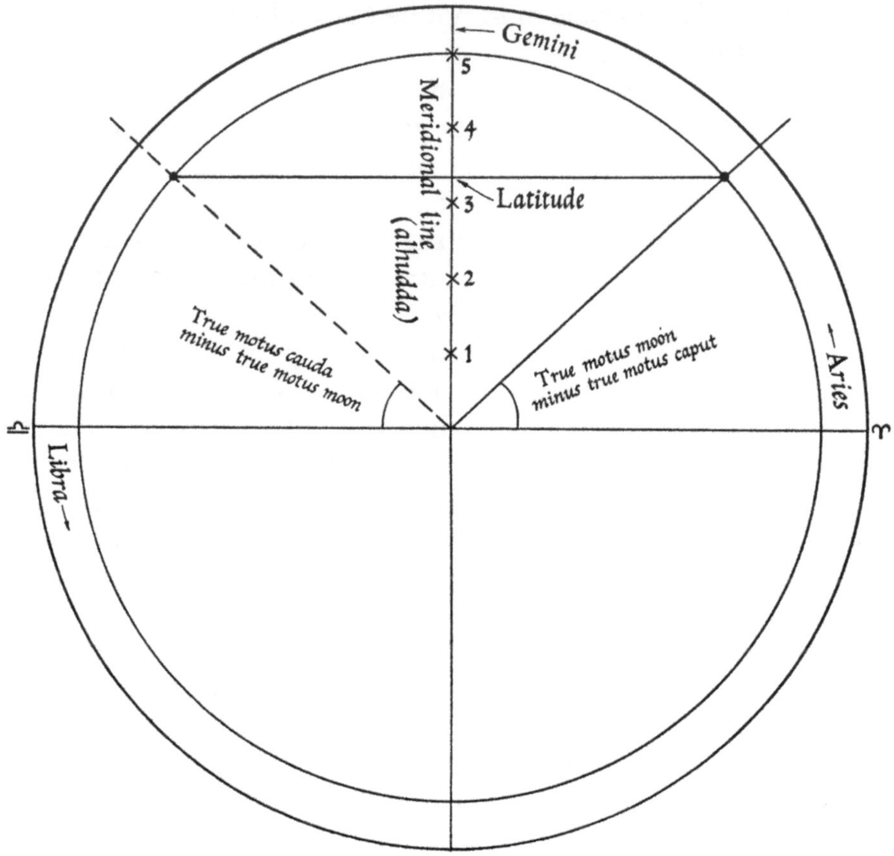


Fig. 4. Use of the meridional line to compute the latitude of the Moon.

If the true *motus* of the Moon is more than six signs from the true *motus* of Caput, then you must subtract the true *motus* of Cauda from the true *motus* of the Moon, and reckon backwards in a clockwise direction beginning at the head of Libra.

If, for instance, one end of the thread was 10° clockwise from the head of Libra, then the other end should lie 10° anticlockwise from the head of Aries.

From the meridional line see how much is the latitude of the Moon, south of the ecliptic. As I have said before, the number of degrees and minutes that the middle of the thread is from centre *aryn* along the meridional line, that number of degrees and minutes is the latitude of the Moon from the ecliptic, no matter whether north or south.

*N.B.* In every case the thread should lie parallel to the diameter that intersects the heads of Aries and Libra.

To help you to understand this explanation, I repeat: when the aforesaid true argument of the Moon is exactly 90° anticlockwise from the head of Aries on the limb, then the Moon has its greatest latitude north (*i.e. ab ecliptica*); M
and if the true argument at all exceeds 90° from the head of Aries, cross the meridional line into the beginning of Cancer, and lay one end of the thread there and the other end on the end of Gemini; and proceeding thus, day after day, descend the meridional line as far as demanded by the reckoning of the true argument until your return to centre *aryn*. For then you will have worked out* the latitude as described, and gone through six signs, and the thread will still be parallel to the diameter that intersects the heads of Aries and Libra.

And always, the number of degrees and minutes that the middle of the thread is from centre *aryn* along the meridional line, so many degrees and minutes is the latitude of the Moon from the ecliptic.

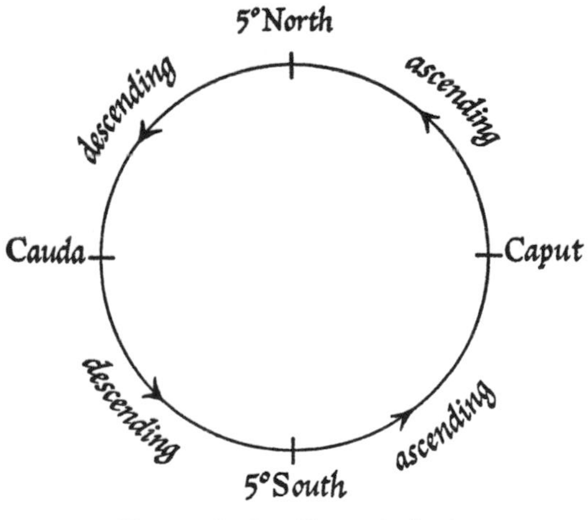


Fig. 5. Cycle of lunar latitude.

When the true argument exceeds six signs employ Cauda, as I taught you, and ascend the meridional line day by day until the end of Gemini on the limb, and thereafter descend again as I stated earlier.

*N.B.* When the Moon is at Caput or Cauda its latitude is zero; and when it goes beyond Caput until it is three signs past Caput (*pro successione signorum*) it is north ascending; at its greatest latitude it is north; and after these three signs until it comes to the opposite of Caput, *i.e.* Cauda Draconis, it is north descending; from Cauda until midway (*in medio*) between Caput and Cauda (it is south descending),† and from there it is south ascending until it returns to Caput.*

† Omitted in the manuscript, but essential to the sense, see p. 72.

17 December 1391.*

Example: The Moon was in 12° 21′ of Virgo, and Caput was in 4° 46′ of Aries; therefore I subtracted the true *motus* of Caput, *i.e.* $0^s$ 4° 46′, from the true *motus* of the Moon, *i.e.* from $5^s$ 12° 21′, and thus found that the true argument of the Moon remained $5^s$ 7° 35′.

Then I reckoned this $5^s$ 7° 35′ anticlockwise from the head of Aries on the limb, and I placed one end of my thread there, and the other end at 22° 35′ of Aries. So the middle of my thread crossed the meridional line at 1° 54′ from centre *aryn*, and I knew from this that the latitude of the Moon was 1° 54′ north, descending from the ecliptic.

N 19 February 1391.

Another example: I found (*scilicet in almenak*)* that the Moon was in 8° 13′ of Virgo, and Caput Draconis at 20° 42′ of Aries. Therefore I subtracted the true *motus* of Caput from the true *motus* of the Moon in the following way: I realized that I could not subtract 20° from 8°, or 42′ from 13′, so I added 30° to the aforesaid 8° of Virgo, and 60′ to the 13′ of Virgo. Then I subtracted the true *motus* of Caput from the true *motus* of the Moon, and there remained the true argument of the latitude of the Moon, *i.e.* $4^s$ 17° 31′.

Therefore I placed one end of my thread at $4^s$ 17° 31′ anticlockwise from the head of Aries on the limb, and the other end lay parallel to the diameter that intersects the heads of Aries and Libra. So I found that the middle of my thread cut the meridional line at 3° 22′ from the centre of the Earth—which is centre *aryn**—and so I knew that the latitude of the Moon was 3° 22′ north, descending from the ecliptic.

23 February 1391.

The third example is this: I found in my almanac* that the true *motus* of the Moon was in 6° 24′ of Scorpio, and the true *motus* of Caput was in 20° 29′ of Aries. Then, because the true *motus* of the Moon exceeded six signs, I had to work with Cauda. Therefore I subtracted the true *motus* of Cauda from the true *motus* of the Moon by this means: I added 30° to the 6° of Scorpio, and 60′ to the 24′ of Scorpio. Thus there remained the true argument of the latitude of the Moon $0^s$ 15° 55′.

I therefore placed one end of the thread at $0^s$ 15° 55′ clockwise from the head of Libra—because Cauda is being used—and the other end of the thread was parallel to the diameter intersecting the heads of Aries and Libra. So I found that the middle of my thread cut the meridional line at 1° 22′ from centre *aryn*, by which I knew that the latitude of the Moon was 1° 22′ north, descending from the ecliptic.

You must proceed upwards, day by day, in this manner from the head of Libra 90° clockwise, *i.e.* to the beginning of Cancer, and then cross the meridional line if the true argument of the latitude of the Moon at all exceeds 90°. And place one end of the thread in Gemini, and the other end in Cancer, and descend, day by day, until your return to centre *aryn*, and then employ Caput as I have already explained.

*N.B.** When any eclipse (*Lune*) occurs in Aries, Taurus, Gemini, Cancer, Leo, Virgo, then the eclipse is at Caput, and all other eclipses are at Cauda.*

# V

# NOTES ON THE TEXT

**A/1.** The opening *bismillah*, 'In the name of God, pitiful and merciful', is typically Arabic in tradition. Compare C. A. Nallino (ed.), *Al-Battānī Opus Astronomicum* (Milan, 1899–1903), Part I, p. 1 and Part II, p. 1, where the Latin translation begins, 'In nomine Dei clementis et misericordis'. For a discussion of the erased word 'Leyk' see p. 165.

**A/4.** The emphasis laid upon large size as a means of obtaining accuracy is more in the Eastern than the Western tradition.[1] During the later Middle Ages, Islamic astronomers often employed gigantic masonry instruments; see the account of the Marāgha instruments used by Nāṣir al-dīn al-Ṭūsī (*c.* 1259) given by G. R. Kaye, *Astronomical Observatories of Jai Singh* (Calcutta, 1918), p. 81. If we assume for purpose of rough estimation that the accuracy of reading a linear scale is of the order of 1 mm., a divided circle 6 ft. in diameter would enable an angular accuracy of about 4 minutes of arc to be obtained. A 'pocket-size' instrument of about 5 in. radius would yield an accuracy of about $\frac{1}{2}^\circ$, comparable to the apparent size of the Sun or Moon. It must be considered doubtful whether large instruments were ever made in metal during the Middle Ages. Suitably uniform metal plates would have been very costly and difficult to obtain in large sizes.

A similar cautionary note on the size and accuracy of a planetary instrument occurs in a commentary on Alcabitius by John of Saxony (Bodleian MS. Digby 97, f. 230v.). The author refers to the *directorium* as a modified form of equatorium for calculating the ecliptic latitudes of the planets; perhaps it is an adaptation from the instrument designed by his teacher, John of Linières (see p. 126). He says, 'Ego autem dico quod instrumentum non est sufficiens nisi sit maxime quantitatis ita quod possit recipere minuta. Tale autem vix posset fieri. In directionibus enim ut plurimum accipitur pro quolibet gradu unus annus. Modo si instrumentum non sit divisum nisi per gradum vix invenitur in eo certitudo usque ad annum. Adhuc si instrumentum sit bene factum ita quod non sit error in dividendo hoc autem non sufficit' (cited from L. Thorndike, *History of Magic and Experimental Science*, vol. III, p. 261). Compare also a remark by Richard of Wallingford, cited from his *Tractatus Albionis* on p. 128 of the present work.

**A/6.** Large inches seem to imply the old Saxon inch equal to 1·1 of our modern inches. These 'modern' inches were first standardized by Edward I in 1305, but the use of the Saxon measures was exceedingly tenacious, and even as late as the sixteenth century various printed 'day-books' and manuscripts purporting to inform merchants on English weights and measures were full of contradictory statements quite at variance with the standards laid down by Statute. Chaucer would have had considerable experience with weights and measures in the course of his official duties at the Customs and as Clerk of Works; perhaps this may be a reason for the unusual care with which the units are stated.

**A/9.** Although 'glewed with perchemyn' probably means that the board is to be covered

[1] But similar instructions about size, material, etc. occur in the *Planispherium* of Ptolemy.

with parchment glued down to the surface, it is perhaps possible to interpret the phrase as a direction that the board should be covered with a size made by boiling parchment until it forms a suitable glutinous paste. Such a size, containing also plaster of Paris, was used for the gesso ground of panel paintings; see D. V. Thompson, *The Materials of Medieval Painting* (London, 1936), p. 34.

**A/12.** This is one of the few places in which the author's amendments by rubbing and scraping can be followed with certainty. Originally he seems to have written, 'tak thanne a cercle of metal þ$^{t}$ be .2. enche of brede / & þ$^{t}$ the hole dyametre contene .72. enches / or .6. fote / & subtili...'. He then saw that this phrasing was ambiguous, for the circle might have been a little tube, 6 ft. in diameter and 2 in. high, instead of the intended annulus. The revised wording completely removes the ambiguity.

**A/13.** The technical term 'lymbe' is explained as 'zodiacus' in the gloss which occurs at G/25. The word is, however, used for the graduated edge of any instrument; not all instruments have this outer edge divided as a zodiac. It might be thought that 'myn equatorie þ$^{t}$ was compowned' is to be taken in the figurative sense of 'my tract on the equatorie, which was written...', but a similar occurrence at I/36 makes it quite clear that the author is referring to an actual instrument in his possession. It is perhaps worth noting that the name *equatorie* bears no direct relation to the celestial *equator* but derives from the use of the instrument to equate (compute) the planetary positions.

**A/14.** The styles of dating used in this manuscript are somewhat confusing, and although the use of 'last meridie' and 'years complete and incomplete' is known from other medieval writers it is desirable to confirm that the usual practice is being followed here. Fortunately the text concludes with three worked examples in which the dates are given in the normal civil form (i.e. 19 February 1391), and these may be used to determine the civil equivalent of the more puzzling astronomical dates (i.e. the last meridie of December, the year of Christ 1392 complete). It is convenient to use for this calculation the zodiacal positions given for Caput Draconis, since this node of the Moon's orbit slowly retrogresses round the zodiac at a uniform rate of about 19° per annum (cf. Fig. 6, p. 73). Using this figure we may first confirm that the date given as 17 December 1391 occurs about ten months after that given as 19 February 1391; this enables us to confirm that the (civil) New Year used here commences with 1 January and not with any of the other dates in use at this time (for a discussion of these other dates see R. L. Poole, 'The beginning of the year in the Middle Ages', *Proc. Brit. Acad.* vol. x, 1921). Having established this we may proceed with a similar calculation which gives

Last meridie of December 1393 complete
=31 December 1393.

Last meridie of February 1393 complete
=28 February 1394.

It is therefore clear that the date given at this point in the text,

Last meridie of December 1392 complete
=31 December 1392.

According to astronomical usage, the day commences at noon, and as a consequence of this the author completes his year at noon on the last day of December; so, for example, 3 p.m. on 31 December on the usual reckoning is already three hours into the New Year. It is also clear from the above equating of dates that the astronomical 'year complete' is *not* taken as commencing in January—if this were the case February of 1393 complete would precede December 1393 complete instead of following it. Since many of the astronomical tables in our volume commence the calendar at 1 March we may presume that this date has been taken here as the beginning of the astronomical year used with the 'years complete'. The correspondence of the two calendars may be shown diagrammatically as follows:

| | | | | |
|---|---|---|---|---|
| *Astronomical* | 1 March | 1393 complete | 1 March | 1394 complete | 1 March |
| *Civil* | 1 January | 1393 | 1 January | 1394 | 1 January |

Commencing the year at 1 March has the great convenience for astronomical tables of making the intercalary day, 29 February, fall at the end of the table; this makes it particularly simple to convert the table for use during leap year. For a similar reason of convenience the radix dates chosen for the tables are those at the end of the year: for tables commencing on 1 January that radix is the last meridie of December; for those which begin with 1 March, the last meridie of February is always given. With such a radix it is easy to reckon how many days have elapsed to any other date—if, for example, a radix date 22 January had been chosen it would be necessary to subtract 22 days during each calculation of days elapsed. In view of this convenience it is suspicious that our author should claim that his instrument was 'Compowned the laste meridie of December the yer of crist 1392 complet'—we must take this as meaning that his instrument was not constructed *on* that date, but rather *for* that date, i.e. using values of auges, etc., valid for that date.

It remains to say something of the origin and significance of the 'years complete and incomplete'. A year complete gives the number of whole years that have elapsed since the date of the Incarnation of Christ (assumed here as 1 March 1 B.C. or civil year 0). Years incomplete, on the other hand, gives the number of the year current during which that date is taken; hence the year 0 complete corresponds to the year 1 current or incomplete. An alternative medieval nomenclature is to be seen in MS. Sloane 407(60), where a date is given as 'A.C. 1467 perfecto et 1468 imperfecto'. This matter is not related in any way to 'Expans and collect years' which refer to tables giving entries for sets of single years and sets of groups of years. One might, for example, have entries for 1, 2, 3, 4, 5, 10, 15, 20, 25 years, etc. By adding the *expans* year 3 entry to that for the *collect* year 20 one may deduce a value for a year 23 (from the radix) which might be sought.

**A/16.** Cf. *Treatise on the Astrolabe*, Part 1, paragraph 5.

**A/17.** Medieval drawing compasses had two points, much like our modern dividers. It is therefore curious that the author is at pains to distinguish the fixed and movable points, as if the instrument were asymmetrical. Perhaps one point was rounded and the other shaped more like a chisel edge for scribing the parchment. The testing of the right angle by 'geometrical conclusion' seems to imply the use of Euclid 1, 11.

**A/22.** The auges go through a small precessional motion of about 1° in a century, and by virtue of this all the centres marked on the brass plate should be turned through 0·6 minute of arc every year. The variation for a single year would be inappreciable, but the cumulative effect in twenty or thirty years would just be noticeable on an instrument of this size.

**A/26.** Although *Aryn*, *Arin* or *Azin* is familiar in medieval geography as the point at the centre of the known or habitable surface of the Earth, the word is used here in a novel, but completely logical sense. It represents the point which is at the centre of the Earth, and therefore at the hub of the geocentric planetary system.

Medieval astronomers had good reason to suppose that the size of the Earth was not insignificant as compared with planetary distances. They were also familiar with the fact that astronomical observations were affected by the location of the observer because of the variation in local time. Cf. the Latin note on f. 7r. of the tables, p. 85.

Arin seems to have been used as the central meridian by Hindu astronomers during the fourth century A.D. The meridian was drawn to pass through places held in the greatest respect by indigenous traditions, notably Ujjain (capital of Malua), Lanka (Ceylon) and Merou (North Pole); see J. Lelewel, *Géographie du Moyen Age* (Brussels, 1852), vol. I, p. xxij. About the eighth century, the Arabs united the Hindu and the older Western systems by regarding the central meridian as lying half-way between the Eternal Islands (occidental limit) and the extremity of China (oriental limit). The name Ujjain or Odjein was corrupted by the translators into Ozein, and since in Arabic there is only a dot to distinguish *r* from *z̄*, the transition to Arin was not difficult. See Nallino (ed.), *Al-Battānī Opus Astronomicum*, p. 165 and J.-J. Sédillot, *Traité des Instruments Astronomiques des Arabes* (Paris, 1835), vol. II, 'Au lecteur' between second title and p. 371.

An interesting contemporary vernacular use of Arin occurs in the *Livre du Ciel et du Monde* (*c.* 1370) of Nicole Oresme: 'Et les astrologiens ymaginent ou milieu de ceste porcion (habitable du monde) une cité il appellent Arim...' and 'Et donques le devant du ciel est sus Arim et le destre du ciel en l'orient d'Arym et le midi d'Arim est le vrai orient immobile et non pas le premier orient de terre habitable...' (quoted from A. D. Menut and A. J. Denomy, *Medieval studies*, vol. IV, 1932, p. 189).

**A/27.** The author is guilty of an oversight in not specifying exactly how much narrower the circle is to be. The complete limb contains five circles enclosing four bands of unspecified width. The outer circle is given as being exactly 6 ft. in diameter, but it is the inner circle (the encloser of the signs) which fixes the scale of the representation. Presumably the inner circle is to be on the inner edge of the 2 in. wide band of the limb, and the four spaces (if equal) are therefore $\frac{1}{2}$ in. each. It might be more convenient to make the spaces of unequal width to suit the various divisions and inscriptions which have to be inserted.

**A/29.** It can hardly be intended that single minutes of arc should be marked on the limb, for there would have to be about 120 such divisions to the inch. Intervals of about 5 minutes are about the limit of fineness which could be engraved on an instrument of this size. Either the author is speaking loosely, or he is being led astray by comparing an ideal scale with that of his own, too small instrument.

**B/2.** Although the author uses ordinary signs of the zodiac (30°) here and elsewhere in the text, confusion arises in the tables because of the sexagesimal division of the circle used in the Alfonsine source. The normal *signa communia* (e.g. $3^s\ 7^\circ\ 13'$) contain 30°, while the Alfonsine *signa physica* (e.g. $1^{ss}\ 37^\circ\ 13'$) are of 60°; in some places the latter are referred to as 'double signs'.

**B/3.** 'Closere of the signes' probably means *encloser*, although *enclosure* is a conceivable alternative. In any case there is no doubt of the general sense, and this is confirmed by the note at E/64 which states that this circle is drawn in red on the diagram on that folio; the innermost circle is indeed drawn in red there.

**B/6.** Again, as in A/29, it is unlikely that individual minutes are to be marked.

**B/7.** Cf. *Treatise on the Astrolabe*, Part I, paragraphs 4 ff., especially paragraph 8. The dorsum of an astrolabe usually contains other markings in addition to the scales of degrees, etc., here prescribed.

**B/8.** This line is to be used later for other purposes, and the gloss directing that it should be drawn from the centre to 30° Gemini ensures that the first point of Aries will then be in its conventional position on the right-hand side (east) of the instrument.

**B/9.** The 32 parts of this radius are constructed so that the eccentric circle of the Sun may have its centre displaced from Arin by a distance equal to one-thirtieth of its radius, this being the value taken for the eccentricity of its apparent orbit round the Earth.

**B/10.** The level of the line drops suddenly by about the height of a letter, commencing with | and nota. This may represent a resumption of work after a lapse of time, or it may simply be due to a change of pen. Cf. H/12 and K/5. It is curious that the line just divided into 32 parts is called 'diametral' instead of 'semydiametral'; either the author has committed an error or he is speaking in very loose terms.

**B/11.** The term 'alhudda' is unquestionably a transliteration from the Arabic; similar forms have been noted by Sédillot, Nallino and Delambre. In all cases the word signifies 'perigee', the nadir of the aux—it bears no relation to the more common *alidada* or *alidade*.

**B/20.** The radius of the circle of the centre deferent of the moon is thus $12° 28'/60°$ of the radius of its deferent circle. This is very nearly equal to the value given by Ptolemy, $10° 19'/49° 41'$ (*Almagest*, book v, section 4) and followed by al-Battānī and al-Kāshī.

**B/22.** Yet again we find a conflict between ideal requirements and the limitations of practical craftsmanship. Drilling 360 holes evenly and precisely distributed around a circle of about 7 in. radius would be an exacting task for a medieval craftsman, and it is very likely that the full number of holes was never actually employed. The use of such circles of holes is a rare and distinctive feature of this instrument; the only other instances of this type of construction which have been noted are the equatorium of al-Kāshī (see p. 131) and an early fifteenth-century volvelle on the back of a physician's quadrant at Merton College, Oxford (see R. T. Gunther, *Early Science in Oxford*, vol. II (Oxford, 1923), plate facing p. 241).

**B/24.** The phrase 'ultimo .10.bre in meridie london' which has been deleted might perhaps have been written before the next couple of lines were added. There could hardly be much point in specifying the full date twice in such a short space. It is perhaps an added indication that part, at least, of the polishing of the text took place during the actual drafting and not at some later date.

**B/26.** A table on f. 6v. gives the aux of Saturn at this date as $4^{ss}\ 12°\ 7'\ 3''\ 1'''\ 47^{iv}$. The last two sexagesimal places, which have been suppressed in the text, are common to all the auges for this date and represent excessive accuracy in the computation of precession.

**B/27.** If the table of auges which is now on f. 6v. is intended to follow the text, the quires have certainly been bound in the wrong order. See p. 14.

**B/29.** Since mention is made of the *minute* where the aux ends, and not the *nearest minute* marked on the limb, it might be taken as an indication that the limb is actually to be divided into single minutes, but as mentioned in the note to A/29 this is more likely to be merely an ideal.

**B/33.** The value of $6° 50'$ for the distance of the centre equant of Saturn, and the corresponding half-value of $3° 25'$ for its centre deferent are the same as used by Ptolemy, al-Kāshī and most of the medieval writers.

**C/4.** The tables in the manuscript contain no list of the distances of centre equant and centre deferent from Arin for the various planets. It would have been a very short table, and of little use once the instrument had been constructed; either it was never inserted, or else it was on one of the folios which have been removed from the volume. See p. 14.

**C/10.** 'an .D.' This would seem to indicate that the name of the letter at this time was pronounced as if beginning with a vowel, i.e. something like 'ed'. But there seems to be no evidence for any such pronunciation in English, and the 'an' may well be an error due to the influence of the preceding 'an E'.

**C/12.** 'in tabulis.' Cf. above note to C/4.

**C/15.** It must be remembered that Mercury is a special case which does not follow the general rule that places the equant at twice the distance of the centre deferent. See p. 101.

**C/21.** It is interesting that the author here confesses that his ideal construction is quite impracticable; even 90 holes would be almost impossible in such a tiny circle. More realistic is the statement at I/36 that it 'hath but 24 holes as in myn instrument'. This tends to confirm the suspicion that the author is working from a (Latin) text which describes the ideal case, but is modifying and adapting it from his own experience with a smaller instrument. We may even use this figure of 24 holes to the circle in order to obtain a rough estimate of the size of the author's instrument. The 'little circle' of Mercury has a diameter which is $\frac{1}{21}$ of the diameter of its deferent and therefore the diameter of the whole instrument. In an instrument 1 ft. in diameter, division of this little circle by 24 holes would place the holes so that they would lie about 12 to the inch. It is unlikely that they could be made finer than this, and, on the other hand, it is equally unlikely that the author would have been satisfied with a division giving only half this number of holes to the inch. It is therefore probable that the author's instrument was between 1 and 2 ft. in diameter—a conveniently portable size, comparable with that frequently found in the larger astrolabes of this period.

**C/24.** The warning about the graduation of the little circle may be expressed more simply in terms of modern procedure; the degrees are to be marked out by means of a protractor placed at the centre of the little circle, *not* at the centre Aryn.

**C/27.** This division of line alhudda into 5° is to be used later for computing the latitude of the Moon, north or south of the ecliptic.

**C/29.** Cf. *Treatise on the Astrolabe*, Part 1, paragraph 4. The reference to 'the tretis of the astrelabie' may be significant, see p. 158.

**C/32.** The division of the midnight line into 9° is devised for computing the trepidation (accession and recession) of the 8th sphere. In addition to the steady precession amounting to one revolution in 49,000 years, it was assumed that the 8th sphere oscillated 9° back and forth with a period of 7000 years. This scale on the midnight line could be used to compute the magnitude of this oscillatory displacement at any given date, the method employed being similar to the calculation of the latitude of the Moon (which is governed by much the same mathematical law). Since, however, both the steady and oscillatory motions of the 8th sphere are so slow, it would be easier to use arithmetic or a short table to find the position. Tables are in fact provided in the volume, but the author says nothing more about the theory of trepidation or its calculation by means of this scale on the midnight line; perhaps this is part of the subject-matter abandoned by the author towards the end of the treatise.

**D/1.** The Equatorie instrument consists of two parts, the solid disc which is the Face, and the ring-shaped Epicycle; the former is here styled the 'visage'—a somewhat curious use of the word. The **Epicycle** part of the instrument must not be confused with the **epicycle** circle of the Ptolemaic planetary theory; in the translation of the text they have been distinguished by capital and small initial letters, as above.

**D/5.** The three lines of writing, D/*a–c*, at the head of this page are preceded by the mark, //.b. They are to be inserted in the text at the end of D/5 where a corresponding symbol .a.// has been added. It is not surprising that the eccentric circle of the Sun is to be omitted from the Epicycle portion—there is no room for it, since the interior of the ring has been cut away.

**D/15.** The coincidence tests whether the two portions of the instrument are of equal radius.

**D/18.** A bar of metal (brass?) 1 in. wide and almost 6 ft. long would have to be very thick to maintain rigidity; this is a further indication that the instrument was never made in the large size called for.

**D/19.** 'Sowded' seems to connote *soldered*, in the modern sense. The operation was known and used in antiquity; Glaucus of Chios is reputed to be the inventor of the art. See A. Neuberger, *The Technical Arts and Sciences of the Ancients* (London, 1930), pp. 45 ff.

**D/24.** It is worth noting that here, unlike D/17, the actual metal 'latoun' (brass) is specified; it is, of course, the most suitable material for the instrument, since it can be worked into thin plates and engraved with ease.

**D/27.** Cf. *Treatise on the Astrolabe*, Part 1, paragraph 13. A contemporary form of the label of an astrolabe is as follows:

See R. T. Gunther (ed.), *Chaucer and Messahalla on the Astrolabe* (Oxford, 1929), p. 25.

**D/35.** This construction shows bad planning or lack of foresight on the part of the deviser of this text. The marks for the epicycles of the planets might have been derived from the line alhudda while it was still divided into 60 parts (B/18, 19); the redivision into 5 parts (C/25) could easily have been deferred until later. The bad arrangement has probably come about through the author's desire to complete the marking out of the Face before beginning the Epicycle.

**D/37.** The value of 6° 30′ for the radius of the epicycle of Saturn is identical with that used by Ptolemy and followed by John of Linières, al-Kāshī, and many other medieval authorities.

**E/5.** Two quite independent schemes are given for calculating the position of the Sun, one using the special eccentric circle drawn on the Face of the instrument, the other employing both Face and Epicycle in a procedure similar to that used for the other planets. In the second method there is, as noted here, to be no epicycle. Perhaps this duplication indicates that either the author's instrument or his source text is a modification of some original design in which only one of the schemes was employed.

**E/6.** The marks on the label trace out all the planetary epicycles in one movement as this label is rotated about its centre. It is worth noting that this mention shows that the instrument is conceived as a model of the Ptolemaic construction, and not just as a mathematical device to solve its equations.

**E/8.** The conventional phrases 'laus deo vero' and 'explicit' indicate that the source text may have ended at this place. A second section (probably from the same source) may begin at G/1, but the intervening comments accompanying the illustrations seem to be original.

**E/13.** Cf. note to E/5. Sparing of metal was of considerable importance owing to the high cost of brass and the difficulty of obtaining large sheets of good quality.

**E/26.** The remedy lies in deforming the Epicycle so that the distance between its centre and the common centre deferent (at its edge) shall be exactly equal to the radius of the Encloser of Signs on the Face. The consequent distortion of the rest of the Epicycle circle would be of little consequence, since it is only used as a sort of protractor for setting the label to its place. There would be considerable chance of 'mishap' because of the difficulty of soldering the parts of the Epicycle and setting it out without straining the metal. Unless this rather telling practical note is derived from the source text, the author would seem to have had practical experience in the manufacture of the instrument.

**E/33.** In numbering the lines of the text, the left-hand column has been taken first. Since this column starts with a reference to the sixth circle, while the right-hand column treats the third and fourth spaces and the Encloser of the Signs, it is quite likely that the right-hand column is intended to be read first, giving the line order: 32, 52–67, 33–51. In any case the change would be of little importance in meaning or style.

**E/44.** The statement that the line is figured 'boistosly' (crudely, roughly) provides a hint that the author has drawn his own diagrams, and that neither they nor the text are the work of a scribe or secretary. This agrees with our hypothesis that this manuscript is a first, rough draft, intended perhaps for later copying. The diagrams, though indeed somewhat rough, are unusually correct in their detail—scribal copies of such technical illustrations are almost invariably lacking in accuracy.

**E/58.** The diagram shows clearly that the equant of Mars (marked *e*) is situated just outside the circle of holes made for the centre deferent of the Moon. Since the tables do not give the distances of the equants from Arin it is hard to tell how far out it actually falls, but almost certainly the inner brass plate has been made large enough to include it (cf. A/20). A difficulty arises here, for the usual Ptolemaic value for the distance of Mars's equant is *less* than the radius taken for the circle of the centre deferent of the Moon. It has already been shown (note on B/20) that the value used in this text for the radius of that circle is substantially the same as the value given by Ptolemy, i.e. approximately 12° 28′. Since the radius of the inner brass plate on the Face of the Equatorie corresponds to 14° 7′, we should expect from the diagram (which seems accurate in other respects) that the distance of Mars's equant was greater than 12° 28′ and less than 14° 7′. The figure given by Ptolemy is, however, 12° 0′, which would lie within the circle of the centre deferent of the Moon. It is unlikely that an error has been made in the diagram, since a similar position is shown in the lower diagram on the next folio, and hence we have an unusual and probably characteristic deviation from the Ptolemaic constants. In this connection, the use of a value 13° 0′ by John of Linières (see p. 126) is of great interest. It seems likely that the constants of orbits have been derived from the works of Linières. This does not necessarily imply a connection between the present instrument and that of Linières, for it is possible that the constants only have been taken from his version of the Alfonsine Tables.

**F/5.** The advice is difficult to interpret; perhaps it means that if the little lip seems to be in the wrong place, or if it breaks in drilling, another might be made at some other point in the circumference of the Epicycle. The signs of the zodiac would then have to start with Aries at some different appropriate place.

**F/26.** The diagram of the complete instrument shows the common centre deferent pinned down to one of the holes in the circle of the centre deferent of the Moon. It is thus in a position for a determination of the ecliptic longitude of the Moon.

**G/d.** There seems to be no reason why this afterthought should be added just here. It is of considerable interest, since it shows explicitly that the writer recognizes that a period of exactly one year enters into the motions of all of the planets; the superior planets are dealt with here, the inferior planets contain the one-year period quite directly in their mean arguments. In the tables, the mean argument of the superior planets has been calculated using the rule expressed here. It has sometimes been suggested that pre-Copernican astronomers did not recognize this common annual period in all planetary motions, and that they therefore missed this strong suggestion of a heliocentric system; other evidence, notably the 'obvious' immobility of the Earth, was the overriding factor in maintaining the geocentric theories.

**G/e.** I can suggest no interpretation for the fragments of erased text here and at the foot of the folio (G/f); the figures for Saturn do not correspond to the aux or to any of the radices cited in the tables.

**G/2.** For a similar direction to use a slate for making rough notes see *Treatise on the Astrolabe*, Part II, paragraph 44.

**G/3.** The gloss 'terre' for 'aryn' is interesting. Cf. note on A/26.

**G/5.** The author appears to use the forms *different* and *defferent* interchangeably for the modern word *different* as well as for the astronomical term *deferent*. See p. 138.

**G/10.** The succession of signs is: Aries, Taurus, Gemini, Cancer, Leo, Virgo, Libra, Scorpio, Sagittarius, Capricornus, Aquarius, Pisces. Conventionally they are arranged in anti-clockwise order on the face of an instrument. The *succession of signs*, the positive direction, is therefore anti-clockwise; the contrary direction, clockwise, is called *against succession of signs*.

**G/25.** The term *equation* is used in two different senses in the text. In general, as 'to find the equation of the Moon', it signifies the position of the Moon in the zodiac. In the more specialized sense, as here, it corresponds approximately to the modern term 'anomaly'. See p. 171.

**G/35.** Tables of the mean arguments of all the planets are included in the volume, but none of these could be described as 'grene'. On f. 64r. there is a table of the declination of the Sun and Ascensions signs which is outlined in (faded) green ink. Perhaps the word is an error for 'grete' (great), the error being the reason for its cancellation.

**G/37.** Unlike most planetary motions, the mean argument of the Moon rotates clockwise; this is a fact which is important in the criticism of the planetary ideas of Plato. See J. L. E. Dreyer, *History of the Planetary System* (Cambridge, 1906), p. 66.

**G/f.** See note on G/e.

**H/1.** Here and at H/25 a word accidentally repeated at the end of one line and the beginning of the next has been expunged.

**H/10.** Of the remaining planets, Mercury and the Moon have movable centres for their deferents, while the Sun has neither epicycle nor equant.

**H/12.** There is a small gap below line H/11, and the new section then begins in the same hand, but considerably more cramped. This may indicate either a fresh start after a lapse of time, or a freshly cut pen.

**H/26.** According to the tables, the true aux of the Sun entered Cancer in July 1378, moving at the rate of approximately 1° per 100 years. By December 1392 it must have been about 0° 9′ of Cancer.

**H/29.** The maximum equation of the Sun is approximately 2°.

**I/whole page.** Throughout this page the paragraph and punctuation signs have been inserted in red ink; there seems to be no special reason for selecting this page for embellishment.

**I/g.** The whole of this page has been deleted by means of four diagonal strokes one way and one crossing in the opposite direction; the strokes are rough and scratchy, perhaps through the author's annoyance at finding such a large section to be wrong. It is, however, difficult to find any error which could not have been remedied by less drastic means. See notes on I/11, I/22 and K/i. Alternatively, one might suggest that the author was not so much concerned with any errors as with the rearrangement of the sections of his treatise; he might have intended to insert a similar passage at some other place.

**I/4.** According to the tables the true aux of Mercury was at 29° 23′ of Libra on 31 December 1392, and moving at a rate of approximately 1° in 100 years.

**I/11.** The reckoning in the little circle should be clockwise, against the succession of the signs. See p. 101.

**I/16.** There is a gap extending for the space of about five letters between 'as' and 'thilk' caused by writing a short correction over a longer erasure; a thin horizontal line has been used to connect the text.

**I/22.** The letters at either end of the diameter of this little circle cannot, of course, indicate the direction in which the circle is to be traversed! A more serious error is that the author insinuates that the equant of Mercury moves so as to be at the nadir of the centre deferent. In Ptolemaic theory this is not so, for the equant is fixed at the point indicated by E, and it is only the centre deferent which rotates. The author may perhaps have been led astray by a fancied analogy with the Moon where, in fact, both E and D do move.

**I/36.** See note on C/21.

**J/5.** The passage which has been cancelled contains the error of supposing that the mean motus of the Moon is always greater than the mean motus of the Sun.

**J/8.** The modern equivalent of 'remenaunt' is 'mean elongation of the Moon'.

**J/14.** The movable equant of the Moon, discussed here, has probably confused the author and led him into the error discussed in the note to I/22. The nadir of a hole is the hole at the other end of the diameter.

**K/i.** Lines 1–20 inclusive have been struck out by a pair of diagonal strokes in a similar fashion to folio I. I can find no serious error in this section.

**K/1–5.** For example, the result of subtracting 50° from 20° would be given as

$$20-50+360=330°.$$

The use of 'theorike' in this context is interesting, for the preceding account is certainly not a *theory* in the modern sense but rather a discussion of a practical topic or technicality. This may affect the interpretation of a passage from the Prologue of the *Treatise on the Astrolabe* (ed. Skeat, line 59) where Chaucer says, 'The .4. partie shal ben a theorik to declare the Moeuynge of the celestial bodies with (þe) causes'. If this is not restricted to a discussion of the *theory* of the planetary motions, it might perhaps refer to an account of the Equatorie instrument which declares (shows) the motions of the planets.

After the word 'mone' the writing suddenly becomes smaller and more cramped. Cf. notes to B/10 and H/12.

**K/6.** The addition 'in voluella' seems to refer to the Epicycle portion of the instrument with its rotating label. The term *volvelle* is now usually applied to the small equatoria made of vellum which are often to be found in the body or on the covers of astronomical manuscripts.

**K/8.** 'Riet' is puzzling. It may have something to do with the sense 'graduated scale' given in *OED* as a meaning for *riete*, though not before the seventeenth century. Or, more probably, it is allied with *rete*, the openwork metal plate on the front of an astrolabe containing pointers indicating the positions of the fixed stars. The author may be thinking of the Epicycle part of the instrument as forming a similar openworked plate.

**K/33.** The meridional line is the former line alhudda, now divided into 5° for use in computing the ecliptic latitude of the Moon.

**L/11.** The approximate value of 5° for the maximum ecliptic latitude of the Moon is that used by Ptolemy; the true value is only slightly greater.

**M/7.** Cf. note on G/25. Here, the meaning of 'mad equacion' is simply 'worked out'.

**M/22.** The sense demands an insertion between lines 22 and 23. A possible form would be:

22 & fro cauda til she come mid wey (in medio) by twix capud & cauda
**Insert** she is meridional descending / & in hir grettest latitude meridional
23 & fro thennes is she meridional assending til she come again at capud /

The form of the insertion is in keeping with the form adopted in the rest of the passage; alternatively, the addition of the words 'she is meridional descending' would be sufficient to preserve the sense and we have made this minimum addition in our translation. Is it significant that the most probable form of the necessary insertion would be of just the right length to occupy a full line of text at this place? Perhaps the author has skipped a line in translating or in working from his own rough notes. For a diagram of the stages in the movement from Caput to Cauda and back see Fig. 5, p. 59.

**M/24.** It may be shown that the actual numbers used in these three dated examples conform to the tables in our volume, and hence the examples are probably quite genuine practical illustrations—unlike some of the examples in the *Treatise on the Astrolabe*, where the dates seem to have been specially chosen so as to yield a simple result (i.e. by taking the date of an equinox or solstice). This check has the incidental result of enabling us to determine the correspondence between the different styles of dating used in our volume (see note to A/14). Because of its slow annual motion (19° retrograde) it is convenient to use the cited positions of Caput Draconis as a means of checking the dates. Taking first the dates 19 February and 23 February we find that Caput has moved from 20° 42′ to 20° 29′ of Aries; this motion of 13′ is only slightly greater than the calculated motion in 4 days. Taking next the date 17 December 1391 and the given position in 4° 46′ of Aries we find that the difference of 15° 56′ corresponds to almost exactly 10 months from 19 February. Having established the variation of Caput it is convenient to examine the positions given by the tables for the various dates cited in the form 'the last meridie of December 1392 complete'. The complete investigation is best illustrated graphically (see Fig. 6)—it will be observed that two of the points lie on a quite different line which corresponds to a motion of 19° per annum in the opposite direction; it seems that the author has committed an error in forgetting that the motion of Caput should be retrograde. A similar investigation may be carried out using the values given for the position of the Moon, but because of its more rapid motion and great irregularity the agreement is not so fine. Returning to the three numerical examples that close this text, it must be noted that the results which the author claims to have found by following the procedure described here are extraordinarily accurate; the differences between the values he gives and those obtained by direct calculation from the quoted figures for the position of the Moon and Caput are as follows:

| Example | Lat. found | Lat. calc. | Error |
|---|---|---|---|
| 1 | 1° 54′ | 1° 55′ 13″ | 1′ 13″ |
| 2 | 3° 22′ | 3° 22′ 37″ | 0′ 37″ |
| 3 | 1° 22′ | 1° 22′ 16″ | 0′ 16″ |

An accuracy of 1′ corresponds to a precision of about $\frac{1}{10}$ in. in finding where the thread cuts the meridional line. With an instrument considerably smaller than the ideally large model described in our text it would have been

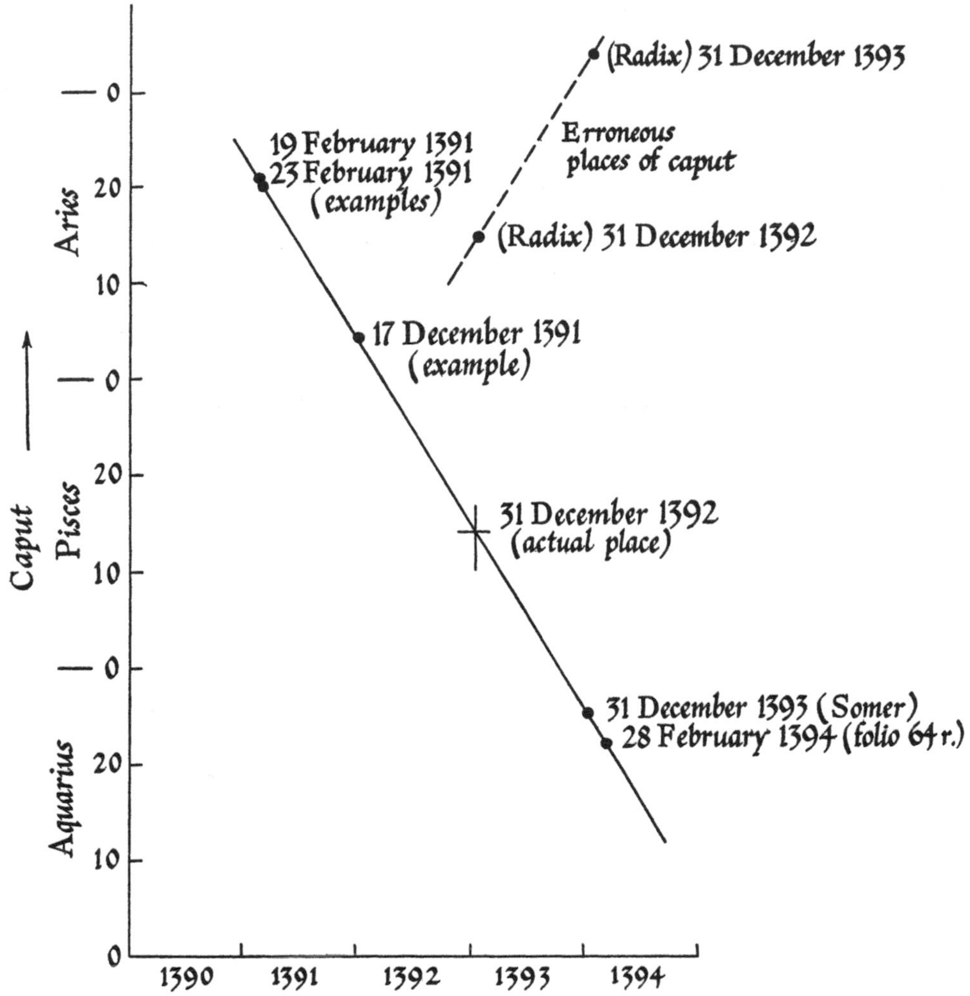


Fig. 6. Positions of Caput Draconis at various dates, as given in text and tables. The time scale shows the normal civil year starting on 1 January.

almost impossible to obtain such accuracy as is shown in these examples. We may conjecture that the examples have, in fact, been solved by direct calculation, or more probably by using a table of the latitude of the Moon such as is given on f. 32r.

**N/1.** Owing to its rapid and irregular motion it was not customary to draw up any almanac showing the true place of the Moon to such a degree of accuracy as is given here. The true place would have to be found from the tables of mean motus and mean argument in the usual fashion.

**N/14.** Once more 'aryn' is glossed as 'Earth'. Cf. G/3 and the note on A/26.

**N/17.** See note on N/1.

**N/29.** A gap of about ten letters in extent has been left between 'mone' and '.1. g^a^', apparently through an erasure. This space has been filled by the insertion of 'was' followed (as in I/16) by a thin horizontal line which connects the text. Cf. N/3.

**N/37.** This closing section is completely false, for it contains the supposition that Caput is restricted to one half of the zodiac, and Cauda to the other half. In point of fact both nodes rotate through the entire zodiac, making a complete revolution in the draconitic period of approximately 18·6 years.

**N/39.** The ending, though somewhat abrupt, and lacking any 'explicit', is sufficiently logical to bring this section to a satisfactory close. If the text were continued on a new quire it might have included a treatment of the trepidation and its computation by means of the 9° graduation of the midnight line. It could also include a treatment of the ecliptic latitude for the several planets, although this is a matter of considerable complication in Ptolemaic theory and leads to great complexity in the equatorium of al-Kāshī. On the whole it seems as if the text has been abandoned at the exact point at which the main features have been sufficiently discussed, and only complex theory remained.

# VI

# THE ASTRONOMICAL TABLES

THE text of the *Equatorie of the Planetis* occupies only eight out of the seventy-eight folios of the manuscript, the remainder containing sets of astronomical tables for use with the instrument described in the text. The tables themselves are of comparatively slight interest since they are a simple modification of the well-known Alfonsine tables, and it is therefore only necessary to indicate their content and the manner in which they have been modified. Of much greater importance to the present study are the notes, comments and fragments of text which are interspersed in these tables and probably represent original work.

Following is a page-by-page analysis of the contents of the volume, in which a short-title description only is cited for the tables and notes, etc., contained on each folio. This is followed by an analysis of the tables themselves and the numerical data from which they have been constructed and modified. In a final section we transcribe and comment upon the additional material which the author has inserted into this predominantly tabular part of the manuscript volume.

## 1. ANALYSIS OF CONTENTS OF THE MANUSCRIPT

Unless otherwise indicated the hand of Tables I (folios 1–13) and the text (folios 71–78) is thought to be that of Chaucer [Hand C], while that in Tables II (folios 14–64), with the exception only of the comments in cipher, is that of some professional scribe [Hand S]. The hand of quire h (folios 65–70) is difficult to identify, but it seems rather similar to Hand C, and may well have been written by the same person. The comments in cipher which occur at six places in Tables II are still more difficult to identify on account of the special symbols employed in the writing; from the few places where unciphered words are included in the notes it seems likely that these again are to be attributed to Hand C.

Special mention must be made of a set of 'notes' occurring throughout Tables I, but not mentioned specifically in the page-by-page analysis. The 'notes' consist of the single word 'exa*minatur*' which is written near almost every separate table in the whole section. Again it is very difficult to be certain of an identification for this hand, but as before similarity to Hand C suggests that the main author is again responsible. If this is the case, the notes imply that the author has followed the medieval practice of collating his copy against some independent copy of the same tables (cf. p. 8). The occurrence of *examinatur* is rather interesting in view of the tradition that William Thynne is said to have edited part of Chaucer's works with the help of a manuscript *in the poet's own hand* which bore on almost every page the phrase *Examinatur Chaucer*. See J. M. Manly and E. Rickert, *The Text of the Canterbury Tales* (Chicago, 1940–3), vol. I, p. 73. The same work notes that at the foot of some of the verses of the Cardigan manuscript of the *Tales* there appears the mark 'ex'—but this is a manuscript of the mid-fifteenth century, so there is no question of it being in holograph.

*Tables I* (*Hand C*)

**1r.** Note giving sexagesimal equivalent of 1392 years, 1393 years. Table for converting numbers of years to numbers of days. Planetary radices (mean motus and mean argument) for 31 December 1392. Table for annual motion of the centre deferent of the Moon; the radix is for the Era of Christ, at London.

**1v.** Tables for annual motion of the mean motus and mean argument of the Moon; the radices are for the Era of Christ, at London. Multiplication table for orders of sexagesimals. Short note in English.

**2r.** Tables for annual motion of the mean motus of Caput and of Saturn; the radices are for the Era of Christ, at London.

**2v.** Tables for annual motion of the mean motus of Jupiter and of Mars; the radices are for the Era of Christ, at London.

**3r.** Tables for annual motion of the mean motus of Venus and Mercury; the radices are for the Era of Christ, at London.

**3v.** Table for annual motion of the 'revolution' or mean motus of the ascendant for latitude 51° 34′; the radix is for 28 February 1393. Note in Latin giving instructions for use of table.

**4r.** Table of the ascensions of signs for latitude 51° 50′.

**4v.** Ascensions of signs, concluded. Table for converting numbers of hours to sexagesimal fractions of a day. Table for converting sexagesimal fractions of a day to hours and minutes.

**5r.** Radices of all tabulated quantities for 31 December 1392 and for the Era of Christ, at London. Radix for the argument of the ascendant for 1393 'incomplete'. A short (Latin) note explains that the argument of the 8th sphere is the same as its accession and recession. The positions of the mean auges of the planets are given for the Era of Christ.

**5v.** (Plate XI). Table for converting numbers of years (up to 800,000) to numbers of days, expressed sexagesimally. The 'radix Chaucer' note gives the number of days in 1392 years; a second note gives the number of days for 1395 years; a third note gives the radix of the ascendant at the time the Sun enters Libra in the year 1393 incomplete. Multiplication table for orders of sexagesimals.

**6r.** Table for diurnal motion of the mean argument of the 8th sphere; the radix is for 31 December 1392.

**6v.** True auges of all the planets for the Era of Christ, for 31 December 1392, and for 31 December 1400. Mean auges for 31 December 1392. A radix is also given for the mean elongation of the Moon, Era of Christ at Toledo.

**7r.** (Plate II). Table for annual motion of the mean motus of the Sun, Mercury and Venus (single table). Table for the annual motion of the argument of the 8th sphere. Table for the annual motion of the mean motus of the auges. Radices for Era of Christ and 31 December 1392, at London. A note in Latin describes the difference between true and local time.

**7v.** Table of the diurnal motion of the mean motus of the Sun, Mercury and Venus (single table); the radix is for 31 December 1392, at London. Table of fractions of a day.

**8r.** Table of the diurnal motion of the mean motus of the Moon; the radix is for 31 December 1392, at London.

**8v.** Table of the diurnal motion of the mean argument of the Moon; the radix is for 31 December 1392, at London.

**9r.** Table for the diurnal motion of the mean motus of Mars; the radix is for 31 December 1392, at London.

**9v.** Table of the diurnal motion of the mean argument of Mars; the radix is for 31 December 1392, at London. A line of numerals has been added to this table in a much later hand, perhaps identical with that of the 1461 note of ownership (?) on f. 74v.

**10r.** Table for the diurnal motion of the mean argument of Saturn; the radix is for 31 December 1392, at London.

**10v.** Table for the diurnal motion of the mean motus of Saturn; the radix is for 31 December 1392, at London.

**11r.** Table for the diurnal motion of the mean argument of Jupiter; the radix is for 31 December 1392, at London.

**11v.** Table for the diurnal motion of the mean motus of Jupiter; the radix is for 31 December 1392, at London.

It is possible that four folios are missing at this point. See p. 14.

**12r.** Table for the diurnal motion of the mean argument of Mercury; the radix is for 31 December 1392, at London.

**12v.** Table for the diurnal motion of the mean argument of Venus; the radix is for 31 December 1392, at London.

It is possible that a single folio is missing at this point. See p. 14.

**13r.** Table for the diurnal motion of the mean motus of Caput Draconis.

**13v.** Table for the diurnal motion of the mean motus of the auges and fixed stars (single table). Mean auges of all the planets for the Era of Christ, at London.

*Tables II* (*Hand S*)

**14r.** Calendar for the mean motus of the auges. Note in cipher.

**14v.** Mean motus of auges, concluded.

**15r.** Calendar for the mean argument of the 8th sphere.

**15v.** Mean argument of the 8th sphere, concluded.

**16r.** Calendar for the equation of the 8th sphere. Explanation of the calendar, written (in Latin) in a formal hand different from S.

**16v.** Equation of the 8th sphere, concluded. Calendar for the mean motus of the Sun.

**17r.** Mean motus of the Sun, continued.

**17v.** Mean motus of the Sun, concluded. Calendar for the mean motus of the Moon.

**18r.** Mean motus of the Moon, concluded.

**18v.** Calendar for the mean argument of the Moon.

**19r.** Mean argument of the Moon, concluded.

**19v.** Calendar for the centre deferent of the Moon.

**20r.** Centre deferent of Moon, concluded. Hand C notes 'Pro instrument' equatorij'.

**20v.** Calendar of the true motus of Caput Draconis.

**21r.** True motus of Caput Draconis, concluded.

**21v.** Calendar for the mean motus of Saturn.

**22r.** Mean motus of Saturn, concluded.

**22v.** Calendar for the mean argument of Saturn.

**23r.** Mean argument of Saturn, previous side corrected and continued.

**23v.** Mean argument of Saturn, concluded.

**24r.** Calendar for the mean motus of Jupiter.

**24v.** Mean motus of Jupiter, concluded.

**25r.** Calendar for the mean argument of Jupiter.

**25v.** Mean argument of Jupiter, concluded.

**26r.** Calendar for the mean motus of Mars.

**26v.** Mean motus of Mars, concluded.

**27r.** Calendar for the mean argument of Mars.

**27v.** Mean argument of Mars, continued.

**28r.** Mean argument of Mars, concluded.

**28v.** Calendar for the mean argument of Venus.

**29r.** Mean argument of Venus, concluded.

**29v.** Calendar for the mean argument of Mercury.

**30r.** Mean argument of Mercury, concluded.

**30v.** (Plate III). Calendar for the mean centre of the Moon; the radix for 28 February 1392 is added in hand C and repeated (in Latin) by a formal hand different from S; it is the same hand as on f. 16r. Note in cipher.

A single folio appears to be lacking at this point, but the table continues without a break. See p. 14.

**31r.** Mean centre of the Moon, continued.

**31v.** Mean centre of the Moon, concluded. Another Latin note in the same formal hand as occurs on f. 16r. and f. 30v.

**32r.** Table for the ecliptic latitude of the Moon.

**32v.** Table for the ecliptic latitude of Saturn.

**33r.** Latitude of Saturn, concluded.

**33v.** Table for the ecliptic latitude of Jupiter.

**34r.** Latitude of Jupiter, concluded.

**34v.** Table for the ecliptic latitude of Mars.

**35r.** Latitude of Mars, concluded.

**35v.** Table for the ecliptic latitude of Venus.

**36r.** Latitude of Venus, continued.

**36v.** Latitude of Venus, concluded.

**37r.** Table for the ecliptic latitude of Mercury.

**37v.** Latitude of Mercury, continued.

**38r.** Latitude of Mercury, concluded.

**38v.** (Plate IV). Tables of proportion, for the multiplication of sexagesimal numbers. Note (in Latin) on the multiplication of sexagesimal units. Note in cipher. The tables continue to the end of f. 44v.

**45r.** Table of the equation of Saturn, continuing to the end of f. 47v. On f. 45r. there is a mark and a (Latin) note referring to the 18th folio following, i.e. f. 62r., where stands a similar mark.

**48r.** Table of the equation of Jupiter, continuing to f. 50v. There is a (Latin) note on f. 49r.

**51r.** Table of the equation of Mars, continuing to f. 54v.

**55r.** Table of the equation of Venus, continuing to f. 58r.

**58v.** Table of the equation of Mercury, continuing to f. 61r.

**61v.** Table of the ascensions of signs and houses for latitude 51° 50′.

**62r.** Precession for the years 1349–1468, at a rate of 1° in 100 years starting from 1349. There is a note that this is the folio referred to at f. 45r.

A different hand continues with a collection of miscellaneous items included in folios 62v.–64v. It may well be the same hand that has written quire h (folios 65–70), and indeed both may be due to the writer of Hand C working at some other time, or using a different pen.

**62v.** (Plate V). Table giving the excess degrees in 1 to 12, and 24 hr., for 600, 624,

PLATE II

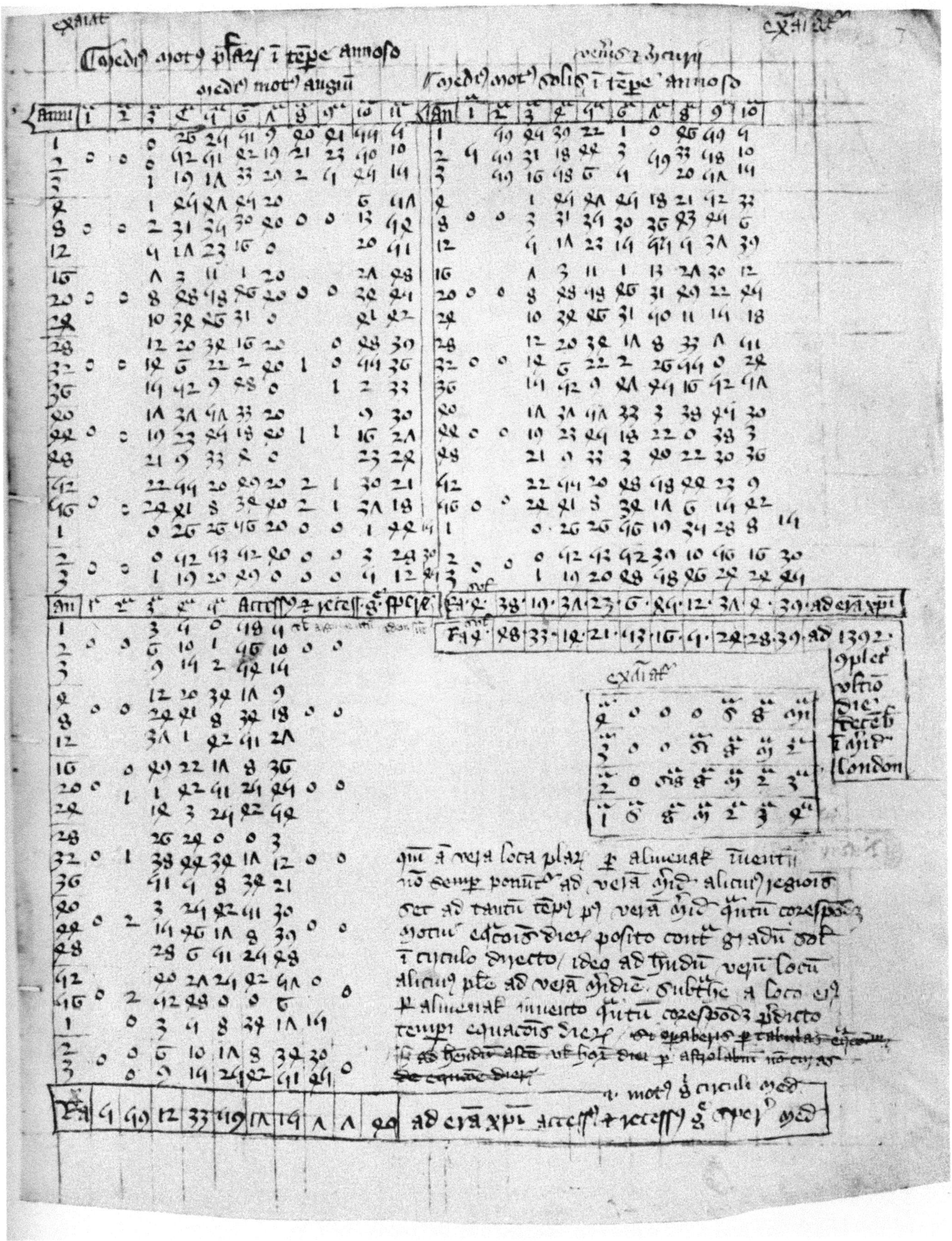

FOLIO 7 RECTO

*see p.* 85

PLATE III

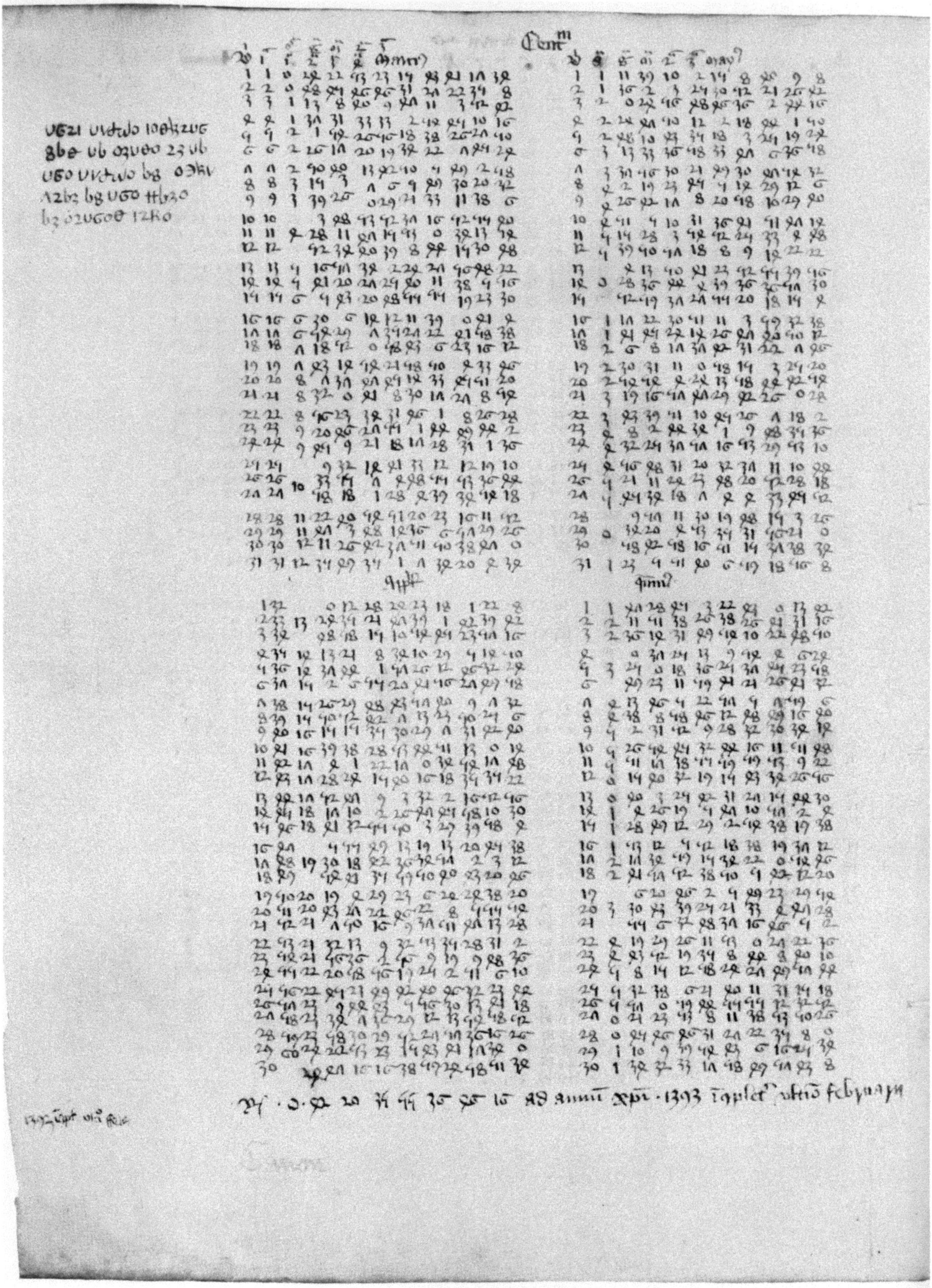

FOLIO 30 VERSO

*see p.* 85

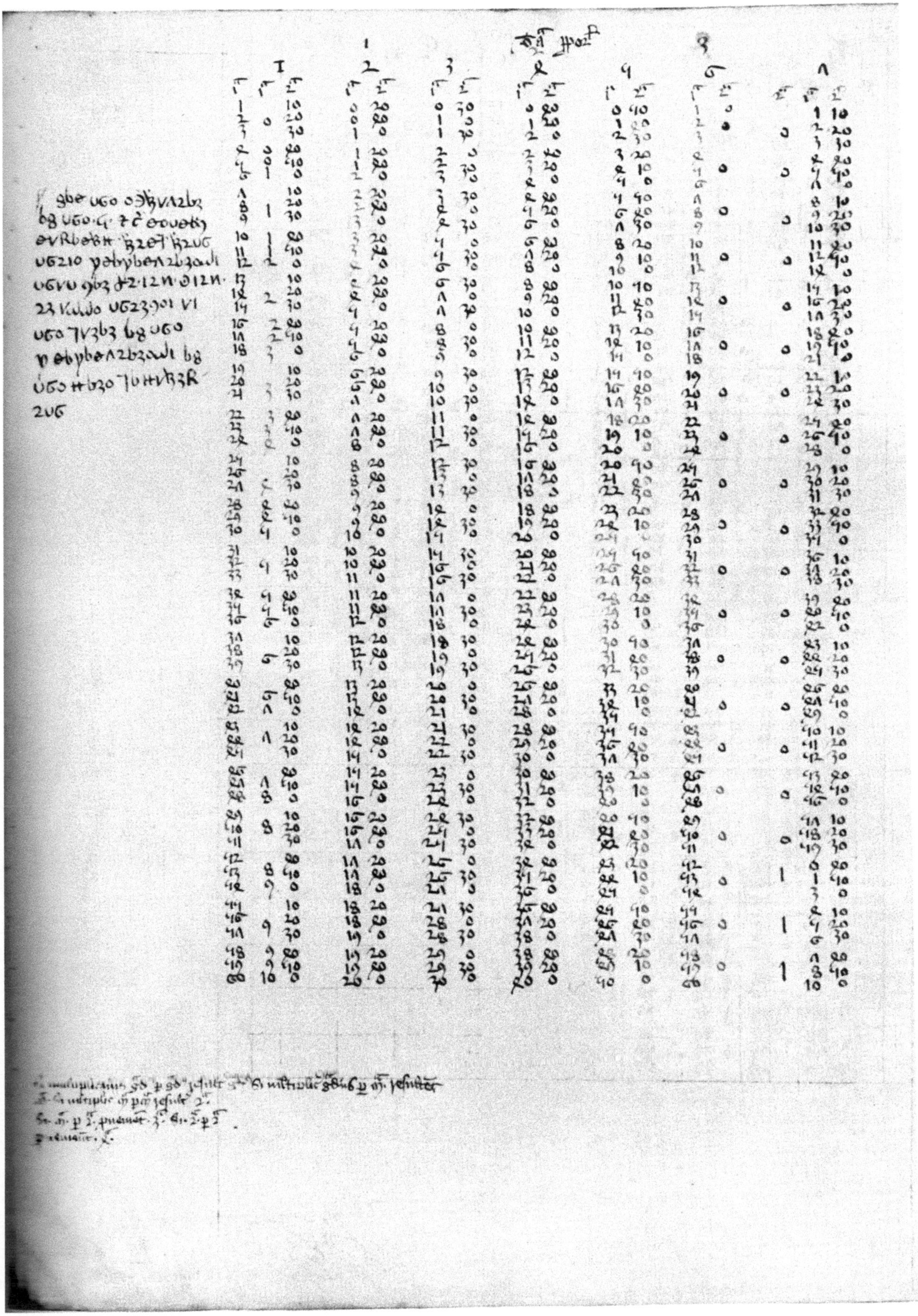

FOLIO 38 VERSO

*see p.* 85

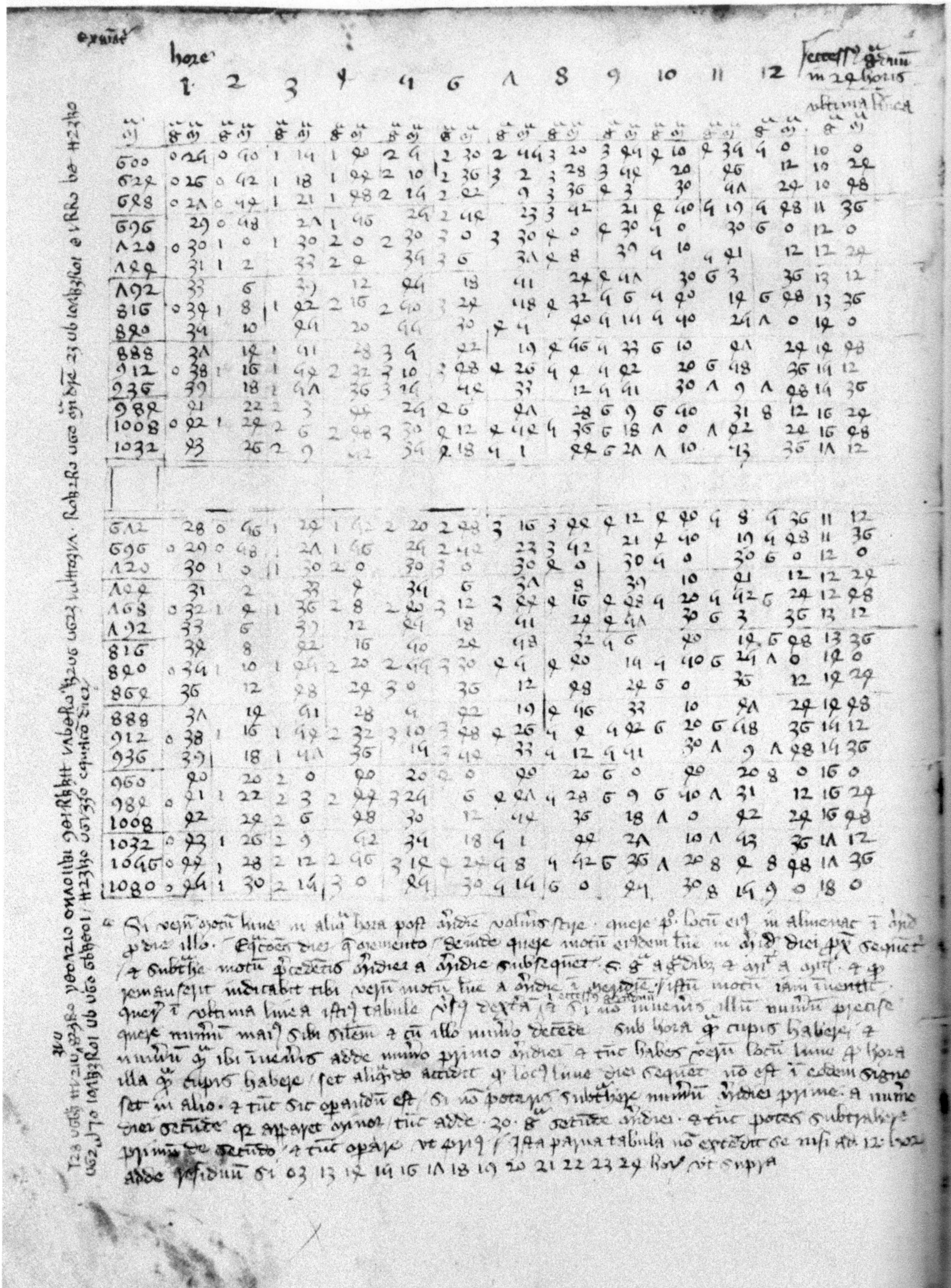

FOLIO 62 VERSO

*see pp.* 85–6

648, ..., 1080 min. There is an explanation in Latin, and a cipher note has been added along the left-hand margin of the page.

**63r.** Clean vacant folio.

**63v.** (Plate VI). Table giving the excess of the longest day, over the equinoctial day, for polar altitudes from 1° to 60°. A cipher note refers to it. List giving the true motus and the ecliptic latitude (in degrees and minutes) on 31 December 1393 for each of the planets; the list is attributed to John Somer of Oxford.

**64r.** (Plate VII). Table of the declination of the Sun, and diverse ascensions of signs, for all the habitable world; it is outlined in (faded) green ink. At the top of the page a quite different hand has added a suggested attribution to Azarquiel. A list, tabulated in a column containing five sections, gives the mean motus, mean argument, mean centre, true auges and 'diversitas' for all the planets on 28 February 1394, at London.

**64v.** (Plate VIII). Latin text accompanying a figure of a horoscope. This leaf is much stained and worn.

This concludes Tables II and quire h follows. It is possible that the first two folios of this quire are missing. See p. 14.

**65r.** (Plate IX). Table of the ascensions of signs for the latitude of Oxford, 51° 50′. The table continues to f. 70v. which is a stained and dirty leaf. On folio 70 *recto* is a (Latin) note referring to Profatius.

---

The two quires containing the text now follow. With the exception only of the note at the foot of f. 74v. this is all in Hand C.

**71r.** (Plate X). An unfinished table of the names and positions of the fixed stars; it is written on a stained and dirty leaf.

**71v. to 74r.** Text of the *Equatorie of the Planetis*, pages A to F.

**74v.** Vacant leaf; at the foot appears an ownership (?) mark dated 19 August 1461.

**75r. to 78v.** Text of the *Equatorie of the Planetis*, pages G to N. Two short lines of writing at the foot of the last side have been erased; they are hardly legible in ultra-violet light, but a possible reading is: '*memorandum quod*...for the additiou*n* of signes.'

## 2. ANALYSIS OF TABLES, NUMERICAL VALUES, ETC.

Although at first sight the two main tabular sections of our manuscript, Tables I and Tables II, appear to be quite different in form, they are both copies from the same recension of the Alfonsine Tables and make no change in the fundamental data contained in the original version of these tables beyond a slight conversion which enables them to be used for the local time of London rather than that of Toledo.

Unfortunately there have been very few mathematical analyses of medieval astronomical codices of tables, and most investigators have had to restrict their means of identification to recognizing some particular radix date or the 'incipit' of a canon of explanation attached to the table. Both these methods involve considerable uncertainty, and neither enables one to tell whether the table represents a new set of calculations or a modification of an earlier work.

Because of such difficulties the identification of the particular version of the Alfonsine Tables is not easy. To make matters worse there is still a confusion, already present in the Middle Ages, between these *Alfonsine Tables* drawn up for Toledo, and the much earlier *Toledo Tables* which they superseded.[1] The original Alfonsine Tables were prepared *c.* 1272 in Toledo under the patronage of Alfonso X (El Sabio), king of Léon and Castile (1252–84); the actual authors of the tables seem to have been two Hispano-Jewish astronomers, Judah ben Moses ha-Kohen and Isaac ibn Sid ha-Hazzan. The original tables are lost, but the introduction has been preserved (*Libro de las taulas Alfonsies*), and thus we have a definite idea of the nature of these tables, and can appreciate their differences from later ones. It is highly probable that the original tables were written in Spanish and used the Roman rather than the Arabic numerals,[2] and in this form the tables could hardly exert any influence outside the School of Toledo.

The Alfonsine Tables owe their tremendous popularity to the Latin versions which seem to have been introduced at Paris by John of Linières and his pupil John Dancowe of Saxony, *c.* 1320, although Dreyer[3] has been able to recognize another group of tables reduced to the meridian of Oxford which seems to be much nearer the original form of the Alfonsine compilation. It is, however, the Linières group which seems to have persisted through various manuscript versions into the first printed edition of 1483 (Ratdolt, Venice), and the many subsequent editions which use the same numerical material with supplementary matter.[4]

The main body of the tables, giving the diurnal and annual motions of the mean motus and argument of the planets, is quite valid for any place of observation, but the radices and auges must be modified to a local time in accordance with the longitude of the place of use. This modification is to be found in numerous manuscripts of the Alfonsine Tables intended for places other than Toledo (e.g. Paris, Oxford), and the tables in our manuscript represent just such an adaptation for the longitude of London. Hence whilst

[1] An excellent account of the Toledo Tables is given by E. Zinner, 'Die Tafeln von Toledo', *Osiris*, vol. I (Bruges, 1936), p. 747. It is quite clear from the constants quoted there (pp. 763–4) that the Toledo Tables are based on numerical data quite different from those of the Alfonsine Tables.

[2] A full discussion of the tables and their authors and patron is given by G. Sarton, *Introduction to the History of Science*, vol. II (Baltimore, 1931), pp. 837 ff. The introduction to the tables has been published by D. Manuel Rico y Sinobas, *Libros del Saber* (Madrid, 1863–7), vol. IV, p. 119; the same volume also gives some 80 pp. of facsimiles of manuscript tables, but these are but fragments of a set of perpetual ephemerides quite different from the planetary tables described in the introduction to the Alfonsine work.

[3] J. L. E. Dreyer, 'On the original form of the Alfonsine Tables', *Mon. Not. R. Astr. Soc.* (January 1920). His identification is based on the fact that these tables use *signa communia* of 30° instead of the *signa physica* of 60° found in the Linières group, and they show no trace of the sexagesimal division of time. See also R. T. Gunther, *Early Science in Oxford*, vol. II (Oxford, 1923), pp. 44 ff.

[4] Later editions: Venice, Hamann, 1492; Venice, 1518, 1521, 1524; Paris, 1545, 1553; Madrid, 1641. In the present investigation we have chiefly used the edition of 1553. For a discussion of the bibliography see A. Wegener, *Bibliotheca Mathematica*, series 3, vol. VI (1905), p. 129.

the constants in Table 1 (below) are identical with those found in the printed editions, the radices and auges in Tables 2 and 3 have been corrected by the subtraction of a quantity corresponding to the motion of the appropriate quantity in a time interval of 33 min. 44 sec., or a longitude 8° 26′ east of Toledo.[1]

This process of adaptation may be illustrated by an example; in the printed version (edition 1553, p. 10) a radix is given for the mean argument of the 8th sphere (i.e. accession and recession) at noon on the day taken as the 'Era of Christ'. This radix value is $5^{ss}\ 59°\ 12'\ 34''$ at the meridian of Toledo. To calculate the correction corresponding to a difference in longitude of 8° 26′ we must first compute a quantity equal to 8° 26′/360° of the diurnal motion of the mean argument of the 8th sphere. This proceeds as follows:

$$8°\ 26'/360° = 0^p 1^1 24^2 20^3\ (=0+1/60+24/60^2+20/60^3).$$

TABLE 1. *Motions*

For an excellent account of the system of sexagesimal arithmetic see O. Neugebauer, *The Exact Sciences in Antiquity* (Oxford, 1951), pp. 16ff.

| | Diurnal motion | Annual motion (365¼ days) |
|---|---|---|
| | **Mean motus** | |
| Sun, Venus and Mercury | 0 00 59 08 19 37 19 13 56 13 | 0 00 00 26 26 56 19 35 28 08 15 |
| Mars | 0 00 31 26 38 40 05 | 3 11 24 56 53 30 26 15 |
| Jupiter | 0 00 04 59 15 27 07 23 50 | 0 30 21 43 48 51 46 50 07 30 |
| Saturn | 0 00 02 00 35 17 40 21 | 0 12 14 04 51 19 52 50 15 |
| Moon | 0 13 10 35 01 15 11 04 35 | 2 12 40 41 22 41 10 39 03 45 |
| Caput Draconis | 0 00 03 10 38 07 14 49 10 | 0 19 20 29 33 36 57 48 07 30 |
| Auges and Stars | 0 00 00 00 04 20 41 17 12 26 37 | 0 00 00 26 26 56 20 00 00 01 44 15 |
| | **Mean argument** | |
| Mercury | 0 03 06 24 07 42 40 52 | 0 54 43 22 56 34 16 33 |
| Venus | 0 00 36 59 27 23 59 31 | 3 45 10 56 32 48 03 27 45 |
| Mars | 0 00 27 41 40 57 14 13 56 13 | * |
| Jupiter | 0 00 54 09 04 10 11 50 06 13 | * |
| Saturn | 0 00 57 07 44 19 38 52 56 13 | * |
| Moon | 0 13 03 53 57 30 21 04 13 | 1 31 59 13 19 00 45 55 08 15 |
| 8th sphere | 0 00 00 00 30 24 49 | 0 00 03 05 08 34 17 15 |
| Centre deferent of Moon | * | 3 47 20 11 31 11 28 31 52 31 |

* These values are not given explicitly in the tables. In all numerical data, as given above, the first (single) numeral gives the number of signs (60°) and this is followed by (double) numerals giving numbers of degrees, minutes, seconds, thirds, etc. It should be noted that in some cases a whole number of complete revolutions has been neglected, e.g. the annual motion of the mean motus of the Sun is $360°\ 0'\ 26''\ 26'''$, etc., but only the excess over 360° is given in Table 1.

[1] London is only about 4° east of Toledo according to modern determinations, but the usual medieval figure for the difference in longitude is 8–9° (cf. Ptolemy, *Geographia*, Toledo 11° E., Oxford 20° E.; cf. also values given by Richard Thorpe, cited p. 157 n.).

Table 1 gives the required diurnal motion as 0 00 00 00 30 24 49, hence the required correction is 0 00 00 00 30 24 49 × 0 01 24 20, i.e.

| | | | | | | | | | | |
|---|---|---|---|---|---|---|---|---|---|---|
| | 0 | 00 | 00 | 00 | 00 | 30 | 24 | 49 | | |
| | | | | | | 12 | 09 | 55 | 36 | |
| | | | | | | | 10 | 08 | 16 | 20 |
| Total | 0 | 00 | 00 | 00 | 00 | 42 | 44 | 52 | 52 | 20 |

Subtracting this from the radix for Toledo, we obtain

| | | | | | | | | | |
|---|---|---|---|---|---|---|---|---|---|
| 5 | 59 | 12 | 34 | | | | | | |
| 0 | 00 | 00 | 00 | 00 | 42 | 44 | 52 | 52 | 20 |
| 5 | 59 | 12 | 33 | 59 | 17 | 15 | 07 | 07 | 40 |

which is exactly the value contained in our manuscript and recorded towards the bottom of the left-hand column in Table 2 (below). A similar procedure has been used to derive all the quantities given for 'the Era of Christ, at London' and noted in the left-hand columns of Tables 2 and 3. Values of radices and auges for other dates *at London* have been calculated from the data for the 'Era of Christ' just found by adding the corresponding motion for the time elapsed. For example, the values for 31 December 1392 are derived by adding the motion for exactly 1392 years to the radix or auge for the Era of Christ.

It is worth remarking that this general process of adaptation leads to a spurious and excessive 'accuracy' in the final figures. The tables have originally been calculated from observations whose accuracy certainly did not exceed several minutes of arc, and a modern computor would not therefore hesitate to discard seconds, thirds, etc., in his final result. It needs considerable mathematical sophistication to discard the accuracy which has been so painstakingly achieved by long calculations, and hence medieval astronomers were loth to neglect any calculated quantity or correction indicated by their theory. In the example worked above, a radix given to seconds of a degree has been 'corrected' by a term amounting to fourths of a degree, and the final answer appears to an accuracy of eighths ($60^{-8}$) of a degree—for comparison, one sexagesimal eighth part of one million years is approximately 5 sec. of time. Although worthless from the astronomical point of view, this excessive accuracy is of great value to a modern investigator of such tables, for he is able to use the unapproximated values to work backwards and ascertain thereby the exact process of computation by which any result has been obtained by the medieval astronomer.

## Table 2. *Radices*

| | Radix Era of Christ | Radix 31 December 1392 | Radix 31 December 1393 |
|---|---|---|---|
| | Mean motus | | |
| Sun, Venus and Mercury | 4 38 19 37 23 06 45 12 37 04 39 | 4 48 33 14 21 53 16 05 24 28 39 | 4 48 18 53 43 |
| Mars | 0 41 24 45 31 12 48 59 38 20 | 1 32 12 38 53 21 48 59 38 20 | 4 43 29 44 06 |
| Jupiter | 3 00 37 13 43 22 36 52 36 10 03 20 | 5 24 45 43 20 41 11 46 36 10 03 20 | 5 55 06 12 19 |
| Saturn | 1 14 05 17 22 30 23 29 37 09 | 3 04 45 56 15 44 13 17 37 09 | 3 16 59 30 58 |
| Moon | 2 02 28 19 04 08 34 09 25 53 28 | 2 10 28 17 23 27 40 24 25 53 28 | 4 19 51 20 00 |
| Caput Draconis | 1 31 55 48 12 03 05 08 50 13 37 | 0 15 21 36 05 36 05 38 50 13 37 | 0 34 41 17 59 |
| | Mean argument | | |
| Mercury | 0 45 19 32 00 05 09 40 33 34 40 | 4 19 47 48 27 17 07 16 34 34 40 | 5 13 44 |
| Venus | 2 09 21 10 56 45 49 16 40 45 40 | 0 23 13 03 55 42 09 04 40 45 40 | 4 08 14 |
| Mars | * | 3 16 20 35 28 31 27 05 46 08 39 | 0 04 49 |
| Jupiter | * | 5 23 47 31 01 12 04 18 48 18 35 40 | 4 53 12 |
| Saturn | * | 1 43 47 18 06 09 02 47 47 19 39 | 1 31 19 |
| Moon | 3 18 41 52 42 26 30 20 23 04 24 | 1 24 38 49 48 11 49 31 47 04 24 | 2 53 22 |
| 8th sphere | 5 59 12 33 59 17 15 07 07 40 | 1 10 47 52 50 45 27 07 07 40 | 1 10 50 |
| Centre deferent of Moon | 1 14 10 55 42 04 56 15 48 15 50 | 1 26 38 11 20 18 51 46 23 03 50 | 5 16 46 27 |

## Table 3. *Auges*

| | Era of Christ | 31 December 1392 | 31 December 1400 |
|---|---|---|---|
| | Mean auges | | |
| Sun and Venus | 1 11 25 22 03 53 53 35 18 04 51 | 1 21 38 59 02 50 | * |
| Mercury | 3 10 39 33 03 53 53 35 18 04 51 | 3 20 53 10 02 50 | * |
| Mars | 1 55 12 13 03 53 53 35 18 08 51 | 2 05 25 50 02 50 | * |
| Jupiter | 2 33 37 00 03 53 53 35 18 08 51 | 2 43 50 37 02 50 | * |
| Saturn | 3 53 23 42 03 53 53 35 18 08 51 | 4 03 37 19 02 50 | * |
| | True auges | | |
| Sun and Venus | 1 11 17 55 11 23 11 45 25 08 55 | 1 30 08 44 01 47 | 1 30 13 34 51 00 |
| Mercury | 3 10 32 06 11 23 11 45 25 08 55 | 3 29 22 54 01 47 | 3 29 27 44 51 00 |
| Mars | 1 55 04 46 11 23 11 45 25 08 55 | 2 13 55 34 01 47 | 2 14 00 24 51 00 |
| Jupiter | 2 33 29 33 11 23 11 45 25 08 55 | 2 52 20 21 01 47 | 2 52 25 11 51 00 |
| Saturn | 3 53 16 15 11 23 11 45 25 08 55 | 4 12 07 03 01 47 | 4 12 11 53 51 00 |

* See note to Table 1.

This method may be used to check some of the data given in Tables 1 and 2; for in Ptolemaic theory,

Centre deferent of Moon = Twice mean motus of Sun − Mean motus of Moon,

and

Mean argument of Mars/Jupiter/Saturn

= Mean motus of Sun − Mean motus of planet.

It is easy to see from the figures cited for motions and radices that these two laws have been followed without deviation. The ease with which these quantities can be calculated from the mean motus values is probably responsible for the absence of certain tables of motions and radices relating to them.

The actual arrangements of the tables using this numerical material may be seen in Plates II and III (folios 7r. and 30v.). For a discussion of the Alfonsine theory of precession (auges, motion of 8th sphere, etc.) see p. 104.

## 3. NOTES AND OTHER MATERIAL INSERTED IN THE TABLES

**1r.** At the head of the page is a note, 'Tempus corespondens an' / 1392 / complet' vltimo die 10bre 23 secunda & 12 primis annorum / & ultimo die 10bre a° 1393 23 secunda & 12 primis &c', so expressing sexagesimally the number of years elapsed which correspond to the dates in question; i.e. A.D. 1392 = 23, 12; and A.D. 1393 = 23, 13, etc. This information is necessary for working out the positions of the planets, starting from tables using a radix given for the Era of Christ. Thorndike[1] has taken this note as providing an 'incipit' for the whole text, but such a mnemonic entry can hardly be considered characteristic enough for the identification of either the 'Equatorie' or the tables.

**1v.** The multiplication table for sexagesimal orders is as follows:

| | | | | | | |
|---|---|---|---|---|---|---|
| $4^{a}$ | o | o | o | $s^{a}$ | $g^{a}d$ | $mi^{a}$ |
| $3^{a}$ | o | o | $s^{a}$ | $g^{a}$ | $m^{a}$ | $2^{a}$ |
| $2^{a}$ | o | $s^{a}$ | $g^{a}$ | $m^{a}$ | $2^{a}$ | $3^{a}$ |
| $1^{a}$ | $s^{a}$ | $g^{a}$ | $m^{a}$ | $2^{a}$ | $3^{a}$ | $4^{a}$ |

The bottom line of this table gives the units (primes), which consist of double signs, degrees, minutes, etc. The line above this gives the result of multiplying each unit by 60, e.g. 60 min. gives 1°. The next line above multiplies the series by a further 60, etc. A note in English, apparently referring to the tables for the mean motus and argument of the Moon, reads, 'Shaltow neu*er* abid þ$^{t}$ yer / þ$^{t}$ $4^{a}$ of yeris wole serue for thyn introit*us*'—the exact meaning of this is not clear.

**3v.** The figure given for the mean motus of the 'revolution' shows that it is intended to give the number of revolutions made by the Sun, with respect to the horizon, in the time of one tropical year. Because of the adoption of the Alfonsine precession, the tables are based on the equation 49,000 solar years = 49,001 tropical years. The 'revolution' is thus (almost) exactly $365\frac{1}{4} \times 360° \times 49{,}000/49{,}001$. In the figure given in the tables the whole number of rotations (365) has been omitted as usual, and the revolution is thus given as 1 27 18 59 42 26 28 28 47 35 12 per annum. The radix for 1393 incomplete is given as 3 15 55 49 12 03 35 08. A note of explanation reads 'Sub-t*ra*he 1393 de nu*mer*o tuo quesit*o* & int*ra* cu*m* residuo i*n* p*ri*ma*m* linea*m* tabule p*re*sent*is* / & adde i*n*uentu*m* i*n*dire*c*to sup*er* radice*m* ꝥ *in* pede istiu*s* tabul*e* / & diuide sig*na* *in*

[1] L. Thorndike, 'Pre-Copernican astronomical activity', *Proc. Amer. Phil. Soc.* vol. XCIV (1950), p. 321, and L. Thorndike and P. Kibre, *A Catalogue of the Incipits of Medieval Scientific Writings in Latin* (Medieval Academy of America, 1937), col. 717. In both places the date has been given as 1329 in error for 1392.

grad*ibus* / q*ui* no*n* potest diuidi i*n* 60 cu*m* minut*is* / quere tu*n*c p*ro*ueniens i*n* tabul*is* asce*nsio*nis circuli tui obliq*ui* / & p*ri*ma linea *illius* tabul*e* o*stend*et *gradum* ascend*entem* / & sig*num* i*n* capit*e* sub q*uo* inue*n*ti nu*merum* tu*um* *est* est signu*m* asce*nsionis* // Incipe i*n* Marci*us* / & no*n* in Januariis'.

**5v.** See p. 159.

**6v.** Radix elongation Moon for Era of Christ at Toledo is given as 3 24 25 49 46 11 56.

**7r.** The note on true and local time reads as follows: '*Quoniam* a*utem* vera loca pla*netarum* p*er* almenak i*n*uentu*m* no*n* [semp*er*][d] ponu*n*t*ur* ad vera*m* m*eridi*e*m* alicui*us* regio*n*is set ad tantu*m* te*mpus* p*ost* vera*m* m*eridiem* . q*ua*ntu*m* corespo*ndet* motui eq*uacio*nis dier*um* posito cont*ra* gradu*m* sol*is* i*n* circulo directo / ideo ad h*abe*ndu*m* veru*m* locu*m* alicui*us* pl*ane*t*e* ad vera*m* m*eridiem* . subt*ra*he a loco ei*us* p*er* almenak inuento q*uantum* corespo*ndet* p*re*dicto temp*or*i equac*io*nis dier*um* [si op*er*aberis p*er* tabulas eq*uacionum* *set* ad h*abe*ndu*m* asce*nsionem* v*el* h*or*a*m* diei p*er* astrolabiu*m* no*n* curas de equac*ione* dier*um*][d].'

*Translation*

In the almanac, the true places of planets do not [always] correspond to true noon in any particular place, but rather to such a time after true noon as will agree with the motus of the equation of days at the Sun's place in the direct circle. Therefore, to find the true place of any planet at the true noon, subtract what corresponds to the aforesaid time of the equation of days from the place that is shown by the almanac. [This is only if the tables of equations are used; but to find the ascension or hour of the day by means of an astrolabe, the equation of days is not employed.]

**14r.** The gloss in cipher reads, '/ if the liketh to knowe the verre auges of pl*anete*s for yeris or montis or daies adde thise mene motes to the rotes of the verre auges of a. ⟨1392⟩. & tak ther the verre aux of thi pl*anete* desired adde a yer for a yer or a dai for a dai'.[1]

**30v.** The gloss in cipher reads, 'this table servith for to entre in to the table of equacion of the mone on either side'.

**38v.** The gloss in cipher reads, 'for the equacion of the ⟨.5. &c⟩ retrog-radorum wirk with thise proporcionels that gon bi six & six in alle thinges as the kanon of the proporcionels of the mone komaundith'. A Latin note expresses verbally the multiplication table for orders of sexagesimals, 'Si multiplicamus gradus per gradus resultent gradus, si multiplicamus gradus per minuta resultent minuta, si multiplicamus minuta per minuta resultent secunda, si minuta per secunda prouenient tertia, si secunda per secunda prouenient quarta'.

**62v.** The gloss in cipher reads, 'yif thou maist ($_\wedge$ nat) finde precise excessus graduum acorde with thin almenac devide the ⟨mi$^{a}$ d*ierem*⟩ in to secundes & adde or minue thilke secundes to the houres / minue thanne ⟨equac*ion* die*rum*⟩'. The Latin note adds an explanation at greater length:

Si veru*m* motu*m* lune in aliq*ua* hora post m*eridiem* volu*eris* scire . quere p*rimo* locu*m* ei*us* in almenac i*n* m*eridie* p*ro* die illo. / Eq*uaciones* die*rum* q*ue* memento / deinde quere

[1] In the transcription of the cipher passages ⟨angle brackets⟩ have been used to denote words and phrases not ciphered in the original but inserted as plain text.

motum eiusdem lune in meridie diei proxime sequentis / & subtrahe motum precedentis meridiei a meridie subsequentis . scilicet . gradus a gradibus & minutas a minutis . & quod remanserit indicabit tibi verum motum lune a meridie in meridiem . / istum motum iam inuentum quere in vltima linea istius tabule versus dexteram (id est eccessus graduum) & si non inueneris illum numerum precise quere numerum maius sibi similem & cum illo numero decende sub hora quam cupis habere / & numerum quam ibi inueneris adde numero primo meridiei & tunc habes verum locum lune pro hora illa quam cupis habere / set aliquando accidit quod locus lune diei sequentis non est in eodem signo set in alio . & tunc sic operandum est / si non poteris subtrahere numerum meridiei prime . a numero diei secunde quia apparet minor / tunc adde .30. gradus secunde meridiei . & tunc potes subtrahere primum de secundo / & tunc operare vt prius // Ista parua tabula non extendit se nisi ad .12. horas adde residuum si 03 / 13 14 15 16 17 18 19 20 21 22 23 24 horas vt supra.

*Translation*

If you wish to know the true motus of the Moon at any hour after noon, first find its place in the almanac for noon of that day; remember also the equation of days. Next look up the motus of the Moon for noon of the day following, and subtract the motus on the first noon from that on the second, i.e. take degrees from degrees, and minutes from minutes, and the remainder will show the true motus of the Moon from noon to noon.

Having found the motus, look at the last line of the same table, towards the right-hand side, i.e. 'eccessus graduum', and if you do not find the exact number, look for the next greater number; and with that number go downwards (in the column) to the hour desired. Add the number which you find there to the first number of noon, and this will then give the true position of the Moon for the hour desired.

But it happens sometimes that the place of the Moon on the following day is not in the same sign, but in another; and in that case you must work in this fashion. If you are not able to subtract the number of the first noon from that of the second, because it is smaller, add 30° to the second noon, and then you will be able to subtract the first from the second and continue as before. This small table does not extend beyond 12 hours; add the remainder if (it is) 13, 14, 15, 16, 17, 18, 19, 20, 21, 22, 23, 24 hours, as above.

**63 v.** True positions and ecliptic latitudes, attributed to John Somer 31 December 1393.

| | | |
|---|---|---|
| Sun | 19° 1′ Capricorn | |
| Moon | 18° 30′ Sagittarius | |
| Saturn | 28° 5′ Libra | 2° 27′ north |
| Jupiter | 15° 46′ Pisces | 1° 13′ south |
| Mars | 11° 56′ Capricorn | 0° 3′ south |
| Venus | 5° 43′ Sagittarius | 2° 51′ south |
| Mercury | 3° 46′ Capricorn | 2° 12′ south |
| Caput | 25° 19′ Aquarius | |

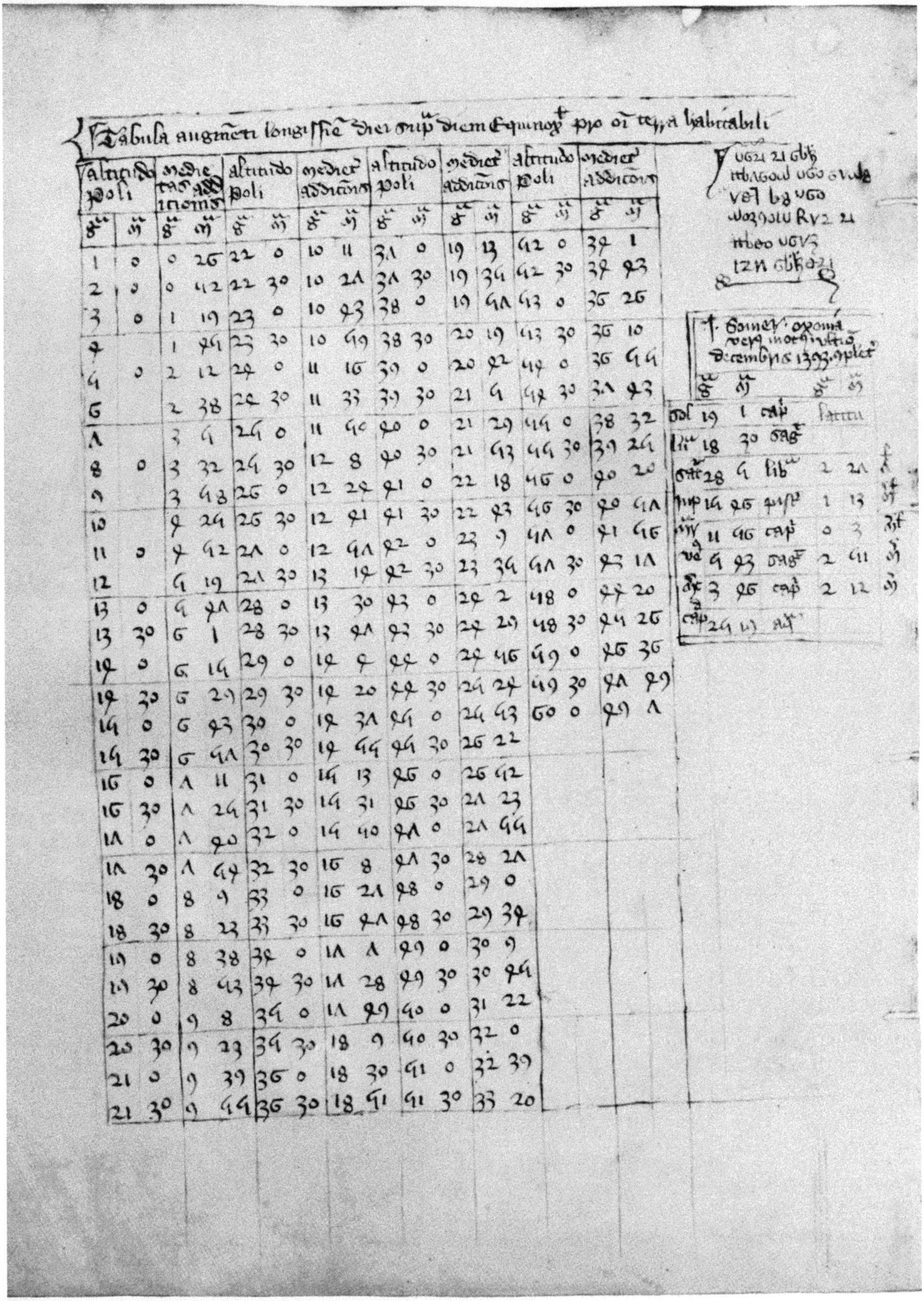

Tabula augmenti longitudinis diei super diem equinoctialem pro omni terra habitabili

| Altitudo poli | | Medietas augmenti | | Altitudo poli | | Medietas augmenti | | Altitudo poli | | Medietas augmenti | | Altitudo poli | | Medietas augmenti | |
|---|---|---|---|---|---|---|---|---|---|---|---|---|---|---|---|
| g | m | g | m | g | m | g | m | g | m | g | m | g | m | g | m |
| 1 | 0 | 0 | 26 | 22 | 0 | 10 | 11 | 37 | 0 | 19 | 13 | 52 | 0 | 34 | 1 |
| 2 | 0 | 0 | 52 | 22 | 30 | 10 | 27 | 37 | 30 | 19 | 35 | 52 | 30 | 34 | 43 |
| 3 | 0 | 1 | 19 | 23 | 0 | 10 | 43 | 38 | 0 | 19 | 57 | 53 | 0 | 35 | 26 |
| 4 | | 1 | 45 | 23 | 30 | 10 | 59 | 38 | 30 | 20 | 19 | 53 | 30 | 36 | 10 |
| 5 | 0 | 2 | 12 | 24 | 0 | 11 | 16 | 39 | 0 | 20 | 42 | 54 | 0 | 36 | 55 |
| 6 | | 2 | 38 | 24 | 30 | 11 | 33 | 39 | 30 | 21 | 5 | 54 | 30 | 37 | 43 |
| 7 | | 3 | 5 | 25 | 0 | 11 | 50 | 40 | 0 | 21 | 29 | 55 | 0 | 38 | 32 |
| 8 | 0 | 3 | 32 | 25 | 30 | 12 | 8 | 40 | 30 | 21 | 53 | 55 | 30 | 39 | 25 |
| 9 | | 3 | 58 | 26 | 0 | 12 | 24 | 41 | 0 | 22 | 18 | 56 | 0 | 40 | 20 |
| 10 | | 4 | 25 | 26 | 30 | 12 | 41 | 41 | 30 | 22 | 43 | 56 | 30 | 40 | 57 |
| 11 | 0 | 4 | 52 | 27 | 0 | 12 | 57 | 42 | 0 | 23 | 9 | 57 | 0 | 41 | 56 |
| 12 | | 5 | 19 | 27 | 30 | 13 | 14 | 42 | 30 | 23 | 35 | 57 | 30 | 43 | 17 |
| 13 | 0 | 5 | 47 | 28 | 0 | 13 | 30 | 43 | 0 | 24 | 2 | 58 | 0 | 44 | 20 |
| 13 | 30 | 6 | 1 | 28 | 30 | 13 | 47 | 43 | 30 | 24 | 29 | 58 | 30 | 45 | 26 |
| 14 | 0 | 6 | 15 | 29 | 0 | 14 | 4 | 44 | 0 | 24 | 56 | 59 | 0 | 46 | 36 |
| 14 | 30 | 6 | 29 | 29 | 30 | 14 | 20 | 44 | 30 | 25 | 24 | 59 | 30 | 47 | 49 |
| 15 | 0 | 6 | 43 | 30 | 0 | 14 | 37 | 45 | 0 | 25 | 53 | 60 | 0 | 49 | 7 |
| 15 | 30 | 6 | 57 | 30 | 30 | 14 | 55 | 45 | 30 | 26 | 22 | | | | |
| 16 | 0 | 7 | 11 | 31 | 0 | 15 | 13 | 46 | 0 | 26 | 52 | | | | |
| 16 | 30 | 7 | 25 | 31 | 30 | 15 | 31 | 46 | 30 | 27 | 23 | | | | |
| 17 | 0 | 7 | 40 | 32 | 0 | 15 | 50 | 47 | 0 | 27 | 55 | | | | |
| 17 | 30 | 7 | 55 | 32 | 30 | 16 | 8 | 47 | 30 | 28 | 27 | | | | |
| 18 | 0 | 8 | 9 | 33 | 0 | 16 | 27 | 48 | 0 | 29 | 0 | | | | |
| 18 | 30 | 8 | 23 | 33 | 30 | 16 | 47 | 48 | 30 | 29 | 34 | | | | |
| 19 | 0 | 8 | 38 | 34 | 0 | 17 | 7 | 49 | 0 | 30 | 9 | | | | |
| 19 | 30 | 8 | 53 | 34 | 30 | 17 | 28 | 49 | 30 | 30 | 45 | | | | |
| 20 | 0 | 9 | 8 | 35 | 0 | 17 | 49 | 50 | 0 | 31 | 22 | | | | |
| 20 | 30 | 9 | 23 | 35 | 30 | 18 | 9 | 50 | 30 | 32 | 0 | | | | |
| 21 | 0 | 9 | 39 | 36 | 0 | 18 | 30 | 51 | 0 | 32 | 39 | | | | |
| 21 | 30 | 9 | 55 | 36 | 30 | 18 | 51 | 51 | 30 | 33 | 20 | | | | |

[illegible] ... Decembris 1393 complet.

| | g | m | | g | m | |
|---|---|---|---|---|---|---|
| sol | 19 | 1 | cap | [illegible] | | |
| lune | 18 | 30 | sag | | | |
| sat | 28 | 5 | lib | 2 | 27 | [illegible] |
| iup | 15 | 46 | pisc | 1 | 13 | [illegible] |
| mar | 11 | 56 | cap | 0 | 3 | [illegible] |
| ven | 5 | 43 | sag | 2 | 51 | [illegible] |
| mer | 3 | 46 | cap | 2 | 12 | [illegible] |
| cap | 25 | 19 | ari | | | |

FOLIO 63 VERSO

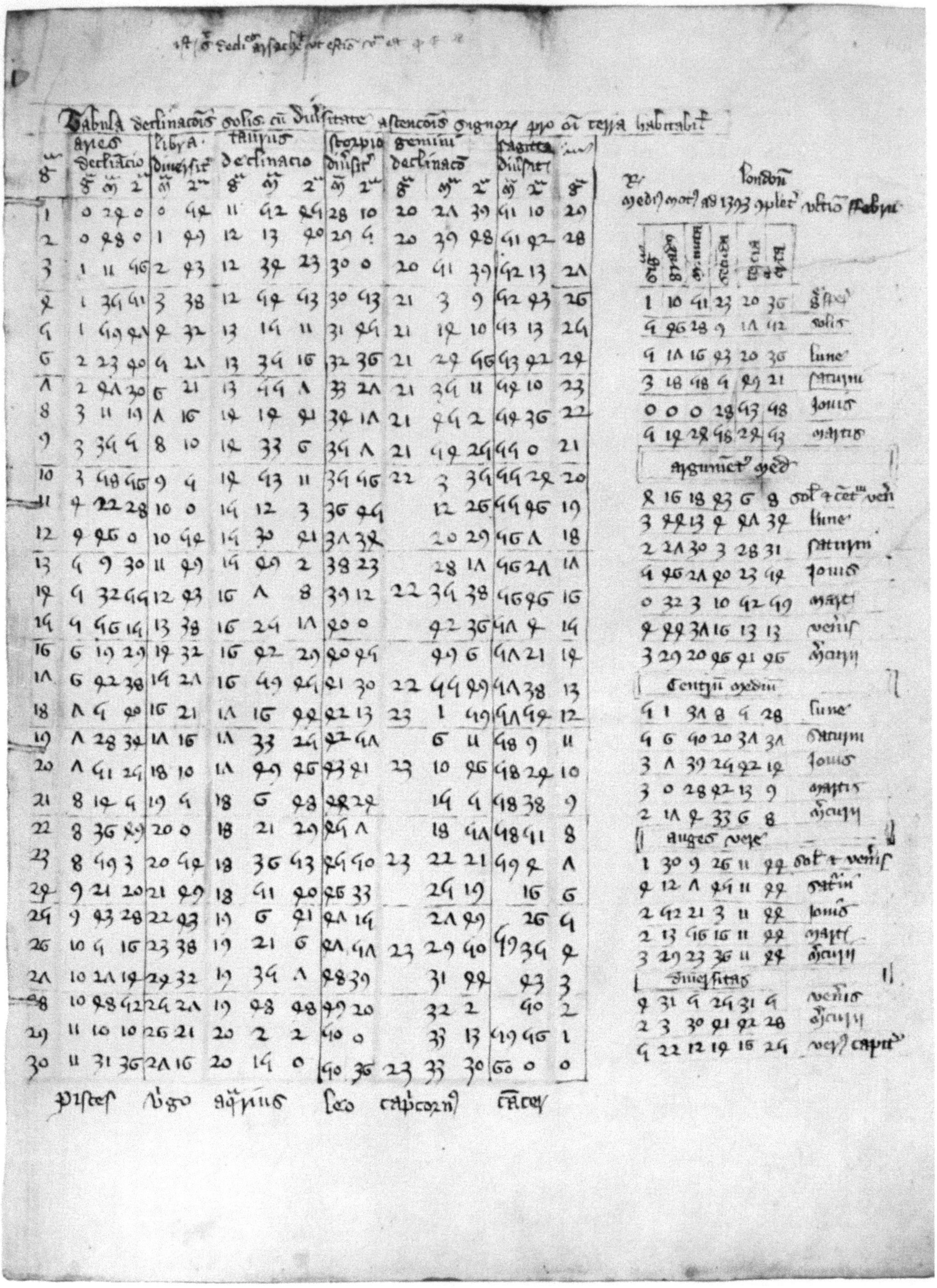

FOLIO 64 RECTO

*see p.* 87

PLATE VIII

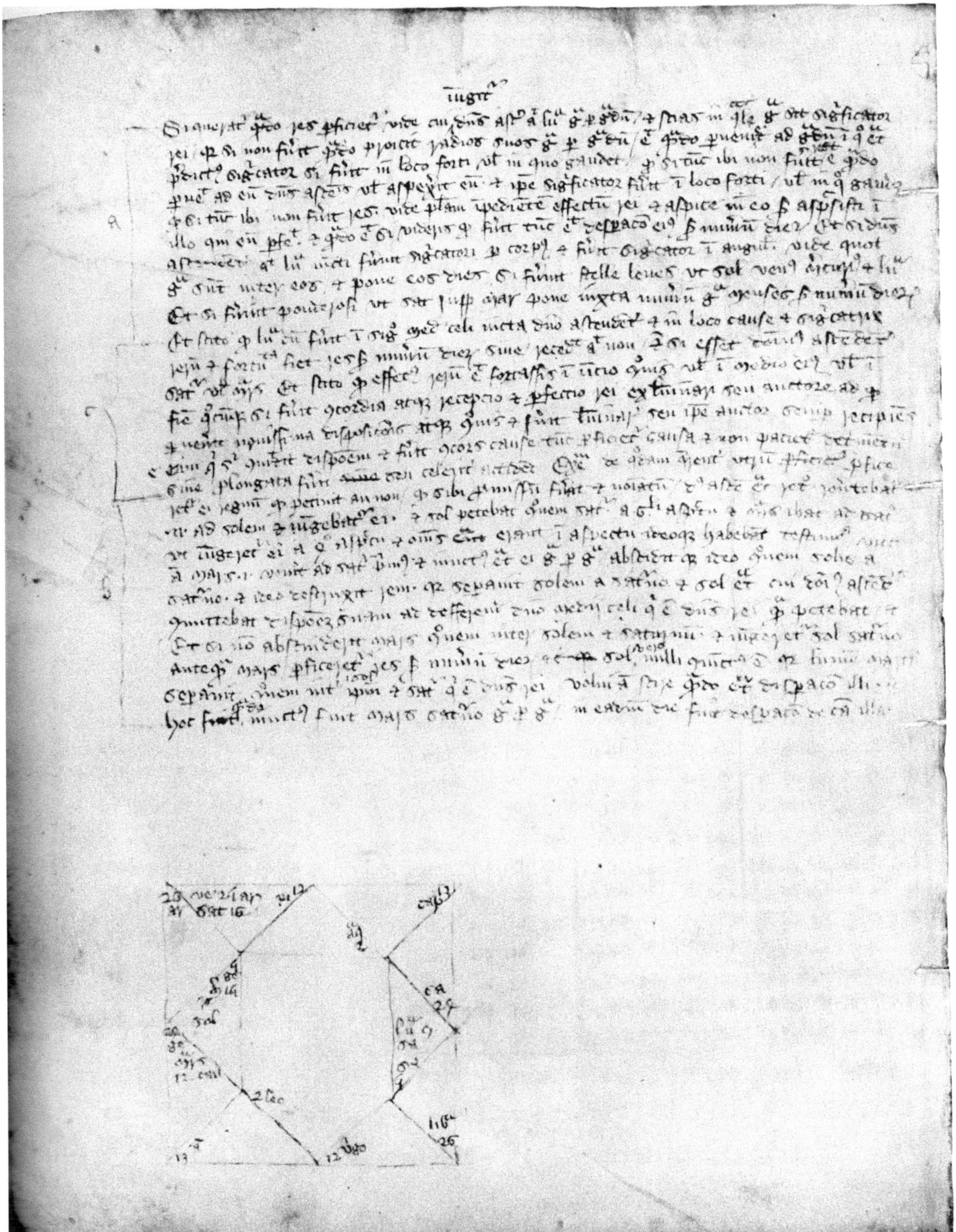

FOLIO 64 VERSO

*see p.* 88

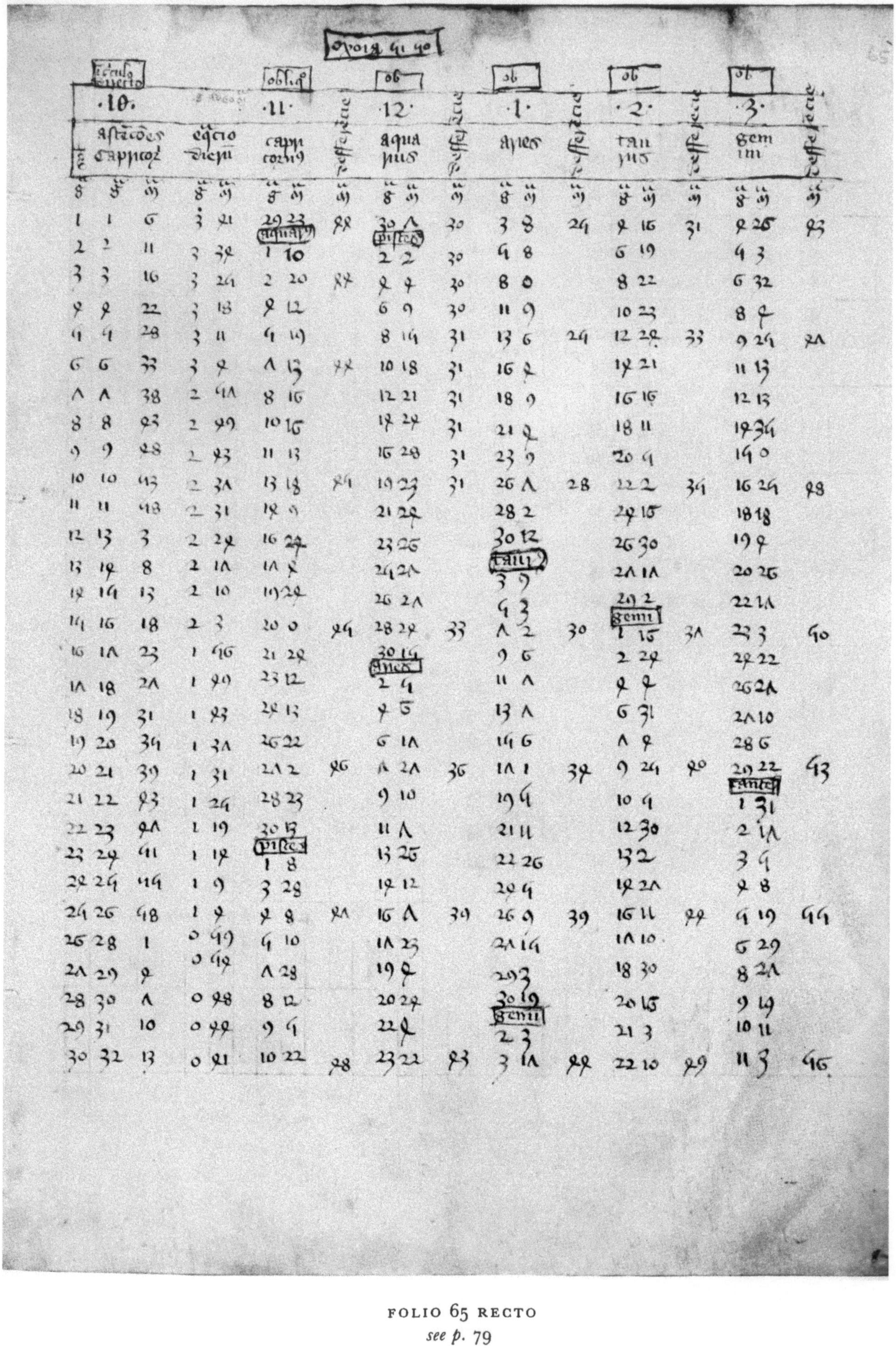

FOLIO 65 RECTO

*see p.* 79

The gloss in cipher reads, 'this is how mochel the half ark of the lengest dai is more than six houris'.

**64r.** Constants for 28 February 1394.

| | | | | | | |
|---|---|---|---|---|---|---|
| **Mean motus** | | | | | | |
| 8th sphere | 1 | 10 | 51 | 23 | 20 | 36 |
| Sun | 5 | 46 | 28 | 09 | 17 | 52 |
| Moon | 5 | 17 | 16 | 43 | 20 | 36 |
| Saturn | 3 | 18 | 58 | 05 | 49 | 21 |
| Jupiter | 0 | 00 | 00 | 28 | 53 | 58 |
| Mars | 5 | 14 | 24 | 58 | 24 | 53 |
| **Mean argument** | | | | | | |
| Sun and Centre Venus | 4 | 16 | 18 | 43 | 06 | 08 |
| Moon | 3 | 44 | 13 | 04 | 47 | 34 |
| Saturn | 2 | 27 | 30 | 03 | 28 | 31 |
| Jupiter | 5 | 46 | 27 | 40 | 23 | 54 |
| Mars | 0 | 32 | 03 | 10 | 52 | 59 |
| Venus | 4 | 44 | 37 | 16 | 13 | 13 |
| Mercury | 3 | 29 | 20 | 46 | 41 | 46 |
| **Mean centre** | | | | | | |
| Moon | 5 | 01 | 37 | 08 | 05 | 28 |
| Saturn | 5 | 06 | 50 | 20 | 37 | 37 |
| Jupiter | 3 | 07 | 39 | 29 | 42 | 14 |
| Mars | 3 | 00 | 28 | 42 | 13 | 09 |
| Mercury | 2 | 17 | 04 | 33 | 06 | 08 |
| **True auges** | | | | | | |
| Sun and Venus | 1 | 30 | 09 | 26 | 11 | 44 |
| Saturn | 4 | 12 | 07 | 45 | 11 | 44 |
| Jupiter | 2 | 52 | 21 | 03 | 11 | 44 |
| Mars | 2 | 13 | 56 | 16 | 11 | 44 |
| Mercury | 3 | 29 | 23 | 36 | 11 | 44 |
| **'Diversitas'** | | | | | | |
| Venus | 4 | 31 | 05 | 25 | 31 | 05 |
| Mercury | 2 | 03 | 30 | 41 | 42 | 28 |
| True Caput | 5 | 22 | 12 | 14 | 16 | 25 |

Above the table of declinations a note has been added in a hand quite different to any appearing elsewhere in the manuscript, 'Ist*e* *sunt* declin*ationes* arsache*lis* ut esti*mo* // *verum* est q*uod* E. B.' The table is based on the value 23° 33′ 30″ for the inclination of the ecliptic, which is near that employed by Azarquiel.[1] It is difficult to suggest who 'E.B.' might be.

[1] The Ptolemaic value is 23° 51′ and that of Azarquiel 23° 33′. The value 23° 33′ 30″ is that actually given by Thabit ibn Qurra but often attributed to Azarquiel. See José Mª Millás Vallicrosa, *El libro de los fundamentos de las Tablas astronomicas* (Madrid–Barcelona, 1947), p. 93 and also J. L. E. Dreyer, *History of the Planetary Systems* (Cambridge, 1906), p. 276.

**64v.** The figure of the horoscope and its accompanying text are as follows:

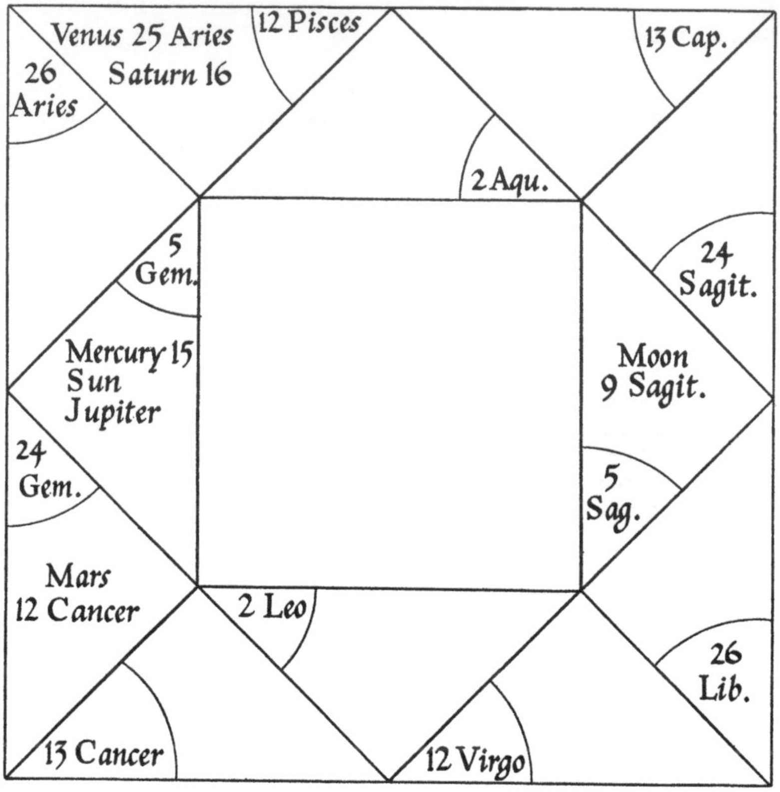


*Horoscope text on f. 64v.*

Si queratur quando res perficietur vide cui (iungitur) dominus ascendentis aut luna gradus per gradum & scias in quale gradu sit significator rei quia si non fuerit quando proicit radios suos gradus per gradum / est quando perueni(e)t ad gradum in quo erat predictus significator si fuerit in loco forti vel in quo gaudet // quod si tunc ibi non fuerit (res) est quando peruenerit ad eum dominus ascendentis vel aspexerit eum. & ipse significator fuerit in loco forti / vel in quo gaudet quod si tunc ibi non fuerit res. vide planetam impedientem effectum rei & aspice in eo secundum aspecsisti in illo qui eum perfecit. & quando est si videris quod fuerit tunc est desperacio eius secundum numerum dierum Et si dominus ascendentis aut luna iuncti fuerint significatori per corpus & fuerit significator in angulis. vide quot gradus sunt inter eos / & pone eos dies si fuerint stelle leues vt sol venus Mercurius & luna Et si fuerint ponderosi vt saturnus juppiter mars pone iuxta numerum gradus menses secundum numerum dierum Et scito quod luna cum fuerit in signo medii celi iuncta domino ascendentis & in loco cause & significatrix rerum & fortunata fiet res secundum numerum dierum siue recedit aut non / $\overline{\&}$ si esset dominus ascendentis saturnus vel mars Et scito quod effetus rerum est fortassis in inicio coniunctionis vel in medio eius vel in fine quocunque si fuerit concordia atque recepcio & perfeccio rei ex luminari seu auctore ad q*uod* peruen*er*it nouissima disposic*io*nis atq*ue* *coniunctio*nis & fu*er*it lu*m*inar*e* seu ip*s*e auctor semp*er* recipie*n*s eum q*ui* *sibi* *com*mittit dispo*sicio*nem & fu*er*it *con*cors cause / tu*n*c *pro*ficietu*r* causa & non pacietu*r* detrime*n*tu*m* siue p*r*olongata fu*er*it [siue] seu celerit*er* accide*n*t Exe*mplum* de q*uo*dam q*ue*renti vtru*m* [p*er*ficietu*r*] p*er*ficeretu*r* ei regnu*m* q*uod* petiuit / an non / q*uod* sibi p*ro*missu*m* fuerat & no*m*inatu*m* / d*ominus* asce*ndentis* *era*t retr*o* reu*er*tebatu*r* *enim* ad solem & iu*n*gebatu*r* ei . & sol petebat *coniunctio*nem satu*rni* a .6. *li* asp*ec*tu & m*a*rs ibat ad satu*rnum* vt iu*n*geretu*r* ei a 4° asp*ec*tu & o*m*nis [*er*ant] erant i*n* aspectu ideoq*ue* habeba*n*t testimo*nia* / vicit (?) *aut* mars . *id est* . venit ad sat*urnum* p*rimus* & iunct*us* *erat* ei *gradus* *per* *gradum* / abscidit q*ue* ideo *coniunctio*nem solis a saturno . & ideo destruxit rem . q*uia* sep*er*auit solem a satu*r*no & sol *er*at cui do*minus* asce*n*de*ntis* *com*mitebat dispo*sicionem* suam ad defferend*um* do*mi*no medii celi q*ui* e*st* do*minus* rei q*uam* petebat / & Et si no*n* abscinderit mars *coniunctio*nem inter solem & saturnu*m* . & iu*n*geret*ur* sol satu*r*no anteq*uam* mars p*er*ficeretu*r* res *secundum* num*er*u*m* die*rum* [&*cetera* q*uia*] sol 'uero' nulli *coniunctus* e*st*. q*uia* lume*n* marti*s* sep*er*auit *coniunctio*nem int*er* ip*s*um [i*d est* sol] & satu*rnum* q*ui* e*st* d*ominus* rei // volui a*utem* scire q*uan*do *er*at disp*er*aci*o* illi*us* & hoc [fuit] 'q*uan*do' iunct*us* fuit mars satu*r*no *gradus* p*er* *gradum* / in eade*m* die fuit desp*er*aci*o* de ca*us*a illa.

*Translation*

If anyone ask when the matter shall be successful, observe the aspects of the Lord of the ascendant or the aspects of the Moon, degree by degree, and you will find which degree contains the indicator of lack of success of the matter. For while the aspects move degree by degree, there would come a time when the aforesaid indicator would be in a certain degree if it is to occupy a strong or favourable position. But if the matter is not then, there will come a time when the Lord of the ascendant will be in conjunction or opposition, and the said indicator will then be in a strong position or a favourable one. But if the matter does not happen then, look for the planet that is hindering success, and examine it in the same way as you have observed the planet that has produced the aspect. And when it comes about, you shall observe what it was, and the number of days determines whether there is no hope. If the Lord of the ascendant or the Moon is in aspect with the indicator, and the indicator is in an angle, see how many degrees separate them, and write it down as days if they are light stars such as the Sun, Venus, Mercury, and the Moon. If they are heavy, such as Saturn, Jupiter, and Mars, take the number of degrees to be months, just as with the days.

And know that when the Moon is in conjunction with the Lord of the ascendant in mid-heaven, it will act as a cause and be a pointer towards events, and the matter will be fortunate according to the number of days, whether it wanes or not, if the Lord of the ascendant is Saturn or Mars.

And know that the success of a matter occurs variously at the beginning, the middle, and the end of a conjunction. And if this agrees with the occurrence and success of the matter as indicated by the luminary or the initiator; that is the aspect or conjunction to which it has most recently come, and if the luminary is the initiator or in good aspect to it, and agrees with the cause, then the cause shall flourish, and shall not suffer damage either swift or prolonged.

An example, concerning someone who inquired whether the kingdom which he sought, and which had been expressly promised to him, was to be secured for him or not.

The Lord of the ascendant was turned away, for it was facing back towards the Sun, and was coming into aspect with it. And the Sun was coming into sextile aspect with Saturn. And Mars was approaching Saturn so as to be in quartile aspect to it. And they were all in aspect, and therefore provided evidence.... Mars came first to Saturn and was joined to it degree by degree, thereby cutting off the aspect of the Sun to Saturn. So it prevented the matter from being successful, because it separated the Sun from Saturn. And it was the Sun to which the Lord of the ascendant was coming into aspect, as Lord of the mid-heaven, which is the Lord of the matter we are investigating. If Mars had not cut off the conjunction between the Sun and Saturn, and the Sun had come into aspect with Saturn before Mars, then the matter would have been successful in the appropriate number of days. But the Sun was not in aspect, because the light of Mars stopped it being in aspect to Saturn (who is Lord of the matter). But I wanted to know when there would be no hope, and this was when Mars was in aspect to Saturn, degree for degree; and on that day there was no hope of success in the matter.

The astrological content of this folio (so far as I can understand the text and the writer's intentions) appears to be quite in keeping with traditional theory, and is not worth detailed comment. It is, however, useful to examine the planetary positions used in the horoscope, and determine whether they relate to any specific date, or whether they are chosen simply to illustrate a point in the theory which the author wished to explain.

The position of the Sun is not given precisely, but it is included between the cusps marked 5° Gemini and 24° Gemini. Since the position of Venus is 25° Aries, and that planet cannot have a greater elongation than about 45°, it follows that the Sun lies between 5° and 10° Gemini, corresponding to a date 17–22 May, the year being indeterminate.

It is most convenient to use the slowest planetary motion to determine the year of calculation. Since Saturn is said to be nearly in sextile aspect with the Sun, and in 16° of either Pisces or Aries, it is clear that it is the latter sign which is intended. From the tables we find that Saturn was near this position at the middle of 1379 and at all times preceding or following this date by a multiple of 29½ years; the beginning of 1350 seems a little too early for associations with this text of 1392, and the end of 1408 seems too far in the future for the casting of such a horoscope. We therefore suggest that the date for which the query is made lies between 17 and 22 May 1379. Unfortunately, the place of Jupiter is not given, and it is difficult to secure any satisfactory check from the positions of Mercury, Venus, Mars and the Moon; the actual date—even the correspondence with any real set of planetary calculations—is a matter of considerable uncertainty.

If the author is Chaucer, and this inquiry is drawn from his experience, it would be of great interest to identify the querent who had been expressly promised a kingdom. It is tempting to think it might be John of Gaunt; certainly in 1379 he had just concluded an alliance with the king of Portugal, whereby he was to be invited to come to Lisbon with a fleet and army in pursuit of the kingdom of Castile to which his marriage

PLATE X

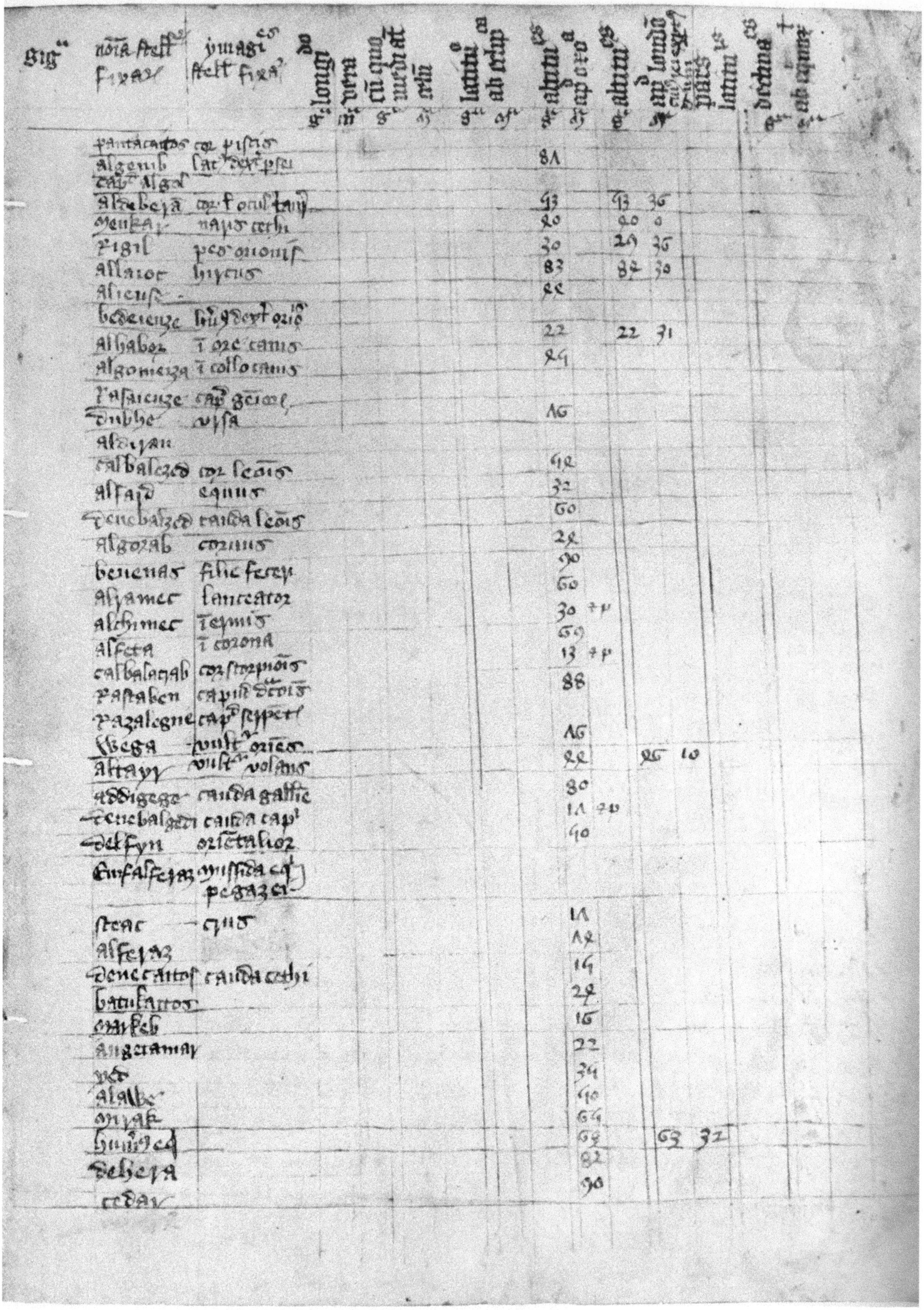

FOLIO 71 RECTO

in 1372 to Constance, daughter of Pedro of Castile, had given him a claim. The case for this identification cannot, however, be pressed on such slight evidence.

**70r.** The note reads, 'Profacius / maiorum equacionem temporis collecti ex diebus inequalibus a tempore in quo est sol in 18 $g^a$ aquarii $9^e$ spere . est cum sol fuerit in 8 & in 9 scorpionis $9^e$ spere / & in 7 $g^a$ & 57 $m^a$ qui valent 31 $mi^a$ hore & 48 secunda'.

**71r.** The list of fixed stars is very similar to that published by Skeat from MS. Ii. 3. 3, f. 70r. in University Library, Cambridge; it gives forty-two out of the forty-nine stars in that list. See W. W. Skeat, *A Treatise on the Astrolabe..., by Geoffrey Chaucer*, E.E.T.S., 1872, p. xxxvii. A comparative table of such 'astrolabe' stars is given by R. T. Gunther, *Early Science in Oxford*, vol. II (Oxford, 1923), pp. 222–5. In the following transcription we append to the name of each star an identification number from Skeat's list.

| Nomina stellarum fixarum | Ymagicorum stellarum fixarum | Altitudo apud Oxon. | Altitudo apud London (cuius rei expertus sum) |
|---|---|---|---|
| Pantacantos (3) | cor piscis | — | — |
| Algenib (7) | latus dexter persei | 87 | — |
| Capud Algol (4) | — | — | — |
| Aldeberan (9) | cor vel oculus tauri | 53 | 53 36 |
| Menkar (6) | naris cethi | 40 | 40 0 |
| Rigil (11) | pes orionis | 30 | 29 36 |
| Allaiot (10) | hircus | 83 | 84 30 |
| Alicuse bedeienze (12) | humerus dexter orionis | 44 | — |
| Alhabor (13) | in ore canis | 22 | 22 31 |
| Algomeiza (15) | in collo canis | 45 | — |
| Rasaicuze (14) | capud geminorum | — | — |
| Dubhe (23) | vrsa | 76 | — |
| Aldiran (18) | — | — | — |
| Calbalczed (20) | cor leonis | 54 | — |
| Alfard (19) | equus | 32 | — |
| Denebalzed (24) | cauda leonis | 60 | — |
| Algorab (25) | coruus | 24 | — |
| Benenas (27) | filie feceri | 90 | — |
| Alramet (28) | lanceator | 60 | — |
| Alchimec (26) | in ermis | 30 & p | — |
| Alfeta (29) | in corona | 69 | — |
| Calbalacrab (32) | cor scorpionis | 13 & p | — |
| Rastaben (34) | capud draconis | 88 | — |
| Razalegne (33) | capud serpentis | — | — |
| Wega (35) | vultur oriens | 76 | — |
| Altayr (36) | vultur volans | 44 | 46 10 |
| Addigege (39) | cauda galline | 80 | — |
| Denebalgedi (44) | cauda capricorni | 17 & p | — |
| Delfyn (37, 41?) | orientalior | 50 | — |
| Enifasteraz (43) | mussida eq pegazei | — | — |
| Steac (45) | crus | 17 | — |
| Alferaz (46) | — | 74 | — |
| Denecaitos (48) | cauda cethi | 15 | — |
| Batukaitos (2) | — | 24 | — |

| Nomina stellarum fixarum | Ymagicorum stellarum fixarum | Altitudo apud Oxon. | Altitudo apud London (cuius rei expertus sum) |
|---|---|---|---|
| Markeb (16) | — | 16 | — |
| Augeramar (17?) | — | 22 | — |
| Yed (31) | — | 35 | — |
| Alawe (30?) | — | 50 | — |
| Mirak (1) | — | 65 | — |
| Humerus equi (47) | — | 63 | 63 32 |
| Dehera (42?) | — | 82 | — |
| Cedar (49) | — | 90 | — |

In the above column giving the meridian altitudes for Oxford, the occasional addition of '& p' after the number of degrees seems to mean *et parte*, i.e. 'slightly more'.

The right-hand column, in which our author gives the meridian altitude at London 'cuius rei expertus sum' (which I have myself measured), is particularly interesting. The altitudes are given to an accuracy of single minutes of arc, for example, *Alhabor* 22° 31′, instead of being expressed to the nearest fraction of a whole degree, as found in most records of actual measurement at this period. To investigate further this unusually precise set of measurements we may compare the seven observations in the above list with the actual places of these stars as calculated from modern observations.[1]

| Star name | Modern name | Altitude at London | Declination at A.D. 1400 | Calculated latitude |
|---|---|---|---|---|
| Aldeberan | α Tauri | 53° 36′ | 15° 7′ N. | 51° 31′ |
| Menkar | α Ceti | 40° 0′ | 1° 37′ N. | 51° 37′ |
| Rigil | β Orionis | 29° 36′ | 9° 4′ S. | 51° 34′ |
| Allaiot | α Aurigae | 84° 30′ | 45° 8′ N. | 51° 38′ |
| Alhabor | α Canis Maj. | 22° 31′ | 16° 3′ S. | 51° 26′ |
| Altayr | α Aquilae | 46° 10′ | 7° 27′ N. | 51° 17′ |
| Humerus Equi | α Andromedae | 63° 32′ | 25° 47′ N. | 51° 15′ |
| | | | Mean calculated latitude | 51° 28′ |
| | | | Average error | ±8′ |

It will thus be seen that the observations do, indeed, show very good precision, the mean calculated latitude agreeing well with the modern determinations of the latitude of London (51° 31′). An average error of only 8 min. indicates that the observing instrument must have had a radius of at least 18 in. (cf. p. 62, note on A/4).

It may also be worth noting that the seven stars observed are such that they could all have been observed in their meridian passages during one night of the year. In this case the night would have to be about the beginning of November, and the stars would be measured, starting with Altair just after sunset, and continuing until the meridian transit of Sirius just before sunrise. There is, of course, no way of determining the actual year of observation.

[1] For this purpose we have used P. V. Neugebauer, *Sterntafeln* (Leipzig, 1912), Table III, entries for A.D. 1400.

# VII

# THE PTOLEMAIC PLANETARY SYSTEM

## I. PREAMBLE

Of all the medieval sciences, astronomy must be accounted by far the most perfectly developed. Indeed, throughout the Middle Ages it must have seemed to those equipped to comprehend the subject almost as if the theory had been completed in all its essentials and needed only slight refinement and a trifling increase of accuracy in its constants. In many ways the situation was similar to that existing in modern science just before the end of the nineteenth century when the theory of evolution, and the discovery of radioactivity, of X-rays and of the electron completely upset the apparent stagnation which had led some people to wonder if there could be more to do than attempt to measure known quantities to more and more places of decimals. But radical change did not occur in astronomy until the times of Copernicus, Tycho Brahe, Kepler and Newton, and in the Western world it was Ptolemaic astronomy which reigned supreme from the period when classical science was being rediscovered through the Arabic in the twelfth century.[1]

It is unfortunate that this peak of medieval scientific thought, second only to Aristotle in influence, has often been held in less esteem than it deserves, partly through the efforts of medieval writers on popular cosmology and partly through the tendency of more modern critics to assume that if Copernican theory is, on the whole, right, then Ptolemaic theory must have been very much in error. The popular cosmologies present the simplified picture of the planets and stars moving on a set of concentric circles about the earth—needless to say this is no more adequate as a statement of Ptolemaic theory than a similar diagram with the Sun in the centre is as a statement of modern astronomical theory. The theory, of course, contains much more than this, and, indeed, it is rather a mathematical device for computing the motions than any cosmological attempt to develop a picture or explanation of the

[1] Modern astronomers should envy one aspect of the medieval situation. A twelfth-century astronomer could rely equally on observations made by his contemporaries and by Ptolemy, 1000 years before. The long time interval must have given him great confidence in his knowledge of the exact size of small secular changes.

system of the universe. Furthermore, the mathematical device is designed to bring order into the movements of the stars and planets as seen from the Earth, and hence, quite apart from the philosophical difficulty of any alternative, a geocentric system was the most logical scheme. It cannot be stated too often that in almost every respect there is strict mathematical equivalence between the geocentric and the heliocentric hypotheses; it matters little for purpose of computation whether it is the Earth or the Sun which is assumed to be at rest, for the lengths and angles are identical and consequently the triangles to be solved will be similar. It is also very misleading to attempt to evaluate the advance made by Copernicus in terms of a reduction in the number of spheres used in the theory. Quite often these spheres would be inserted by cosmologers for physical rather than mathematical reasons, and if we only admit those spheres which represent a parameter of the theory then it can be shown that Copernicus reduced the number by one only for each planet (with the exception of the Sun and Moon which remained unchanged), and one might even maintain that modern theory, by its recognition and evaluation of many extra terms, has effectively increased the 'number of spheres' far beyond even the wildest estimates of Ptolemaic theory.

The theory with which we are concerned takes its name from its author Claudius Ptolemaeus [i.e. Πτολεμαῖος], who was born in Egypt and flourished in Alexandria in the second quarter of the second century, dying after 161. His astronomical work has come down to us in the great Mathematical Treatise, commonly called the *Almagest*,[1] and although it remains a matter of conjecture how much Ptolemy drew on previous observations and theories (in particular those of Hipparchus) this encyclopaedic work remained authoritative until the sixteenth century. The *Almagest* was translated into Arabic under al-Ma'mūn at the beginning of the ninth century and revised by Thābit ibn Qurra at the end of the same century. It was known to medieval scholars[2] through the translation from Arabic into Latin made by Gerard of Cremona in 1175, and also by a number of epitomes and commentaries.[3]

[1] The full title is ἡ μαθηματικὴ σύνταξις or μεγάλη σύνταξις τῆς ἀστρονομίας or, as found by O. Neugebauer, μεγίστη σύνταξις, from which was derived *Almagest* = al-megiste.

[2] It may be worth pointing out that Chaucer's references to Ptolemy and the *Almagest* in the *Wife of Bath's Tale* [D. 180–3, 323–7] are not erroneous as once suggested, but drawn from an introduction to this translation by Gerard of Cremona; this was printed in the 1515 edition at Venice. See R. Steele, *Chaucer and the Almagest* (London, 1920) and F. M. Grimm, *Astronomical Lore in Chaucer* (University of Nebraska Studies in Language, Literature and Criticism, 1919), p. 10.

[3] For modern translations there is one in French by Abbé Halma (2 vols., Paris, 1816) and an excellent German translation by Karl Manitius (2 vols., Leipzig, 1912–13), which is based on the Greek text published by J. L. Heiberg (2 vols., Leipzig, 1898, 1903). The lack of a good English translation has long been a hindrance to scholars, and I must acknowledge my gratitude to Miss Elisabeth Williamson for allowing me access to a translation in manuscript which she has prepared. Another translation has

## 2. GENERAL FOUNDATIONS OF THE THEORY

It is an important feature of the Ptolemaic theory that it consists of a number of quite separate mathematical schemes, one for each planet or detail, rather than an integrated system for all the phenomena. These schemes are all of the same general pattern, but there is no connection between any two of them such as occurs in the Copernican system. To put this another way, each scheme gives an account of the way in which that particular planet appears to move, quite independently of the motion of any other planet. Ptolemy was not concerned with the distances between the planets (i.e. the size of their spheres), and even their order was of little importance to him: 'It is impossible to come to a decision on this point, since none of the planets has a sensible parallax, and this is the only phenomenon which would be capable of yielding the distances. It would appear better to retain the order of the ancients as being more probable' (Book IX, Ch. I).

This order of the ancients was as follows, proceeding from the central Earth to the outermost sphere of the fixed stars: Moon, Mercury, Venus, Sun, Mars, Jupiter, Saturn, Fixed Stars. The actual sizes of the spheres was hence a matter of philosophical speculation rather than scientific inquiry throughout the Middle Ages, and to answer the difficulty it was proposed from time to time that the sizes ranged in an orderly arithmetical or geometrical progression, or were proportional to the lengths of strings giving a set of musical notes (music of the spheres), or the more physical explanation that the spheres were packed together as tightly as possible. This last suggestion arose indirectly from the Ptolemaic theory itself,[1] for the mathematical scheme is such that each planet, instead of being confined to the wall of some sphere, oscillates back and forth so as to take up an appreciable thickness around this wall. Since collisions of planets did not occur, it was at least likely that Mercury, for example, when at its farthest distance from the Earth, did not trespass on the nearest part of the territory allowed to Venus; perhaps it also seemed unlikely that space would be wasted between these spheres.

Since Ptolemy is not concerned with planetary distances, his theory must be understood as referring only to the angles at which planets are observed, that is, their angular motion as seen projected against the background of the fixed stars. This emphasizes the characteristic that the theory is a mathematical

become available while this work was in progress, *Ptolemy, Copernicus, Kepler*, Great Books of the Western World, no. 16, *Encyclopaedia Britannica*, Inc., 1952. The translation of the *Almagest* is by R. Catesby Taliaferro, and I am indebted to the publishers for their kindness in permitting me to examine a copy.

[1] An explicit arrangement of contiguous spheres is given by Ptolemy in the CANOBIC Inscription and in the 'Planet Theory'.

'saving of appearances' rather than an attempt to give a picture of the orbit in space of any one planet. It seems likely that it was not until the time of Kepler that the astronomer pictured the planet as actually moving in the path prescribed by the theory. The path was there purely as a geometrical device to express the mathematical analysis, since algebraic methods were still far too

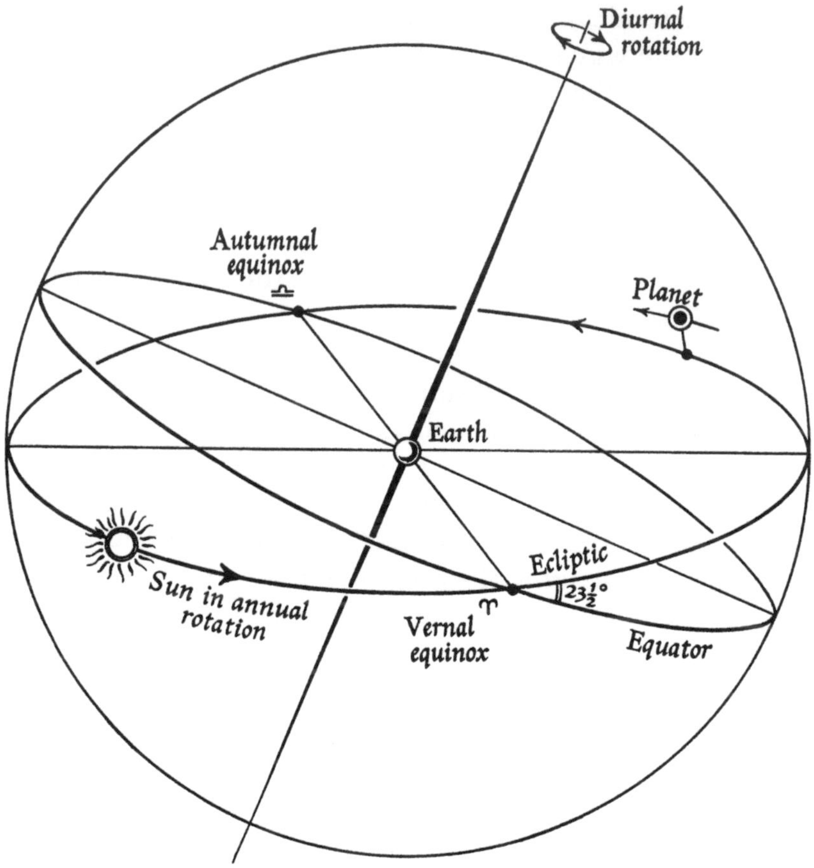


Fig. 7. The diurnal rotation of the celestial sphere, and the annual motion of the Sun.

crude for the purposes. The employment of a complex epicycloidal orbit must not be taken as a physical reality any more than the use of a wave equation for light implies that something is moving back and forth.

Since all the planets seem to move approximately in one great circle in the heavens, it is natural to choose this circle as a means of reference. The Sun, by reason of its obvious importance and most regular motion, is taken as defining this circle which is called the Ecliptic, and fixed points in the circle are provided by its intersections with the celestial equator at the two equinoxes, the vernal equinox being taken as the origin. The longitudes of the planets are measured in degrees from this origin, the direction of motion of the Sun throughout the year being taken as positive. The straying of planets north or south of the ecliptic is measured by their latitude taken in degrees along an arc perpendicular to the ecliptic. It is hardly necessary to mention that the ecliptic is divided by tradition into the twelve signs of the zodiac, starting at the vernal equinox with the first point (degree) of Aries and changing every 30° into a new sign.

## 3. THE SUN

Since the Sun is taken as a reference body for defining the plane of the ecliptic and reckoning the positions of the other planets, its theory is considerably more simple than that of any of the others. Since also the mathematical theory is the same whether we consider it is the Earth or the Sun which is revolving round the other, it follows that the theory of the Sun in Ptolemaic theory must be practically identical with the theory of the Earth's motion in modern theory.

Although the Sun is seen to move through the twelve signs of the zodiac in one solar year, its motion is not exactly uniform, since the path from vernal to autumnal equinox takes a little over one week more than the subsequent path from autumnal back to vernal equinox. It is this slight inequality that the theory has to present, and it does so by the use of a simple eccentric circle round which the Sun moves at a completely uniform rate of one revolution per year. The centre of this eccentric circle is displaced from the Earth by a relatively small amount in a direction which is termed the *line of the Aux* of the Sun, and occupies a position in the zodiac which was considered almost fixed or participating in the general precessional motion of the order of one degree in a century. The amount of eccentricity is determined by the distance between the Earth and the centre of the eccentric circle, and may be expressed as a fraction (about $\frac{1}{30}$) of the radius of that circle. It should be noted that provided this fraction is preserved, the actual radius of the eccentric circle can make no difference to any angles measured on the outer circle of the zodiac.

It is an important feature of Ptolemaic theory that all angles must be measured from the Earth in the first place; it is therefore necessary to construct the uniform motion of the Sun round the eccentric circle by means of a 'protractor' based on the Earth. This is effected by introducing a point called the Mean Motus which moves round the ecliptic at the same uniform rate. The position of the mean motus is always reckoned as so many signs, degrees, minutes, etc., starting from the first point of Aries—*not* from the Aux. In modern terminology this uniformly moving mean motus corresponds to the *Dynamical*[1] *Mean Sun*, and the True Sun exceeds or falls short of its position by a maximum of about $2\frac{1}{2}°$, the actual amount being called by Ptolemy the *Equation of Centre*. Having first found the direction of the mean motus as seen

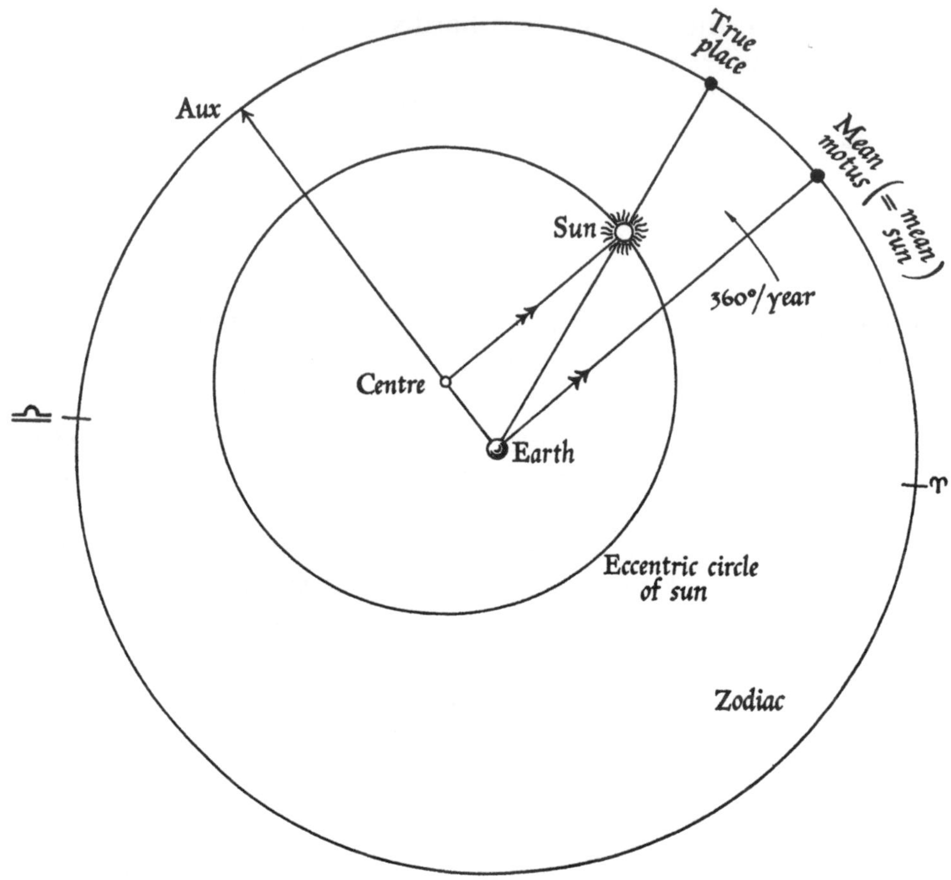


Fig. 8. Ptolemaic construction for the position of the Sun. The eccentricity (distance between the Earth and the centre of the eccentric circle) has been considerably exaggerated (× 11) for the sake of clarity.

from the Earth, a line parallel to this is drawn from the centre of the eccentric circle of the Sun; its intersection with this circle serves to locate the Sun whose true place may finally be found by means of a line drawn from the Earth, through this locating point, and out to the circle of the zodiac.

[1] This must be distinguished from the Astronomical Mean Sun which is assumed to move round the Celestial Equator (not the Ecliptic) at uniform rate.

## 4. VENUS, MARS, JUPITER AND SATURN

Ptolemy considers the motions of these four planets by using the same mathematical device for each. It may be considered the basic pattern of his system, since the special devices used for the Moon and for Mercury are obviously modifications of the original device. The theory contains two features which are not necessary for dealing with the Sun: the introduction of an Equant and an epicycle.

The equant is a modification of the simple eccentric circle used with the Sun. The Earth is displaced from the centre of the circle as before, but the point *P* does not move round uniformly as seen from the centre. Instead, we introduce an Equant so that the centre is half-way between the Earth and the equant, and then *P* moves round uniformly as seen from this equant. In the same way as before a mean motus is used to provide a protractor mechanism, the line from the equant to *P* being drawn parallel to the mean motus.

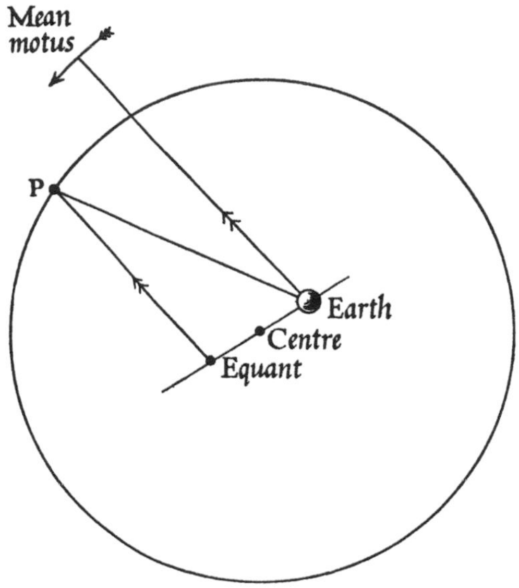


Fig. 9. Ptolemaic use of an equant and a uniformly rotating line to produce non-uniform rotation about the Earth.

In addition to this somewhat slight modification, Ptolemy introduces a second, smaller circle which rides on the back of the main eccentric circle and carries the planet. This is necessary to account for the fact that each of the planets seems to have two separate periodicities—in modern theory these correspond to one periodicity of the planet in its own orbit round the Sun, and a second due to the annual motion of the Earth (from which the planet is being viewed against the background of the heavens). In the case of the outer planets (Mars, Jupiter, Saturn) the epicycle corresponds to the motion of the Earth, and the eccentric main circle to that of the planet concerned. For the inner planets (Mercury, Venus) this situation is reversed, and it is the main circle that contains the annual revolution.

Ptolemy did not complicate the epicycle by making it eccentric or introducing an equant, and the planet is assumed to move round the epicycle at constant angular velocity.

The uniform revolution in the epicycle is governed by the angle termed the

Mean Argument—a quantity which can conveniently be tabulated in the same way as the mean motus which governs motion in the main eccentric circle. The motion in the epicycle is, however, not quite so regular as might appear at first glance, for the mean argument is to be laid off starting from the line drawn from the Equant to the centre of the epicycle (this is the line which

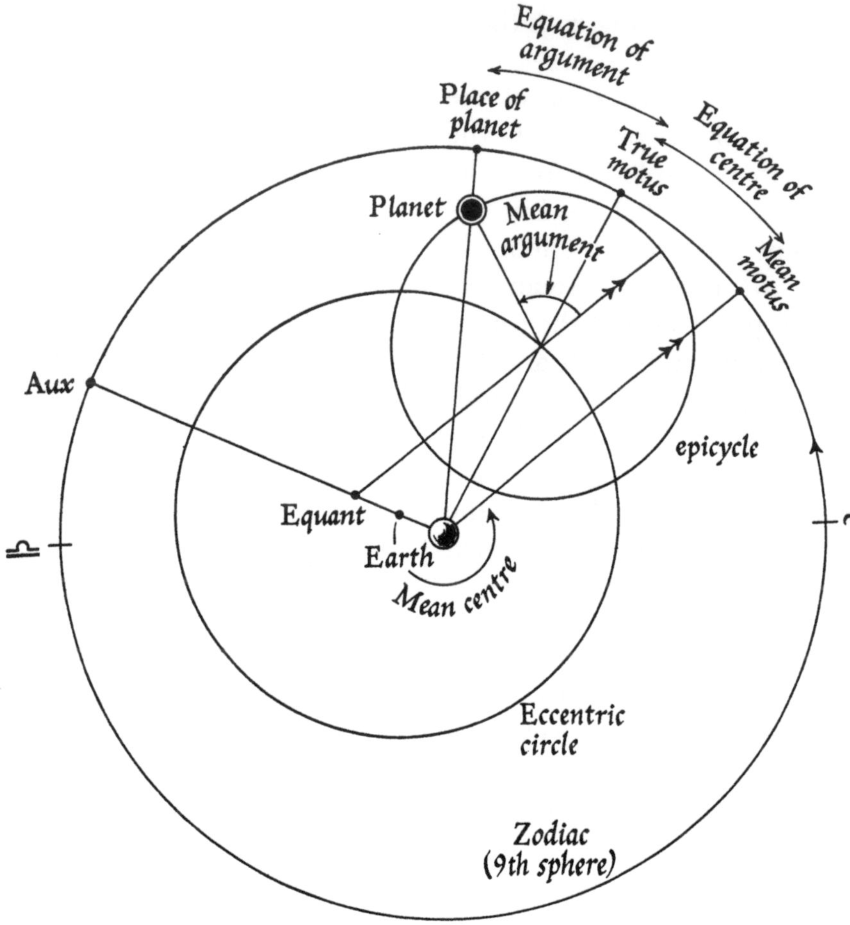


Fig. 10. Complete Ptolemaic construction for Venus, Mars, Jupiter and Saturn.

is drawn parallel to the mean motus line). The angle in the epicycle, when measured with respect to the line running from the Earth to the centre of the epicycle (this line is the line of the true motus), is called the True Argument of the planet, and does *not* increase at a uniform rate.

To sum up the technical terms of Ptolemaic theory, as rendered in English in the Peterhouse manuscript, we have two main angles, the *motus* which governs position in the main eccentric circle, and the *argument* which governs the position of the planet in the epicycle. The *mean motus* and *mean argument* are angles which increase uniformly with time and can therefore be tabulated for general use in planetary computation. When these angles are 'corrected' by the use of a line drawn from the equant, the corresponding angles are known as *true motus* and *true argument* respectively.

The general equivalence of the geocentric and heliocentric hypothesis can be seen from Fig. 16. For the outer planets it will be observed that their mean arguments are not independent quantities, but correspond to the position of the Earth in its (Copernican) orbit; in fact one has in such a case

Mean argument of planet = Mean motus of Sun − Mean motus of planet.

For the inner planets the situation is even more simple, since the geometrical equivalence in this case indicates that the centre of the epicycle corresponds to the apparent position of the Sun as seen from the Earth—it is, in fact, the mean Sun which is taken, and hence in Ptolemaic theory the Sun, Mercury and Venus all have the same mean motus.

The independent periodicities are therefore the mean motus of each of the three outer planets, the mean argument of both the inner planets, and the mean motus of the Sun.

## 5. MERCURY

The mathematical device used for Mercury is constructed in the same manner as for the three outer planets and Venus, but adds a further correction. The Earth and Equant are fixed as before, but the centre deferent (centre of the eccentric circle), instead of being midway between them, moves on a circle as shown in Fig. 11. As the line from the Equant moves round parallel to the line from the Earth indicating the mean motus, a radius moves round the little circle of the centre deferent at an equal rate but in the opposite direction, the sense of both lines being the same when they are in the direction of the aux.

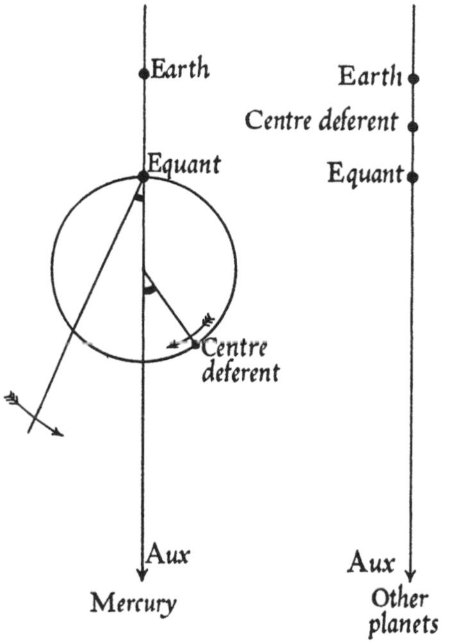


Fig. 11. The placing of Earth, equant and centre deferent in the theory of Mercury and in the other planets.

The effect of this 'wobbling' of the centre deferent is effectively to yield a non-circular deferent; if the actual path of the centre of the epicycle is traced round a complete circle using this movable centre for the deferent, the path will be found to consist of an elongated oval figure. Dreyer[1] has cited a number of examples of the use of non-circular paths for the deferent of Mercury. Perhaps it should also be pointed out that the oval shape does not correspond to the actual (heliocentric) highly elliptical orbit of Mercury. The deferent is associated with the annual periodicity, and it is the epicycle which corresponds to the motion of Mercury in its orbit round the Sun. The effect of the oval is

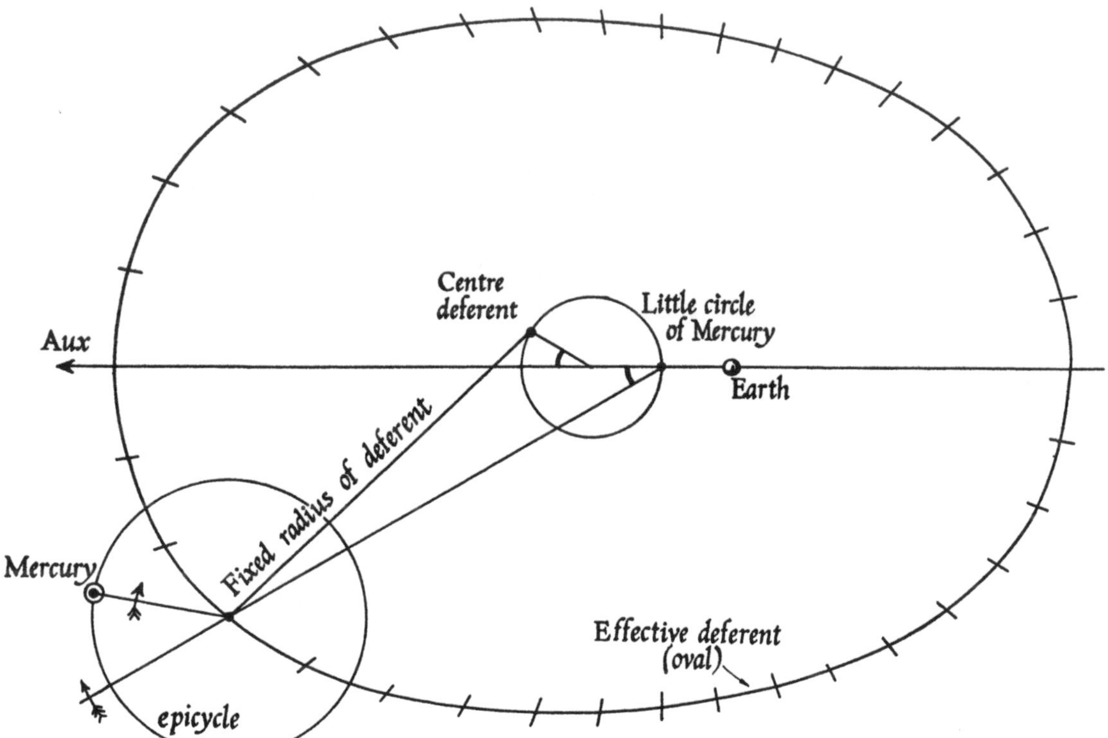


Fig. 12. Ptolemaic construction for Mercury. To show the oval form of the effective deferent, it has been exaggerated by taking $e$ as 10° instead of about 3°.

to make the angle subtended by the epicycle larger and smaller. This increases and decreases the maximum elongation of the planet from the mean centre of Mercury's orbit. Since the planet can only be observed when it is not near the Sun, this theory manages to account satisfactorily for all naked-eye observations.

[1] J. L. E. Dreyer, *History of the Planetary Systems* (Cambridge, 1906), p. 274.

## 6. THE MOON

Ptolemy's theory of the Moon is similar in many respects to that of Mercury which has just been discussed. The centre of the deferent and the equant are here both movable, being situated at opposite ends of a rotating diameter of a circle whose centre is the Earth. This diameter moves as shown in Fig. 13*a*; since MM (i.e. mean motus) Moon is greater than MM Sun the sense of rotation is opposite to that of MM Moon. Once the centre deferent and the equant have been found from the values of MM Moon and MM Sun, the procedure is exactly the same as for any of the other planets.

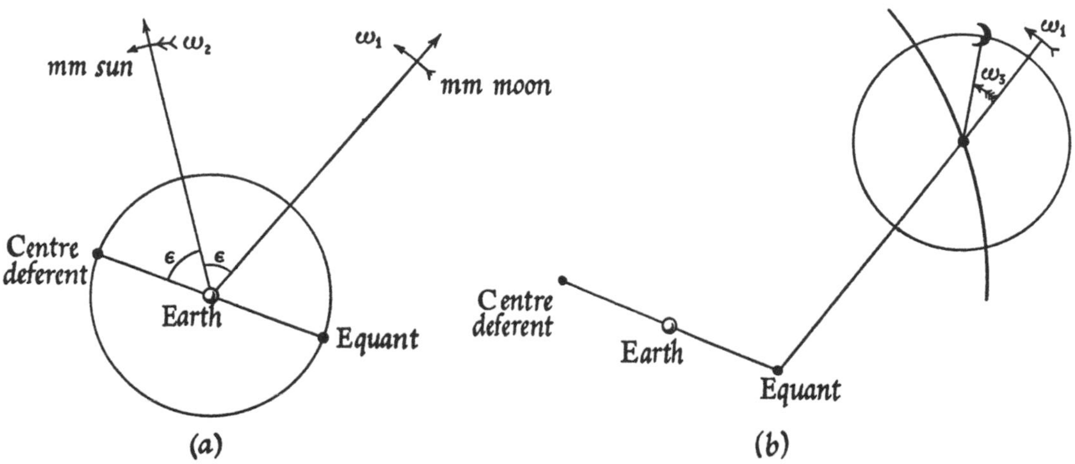


Fig. 13. (*a*) Location of the equant and centre deferent of the Moon.
(*b*) Ptolemaic construction for lunar longitude, after location of equant and centre deferent.

Since the Moon is in fact moving in an orbit around the Earth, it follows that Ptolemy's theory did not require inversion by Copernicus. The movable centre and equant in this device were a particularly great achievement, since by employing them Ptolemy successfully allowed for the third[1] anomaly in the Moon's motion, but for this he had to pay by a gross error in the magnitude of the Moon's variation in distance from the Earth.

The latitude of the Moon was of considerably greater importance to the medieval astronomer than the theory of latitudes of any of the other planets, since it was necessary to compute it for the prediction of eclipses. The Moon's orbit is inclined at approximately 5° to the ecliptic, and unless the Moon was near a point where this orbit intersected the ecliptic plane, no eclipse could

[1] The first anomaly is that represented by the eccentric circle and the equant; the second is represented by the epicycle.

take place. The two points of intersection (or rather, their places in the zodiac) were called the Head and the Tail of the Dragon,[1] and they were taken to move round the zodiac in a direction contrary to that taken by all the planets; the time for a complete rotation is approximately 18½ years. Since these nodes lie in a straight line through the Earth they are always 180° apart, and only one of them, usually the Head of the Dragon (Caput) needs to be tabulated. If Caput is treated as a 'planet' then its theory could hardly be more simple—uniform rotation in a circle with the Earth as centre, i.e. the centre deferent, equant and Earth are coincident. The latitude of the Moon varies approximately as the natural sine of the angular distance from Caput, rising to extreme values of $\pm 5°$ when the Moon is at right angles to the line joining Caput and Cauda.

## 7. THE ALFONSINE 'PRECESSION'

It is impossible to assess the accuracy of the Alfonsine Tables or make any comparison with modern numerical values without a full understanding of the reference system composed of the 8th, 9th and 10th spheres to which all other movements of planets are 'tied'. Medieval astronomers were aware of two examples of exceedingly slow variation in the heavenly motions, the gradual change of the date of the vernal equinox, and a small (precessional) motion of the sphere of the fixed stars; one slowly rotating sphere was used by the Alfonsine astronomers to account for each of these variations.

Modern practice is to consider the fixed stars as providing (apart from their negligible proper motions) an immutable frame of reference for the measurement of all angles and directions in our heliocentric planetary system, and we thus have to measure the time taken for one revolution of the Earth about the Sun, and the rate of movement of the vernal equinox, both with respect (ultimately) to the fixed stars. Medieval practice, on the other hand, was geocentric, not only in the sense of placing the Earth at the centre, but also in taking the horizon of the place of observation as the immutable direction to which all movements must be referred.

With respect to this horizon the Sun performed its daily revolution, so defining the solar day as the interval between two successive meridian transits of the Sun. In addition to this motion the Sun had its annual revolution round the heavens in approximately 365¼ solar days; the Alfonsine astronomers chose to regard this period of exactly 365¼ days as a natural fundamental quantity, and they postulated the existence of a sphere (the Crystalline) in which the

[1] Caput Draconis = Ascending Node, Cauda Draconis = Descending Node. The dragon is, of course, the legendary beast which swallows the luminary and causes eclipses.

Sun made precisely one complete revolution in that period of one mean calendar year of 365¼ solar days.

For astrological purposes it was necessary to refer the motion of the Sun to the band of the zodiac, or rather to its initial point—the first point of Aries or the vernal equinox (see Fig. 7). Since the Sun makes one tropical revolution, returning to the vernal equinox, in slightly less than 365¼ solar days, it will be apparent that for each complete revolution of the Sun round the Crystalline it was exceeding farther and farther beyond the vernal equinox,

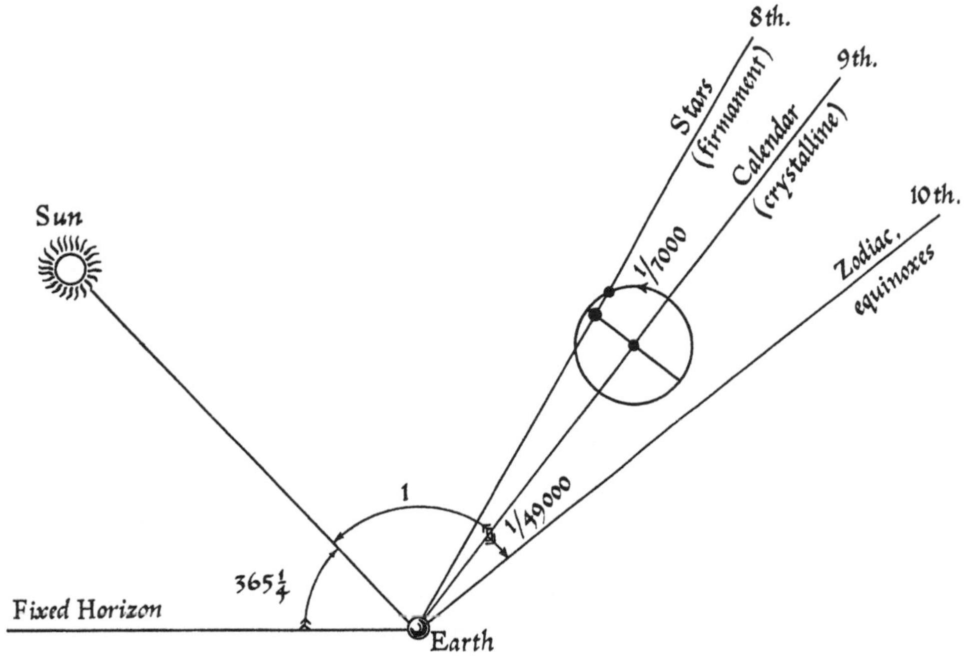


Fig. 14. Alfonsine-Ptolemaic system of reference. The numbered arrows show the number of revolutions made in 365¼ solar days.

or to put matters slightly differently, the Crystalline was edging slowly round the zodiac and vernal equinox. This slow shift was sensible as a gradual and steady decrease of the date of the vernal equinox by about 3 days in every 400 years. The Alfonsine astronomers, partly for mystical reasons,[1] took this decrease as exactly one revolution (or 365¼ days) in 49,000 years—it is the constant from which are derived the 'Mean motus of auges and stars' given in Table 1 (p. 81).

The 49,000-year period compares very well with our modern practice of rectifying the calendar by omitting three out of every four intercalary days

[1] See José Mª Millás Vallicrosa, *Estudios sobre Azarquiel* (Madrid, 1950), pp. 410–11.

at the beginning of the centuries. The 3 days in 400 years is equivalent to $365\frac{1}{4}$ days in exactly 48,700 years. Alternatively, since the Sun is assumed to make 49,001 tropical revolutions in $49{,}000 \times 365\frac{1}{4}$ days, we have the length of a tropical year as 365·24254 solar days, as against the modern determination as 365·24224 solar days. The error corresponds to a difference of about 7 hr. in the time of the vernal equinox over a period of 1000 years; in view of the considerable difficulty of determining the exact time of the equinox this is a very commendable result, and supports the conjecture that medieval observations with the naked eye were by no means inferior to similar measurements by any modern trained observer without a telescope.

It is perhaps worth noting here one of the more obvious errors found in medieval and even modern accounts of the geocentric planetary system. In many of these works it is stated that the *Primum Mobile* or 10th sphere, carrying the zodiac and equinoxes, moves round the Earth once every 24 hr., carrying the stars and the planets with it. The 10th sphere certainly does carry the Sun and planets with it, but its period cannot be exactly the same as the solar day; even without 'calendrical precession' the primum mobile must make $366\frac{1}{4}$ revolutions with respect to the horizon in the time of $365\frac{1}{4}$ days; with the Alfonsine precession it must make altogether $366\frac{1}{4}+\frac{1}{49000}$ revolutions in that same time.

It is well known from modern determinations that because of the 'precession of the equinoxes' the first point of Aries (vernal equinox) is retrogressing through the fixed stars at the rate of a little more than 50 sec. per year, or about one complete rotation in 26,000 years—roughly twice the rate of the Alfonsine calendrical precession. To the Alfonsine astronomers this precession was apparent as a slow drift of the fixed stars with respect to the Crystalline (9th sphere, Calendar) which had already been taken as standard for the consideration of the calendrical precession. They therefore postulated the existence of a sphere containing the fixed stars (and also the auges of the planets—since these were treated as fictitious stars) and arranged for this 8th sphere—the Firmament—to move slowly with respect to the 9th or Crystalline sphere.

For some reason, again probably mystical, they did not use the idea of a steady rotation here, but preferred to make the Firmament oscillate back and forth through an angle of 9° on either side of the Crystalline; the period of one complete double oscillation being taken as exactly 7000 years, and the era of coincidence of the Firmament with the Crystalline being set as 17 May, A.D. 15. Using modern terminology, we have

$$\theta_{(\text{Firmament})}-\theta_{(\text{Crystalline})}=9^\circ \sin\frac{360^\circ\,(t-17\text{ May, A.D. }15)}{7000},$$

where $t$ is the date in years A.D. If we call the right-hand side of this equation $9° \sin \phi$, then $\phi$ is the angle tabulated as 'Mean argument of the 8th sphere' and recorded in Table 1, p. 81. The radix for the Era of Christ, which is given in Table 2, has been calculated from the date 17 May, A.D. 15, when the mean argument should be zero.

To compare the actual amount of precession given by this device we calculate the calendrical and ordinary precession in the interval from A.D. 15 to A.D. 1272 (date of Alfonsine tables). The precession, in the modern sense, during those 1257 years is approximately 17° 30′, and the Alfonsine calendrical precession accounts for 9° 12′ of this, leaving a displacement of the sphere of fixed stars amounting to 8° 18′. By calculation we find that

$$9° \sin (1257 \times 360°/7000) = 8° \ 7',$$

and hence the Alfonsine equation underestimates the precession by 11′ of arc or about 2% of the displacement—again, considering the difficulties of the measurements, this is a very creditable approximation to the truth.

It should be remarked that the medieval astronomers considered the auges of the planets as fixed in the sphere of the Firmament; thereby they neglected a true motion of the apogees amounting, in the use of the Earth's orbit round the Sun, to a shift of about 1° in 300 years. Since the apogee only determines the phase of the first anomaly of the planet, this neglect can only have a negligible second order effect in the calculation of planetary positions.

## 8. THE TECHNICAL TERMS OF PTOLEMAIC ASTRONOMY, AS FOUND IN THE TEXT

It is of the greatest importance in any discussion of medieval astronomy that the technical terms should be clearly defined and, when necessary, translated. Throughout this edition we have followed the usage of the text of the *Equatorie* and, in general, this may be regarded as establishing a very useful convention for the use of these technical terms in the English language.

Most of the technical terms are illustrated in Fig. 15, which shows the equatorie instrument set in position for computing the ecliptic longitude of Venus, Mars, Jupiter or Saturn. In the case of Mercury it is first necessary to locate the variable centre deferent; in the case of the Moon both centre deferent and equant must first be found.

In general the adjective 'mean' is used to refer to a point moving uniformly (in a circle) or to an angle increasing at a constant rate. The corresponding

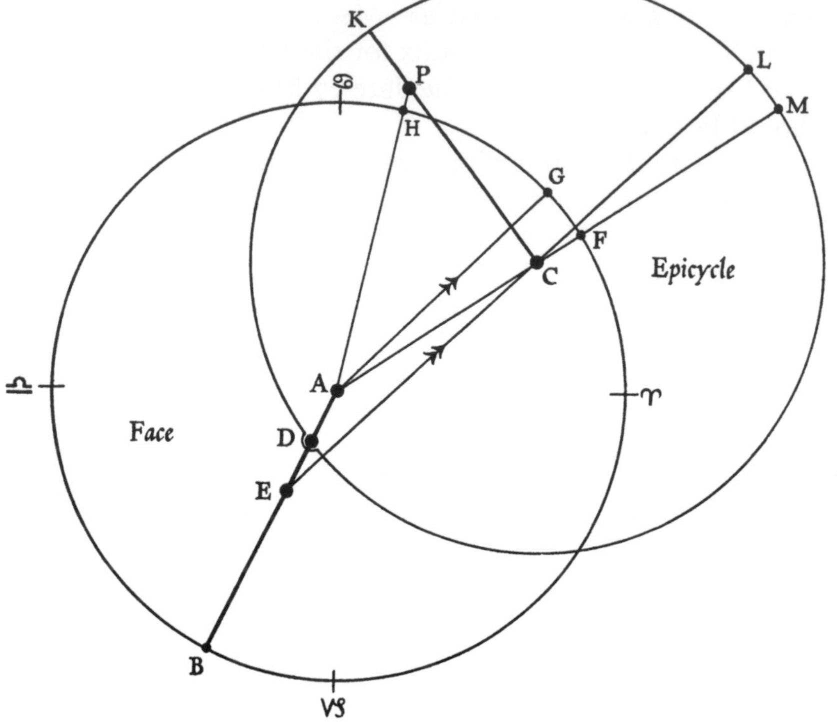


Fig. 15. Simulation of Ptolemaic theory by the Equatorie instrument.

*Key to figure*

| | | | | |
|---|---|---|---|---|
| Points: | ♈ | First point of Aries. | F | True motus of planet (in zodiac). |
| | A | Aryn = Earth. | G | Mean motus of planet (in zodiac). |
| | B | Aux of planet. | H | True place of planet (in zodiac). |
| | C | Centre of Epicycle, pole. | K | Mean argument of planet (on Epicycle). |
| | D | Centre deferent of planet, attached to common centre deferent of Epicycle. | L | Mean aux of planet (on Epicycle). |
| | | | M | True aux of planet (on Epicycle). |
| | | | P | Mark of planet (on label). |
| | E | Centre equant of planet. | | |
| Lines: | AB | Line of aux of planet. | CK | Label. |
| | EL | White thread, fixed to equant. | | |
| | AG, AM, AH | First, second and third positions respectively of black thread fixed to Aryn. | | |
| Angles: | ♈AG | Mean motus. | MCK | True argument. |
| | LCK | Mean argument. | FAG | Equation of centre. |
| | BAG | Mean centre. | FAH = MCL | Equation of argument. |
| | ♈AF | True motus. | ♈AB | Aux. |
| | ♈AH | True place. | | |
| Distances: | CD | Radius of deferent. | CP | Radius of epicycle. |
| | AE = 2AD | Eccentricity of deferent. | | |

'true' quantity is the result of applying a small correction to this uniform motion; the correction itself is called the 'equation'. We thus have:

Mean motus − Equation of centre = True motus,

Mean argument − Equation of centre = True argument,

Mean motus − Equation of centre + Equation of argument = True place,

where the 'Equation of centre' is the correction introduced by virtue of the eccentricity of the deferent circle and the use of an equant, and the 'Equation of argument' is the correction introduced by the existence of the epicycle.

There is a slight ambiguity in that part of the nomenclature which concerns angles measured from ♈, the vernal equinox. Any such angle defines the position of a certain point in the zodiac and therefore the same term may be used either for the angle or for the point itself. Thus we have, on the diagram, a point F and an angle ♈AF, both of which may be called the 'mean motus' of the planet at that particular time.

It would be pointless, perhaps misleading, to attempt to translate the technical terms into modern heliocentric equivalents; a few terms could be taken over unchanged, but in many cases there are slight differences of definition which would be fatal to any comparison of numerical data. For example, 'Mean centre' corresponds to the modern *Mean Anomaly*, and the geocentric 'Aux' is similar (but not equal) to the heliocentric *Apogee*, but the modern term *True Longitude of the Planet* is not equivalent to the medieval 'True place'.[1]

It is much to be regretted that many modern writers have seen fit to translate 'medius motus' as 'mean *motion*' instead of 'mean motus' as found in our text; it refers to an angle or the position of a point in the zodiac rather than to any rate of movement; worse still, it is not the position of any part of the planetary theory, but rather a point or angle which *when corrected* (by the equation of centre) shows the position of the centre of the epicycle. On the whole it is far better to preserve the original term.

In one case only the author has used his terms somewhat ambiguously, so that a certain amount of caution must be used. In the description of the Equatorie instrument he writes of the *mean aux* and *true aux in the Epicycle* (points L and M respectively in Fig. 15) which serve as the datum lines for laying off the true argument and mean argument in the epicycle. In the astronomical tables giving the positions of the auges, the mean aux is the value found by correcting the 'Era of Christ' radix of the aux by the addition of the steady precession at a rate of one revolution in 49,000 years. The true aux

[1] The true longitude is measured partly in the plane of the ecliptic and partly in the plane of the planet's orbit; the true place is measured wholly in the plane of the ecliptic.

is then derived from this by allowing for the effect of the accession and recession of the 8th sphere. It would have been better to retain the conventional latter use, and change the former special application to the epicycle into *mean* and *true starting-point* or some similar pair of terms.

## 9. ACCURACY OF THE THEORY AND THE EQUATORIE

The Equatorie instrument described in our text faithfully reproduces the geometrical construction demanded by Ptolemaic planetary theory. Since it makes no approximations to this theory it will compute planetary positions to the same accuracy as that theory, limited only by the accuracy with which the scales on that instrument may be set and read. It will now be shown that an Equatorie would have to be more than 9 ft. in diameter to enable the maximum discrepancy of the theory to be detected.

To simplify matters we shall exclude the theory of the Moon from this discussion. Ptolemy certainly made a great advance by discovering the Moon's second inequality or *evection* and fixing its amount with considerable accuracy, but in other ways his theory was not so satisfactory. Since the motion of the Moon is 'geocentric' in any case, Copernicus could add nothing fundamentally new to the theory, and, for our present purpose, the discussion of this special case is not material. We shall also omit consideration of the special case of Mercury since Ptolemy had to modify his theory considerably to deal with the *inner* planet whose eccentricity could not be considered negligible.

Limiting the discussion then to the cases of Venus, Mars, Jupiter and Saturn, we may first show that the geocentric system of eccentric circle and epicycle is exactly equivalent to a heliocentric system with two circular orbits, one central and one eccentric. This may be seen in Fig. 16, where diagram *a* shows the geocentric system, *b* the heliocentric equivalent for an outer planet, and *c* that for an inner planet. It will be seen that in all cases the angle and distance of the planet from the Earth is identical.

From this it follows that the Ptolemaic arrangement is equivalent, in the case of an outer planet, to the use of an eccentric circle (instead of an ellipse) for the planet's true orbit, and to the use of rotation in a central circle to represent the motion of the Earth round the Sun. In the case of an inner planet the situation is reversed and the planet is given a central circular orbit, while the Earth is permitted to have an eccentricity.

At first sight this should introduce a considerable error into the theory. For the outer planets it is tantamount to neglecting the equation of centre for the Earth, and since this has a greatest value of about 2° 30′ it would cause

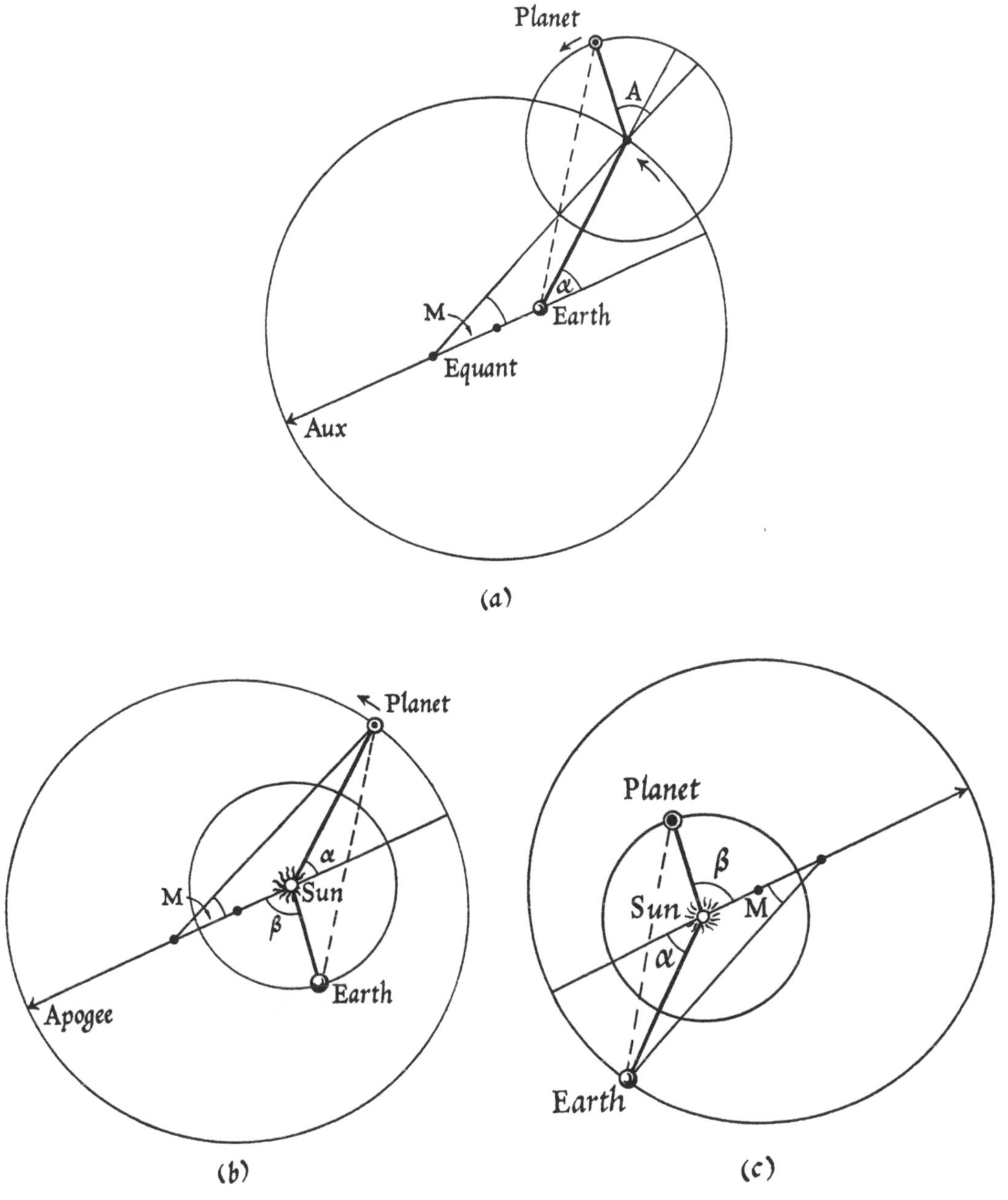


Fig. 16. Equivalence of the Ptolemaic and Copernican hypotheses; the diagram illustrates the two possible cases. (*a*) Ptolemaic construction. (*b*) Heliocentric construction for Mars, Jupiter and Saturn. (*c*) Heliocentric construction for Mercury and Venus.

M = Mean centre, Mean anomaly. A = Mean argument. $\beta = A + M$.

a maximum error of roughly 1° 30′ for Mars, 30′ for Jupiter and 15′ for Saturn (corresponding to epicycles of 40°, 12° and 6°). In the case of Venus, the theory would allow the correct eccentricity for the Earth but none at all for the planet; this would be tantamount to neglecting a greatest equation of centre of about 50′ for that planet. A similar arrangement for Mercury would result in an error of about 7°, hence the need for a special treatment of this case.

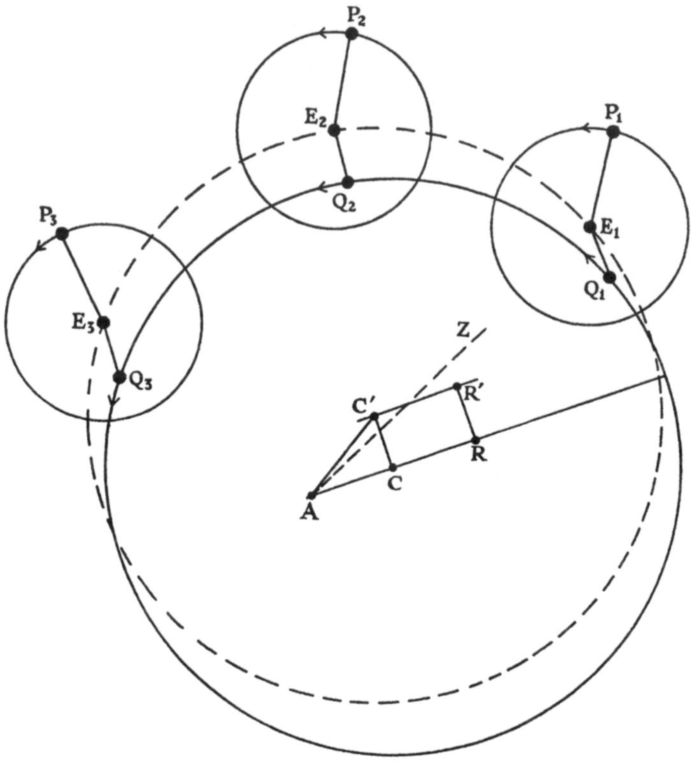


Fig. 17. Incorporation of an eccentric epicycle into an eccentric deferent. AC=CR=C′R′; CC′=RR′=$E_1Q_1$=$E_2Q_2$=$E_3Q_3$. The eccentricity in both deferent and epicycle has been exaggerated to show the effect clearly.

Fortunately for Ptolemy the situation is not so bad as this, since it is possible to use the deferent circle to introduce an eccentricity into the epicycle. In Fig. 17 the full curve shows an eccentric epicycle moving on an eccentric deferent ($Q_1Q_2Q_3$) and maintaining the direction of its eccentricity ($E_1Q_1$, $E_2Q_2$, $E_3Q_3$). It can be seen that this is equivalent to a central epicycle moving on the dotted deferent circle ($E_1E_2E_3$) which has C′ and R′ as its centre deferent and equant instead of C and R. Such a change would mean that the centre deferent and equant were no longer on a straight line through the Earth (A), but it is always possible to find a line AZ which gives a good approximation to this model. The effect of this alteration is to reduce the error to less than one-tenth of the values quoted above, physically it is equivalent to considering the planets as moving in eccentric circles about the empty focus of the Earth's orbit round the Sun. Ptolemy, of course, would not recognize this modern interpretation, he would only notice that his simple scheme using an eccentric circle for the deferent and a central circle for the epicycle gave results agreeing with the observations within the limit of accuracy of his measurements. It should be noted that this deviation from the strict equivalence of the geocentric and heliocentric schemes means that the Ptolemaic aux and eccentricity must be slightly different from the heliocentric apogee and eccentricity even when these are calculated from identical observations. One cannot help remarking also that the extraordinary success of the Ptolemaic theory is due to a large extent to the smallness of the Earth's eccentricity of orbit. A similar astronomer on a planet like Mercury would have to produce a heliocentric theory of the planets or fail hopelessly to account for even the crudest observations.

We must now consider the manner in which Ptolemy makes use of the equant and eccentric circle, and see how far it corresponds with the Kepler ellipse of modern theory. The agreement is surprisingly good for ellipses of small eccentricity, and well within the errors of the observing instruments which were available. For an elliptical orbit of eccentricity $e$, it may be shown[1] that the equivalent eccentric circle gives the angular variation with an error of approximately $\frac{1}{4}e^2 \sin 2M$, and the radius vector with an error of approximately $\frac{1}{2}e^2 \sin^2 M$. The greatest error, that for Mars ($e=0{\cdot}1$), is about 9′ of arc in direction, and 0·5 % in radius vector.

Although not part of the geometrical machinery, the constants used in the theory must accord with fact if the results of computation are to agree with observations. The fundamental constants determining the geometry of the 'orbits' are the values of eccentricity and epicycle radius used for the several planets corresponding in modern theory to the greatest equation of centre and the ratio of mean distances of orbit. In the case of outer planets the epicycle radius is equal to the Earth's mean distance from the Sun divided by the planet's mean distance; for an inner planet this ratio must be inverted—the

[1] The eccentric circle is a good approximation to the better approximation of Ishmael Boulliaud (Bullialdus), whereby the motion of a planet in a Kepler ellipse is taken so that the angular velocity about the empty focus (the *kenofocus*) is uniform. See J. B. Lindsay, *The Chrono-Astrolabe* (Dundee, 1858), p. 40—this is an extraordinary, but nevertheless valuable, discussion of ancient astronomy and the accuracy of observations.

epicycle is always smaller than the deferent. It will be seen from Table 4 that Ptolemaic and medieval values for these constants are in tolerable agreement with modern determinations.

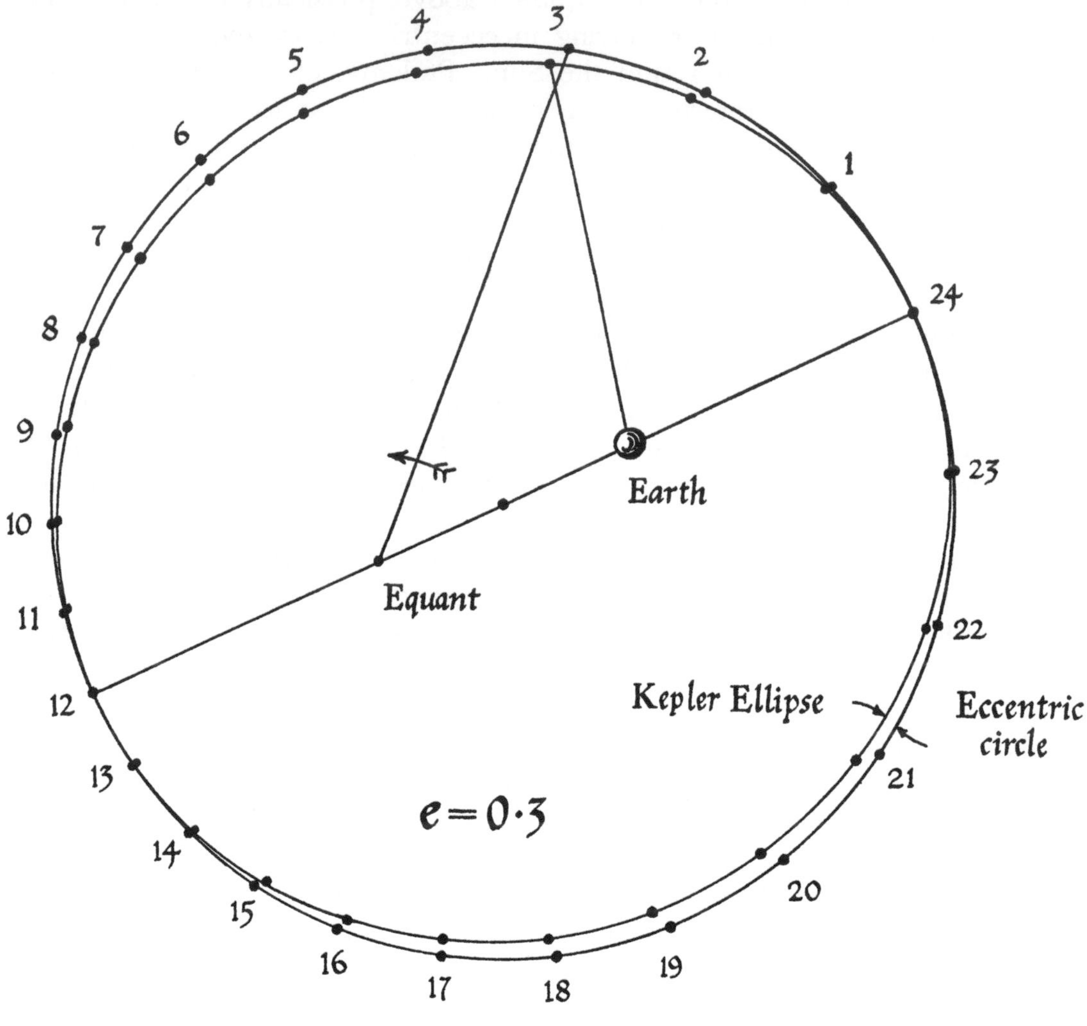


Fig. 18. Equivalence of a Kepler Ellipse and an eccentric circle. The error has been exaggerated by taking $e = 0{\cdot}3$. Since the errors in angle and radius vector are proportional to $e^2$, the orbits and positions would be scarcely distinguishable on a diagram this size constructed for Mars ($e = 0{\cdot}1$).

It is also of interest to compare the radix values of the parameters (mean motus, mean argument, true aux) with the actual mean longitudes for the date in question. Here, however, the comparison is a little different, for any error represents the cumulative effect of the years elapsed between the date of observations from which the radix is calculated and the date of the radix which is being used. For the Alfonsine Tables, the original date of observation is *c.* 1272 and the date of radix is 31 December 1392; the error therefore corresponds to an interval of about 120 years (see Table 5).

TABLE 4. *Fundamental geometrical constants of Ptolemaic orbits*

*Eccentricities (greatest equation of centre, Earth-equant distance)*

| | Sun | Mercury | Venus | Mars | Jupiter | Saturn |
|---|---|---|---|---|---|---|
| Ptolemy | 2, 30 | 3, 00 | 2, 30 | 12, 00 | 5, 30 | 6, 50 |
| al-Kāshī | 2, 04 | 3, 00 | 2, 06 | 12, 00 | 5, 28 | 6, 50 |
| al-Biruni | [2, 04¾] | 3, 10 | 2, 05 | 12, 00 | 5, 30 | 6, 45 |
| al-Khwarizmi | 2, 14 | 4, 02 | 2, 14 | 11, 13 | 5, 06 | 8, 37 |
| Toledo Tables | 1, 59 | 3, 02 | 1, 59 | 11, 24 | 5, 15 | 6, 31 |
| Alfonsine Tables | 2, 10 | 3, 02 | 2, 10 | 11, 24 | 5, 57 | 6, 31 |
| John of Linières | 2, 30 | 3, 00 | 2, 30 | 13, 00 | 5, 30 | 6, 50 |
| Modern | 2, 10 | (23, 24) | (0, 50) | 10, 44 | 5, 26 | 6, 36 |

*Epicycles (ratio of mean distances, Earth and planet)*

| | Sun | Mercury | Venus | Mars | Jupiter | Saturn |
|---|---|---|---|---|---|---|
| Ptolemy | No epicycle | 22, 30 | 43, 10 | 39, 30 | 11, 30 | 6, 30 |
| al-Kāshī | | 22, 27 | 43, 10 | 40, 18 | 11, 30 | 6, 30 |
| al-Battani | | 22, 30½ | 43, 09 | 39, 27 | 11, 30 | 6, 29, 50 |
| al-Khwarizmi | | 21, 30 | 47, 10 | 40, 31 | 10, 52 | 5, 44 |
| Toledo Tables | | 22, 02 | 45, 59 | 41, 09 | 11, 03 | 6, 13 |
| Alfonsine Tables | | 22, 02 | 45, 59 | 41, 10 | 11, 03 | 6, 13 |
| John of Linières | | 22, 30 | 43, 10 | 39, 30 | 11, 30 | 6, 30 |
| Modern | | 23, 02 | 43, 38 | 39, 22 | 11, 32 | 6, 17 |

[ ] From al-Battani.

( ) Eccentricity of deferent of Mercury and Venus should be almost the same as that of the Sun i.e. 2, 10.

TABLE 5. *Comparison of radices with actual positions of planets on 31 December 1392*

| | Mercury | Venus | Sun (Earth) | Mars | Jupiter | Saturn |
|---|---|---|---|---|---|---|
| | | | Mean longitude | | | |
| Radix | 188° 21′ | 311° 46′ | 288° 33′ | 92° 13′ | 324° 46′ | 184° 46′ |
| Calculated | (196° 49′) | 310° 47′ | 288° 22′ | 92° 02′ | 325° 04′ | 183° 48′ |
| Error | Not comparable | +0° 59′ | +0° 11′ | +0° 11′ | −0° 18′ | +0° 58′ |
| | | | Aux (Aphelion) | | | |
| Radix | (209° 33′) | None | 90° 09′ | 133° 56′ | 172° 20′ | 252° 07′ |
| Calculated | 248° 00′ | 303° 03′ | 92° 30′ | 144° 48′ | 184° 43′ | 261° 36′ |
| Error | Not comparable | | −2° 21′ | −10° 52′ | −12° 23′ | −9° 29′ |

Average error in mean longitude: 0° 32′ in *c.* 120 years.

Average error in aux (aphelion): 8° 46′ (producing only a second-order effect on position of planets).

The radix values are taken from Tables 2 and 3 (p. 83), and the modern values of mean longitude were kindly calculated for me by Her Majesty's Nautical Almanac Office, using the theories of Le Verrier and Newcomb; they agree with the tables of P. V. Neugebauer. The aux values are from J. B. Lindsay, *The Chrono-Astrolabe* (Dundee, 1858).

## 10. THE CALCULATION OF PLANETARY POSITIONS

Having examined the general outlines of Ptolemaic planetary theory and of the Alfonsine Tables, we must now turn our attention to the manner in which planetary positions may be calculated by those means.

The skeleton construction demanded by the theory is illustrated in Fig. 19, which shows one particular position of any outer planet or Venus. For purpose of this general computation we may consider that the eccentricity ($e$), the

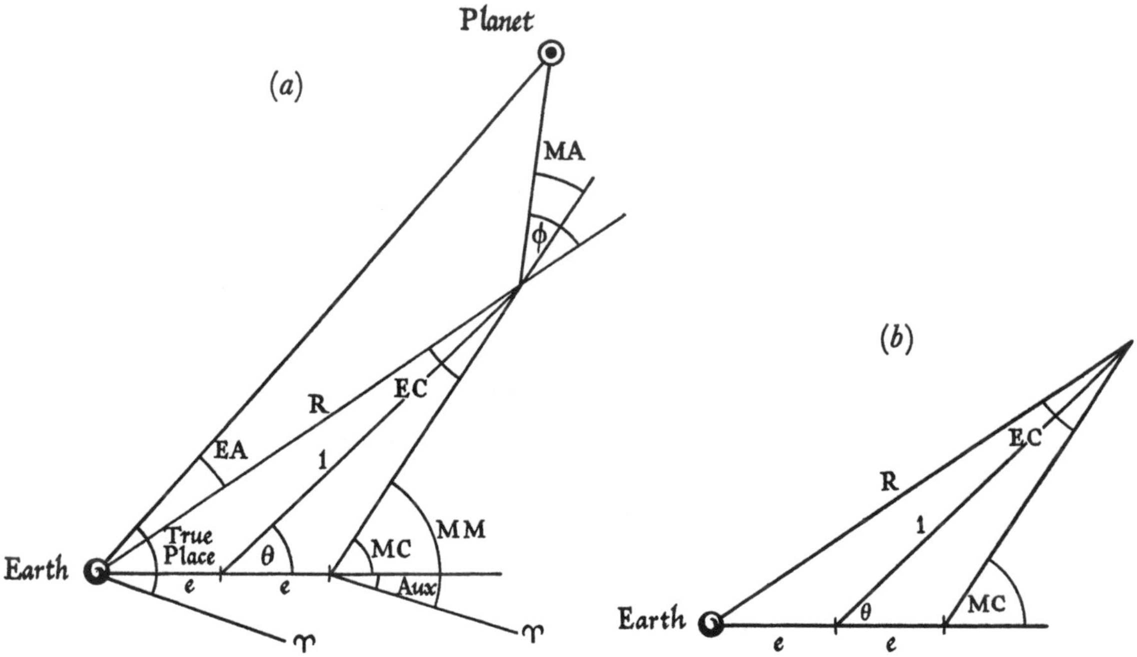


Fig. 19. Skeleton geometrical construction for longitude of Venus, Mars, Jupiter or Saturn.
(*a*) Complete construction. (*b*) Deferent portion only.
MC = Mean centre = Mean motus − Aux. EC = Equation of centre.
MA = Mean argument. EA = Equation of argument.

radius of the epicycle ($r$) and the position of the aux are given as constants, and the angles corresponding to the mean motus and mean argument are the two parameters from which it is required to calculate the true position of the planet.

Considering at first only that part of the figure which is due to the eccentricity in the deferent, we have

$$\tan \text{MC} = \sin \theta / (\cos \theta - e), \quad \text{(I)}$$

$$\tan \text{EC} = 2e \sin \theta / (1 - e^2), \quad \text{(II)}$$

$$\text{R}^2 = 1 + e^2 + 2e \cos \theta. \quad \text{(III)}$$

Eliminating $\theta$ we obtain

$$\tan \mathrm{EC} = \frac{2e \sin \mathrm{MC}}{1-e^2} [\sqrt{(1-e^2 \sin^2 \mathrm{MC})} - e \cos \mathrm{MC}], \tag{IV}$$

$$\mathrm{R}^2 = 1 + 2e \cos \mathrm{MC} \sqrt{(1-e^2 \sin^2 \mathrm{MC})} + e^2 (1 + 2 \sin^2 \mathrm{MC}). \tag{V}$$

Having thus obtained EC and R in terms of MC, we may proceed to the calculation of the equation of argument, for setting $\mathrm{MA}+\mathrm{EC}=\phi$,

$$\tan \mathrm{EA} = r \sin \phi / (r \cos \phi + \mathrm{R}). \tag{VI}$$

The true place of the planet is then given by

$$\text{True place} = \mathrm{MM} - \mathrm{EC} + \mathrm{EA}, \tag{VII}$$

which therefore gives the required solution by direct trigonometrical calculation. It is easily seen that this method is somewhat cumbrous and ill-adapted for routine purposes, and as Ptolemy himself suggests (*Almagest*, XI, 10) it is far more convenient to construct tables for values of $e$ and $r$ corresponding to each of the planets. Such a table may readily be drawn up for EC using equation (IV), where the right-hand side is a function of MC only. For EA we cannot make use of equation (VI) as it stands, for the right-hand side is a function of both $\phi$ and MC (through R).

We therefore define EA′ by

$$\tan \mathrm{EA}' = r \sin \phi / (r \cos \phi + 1), \tag{VIII}$$

and from this and (VI) it follows that

$$\tan (\mathrm{EA} - \mathrm{EA}') = (1 - \mathrm{R}) \frac{1 + r \cos \phi}{\mathrm{R} + (1+\mathrm{R}) r \cos \phi + r^2} \tan \mathrm{EA}'. \tag{IX}$$

Since $(1-\mathrm{R})$ is a small quantity we may take as a first approximation

$$\mathrm{EA} - \mathrm{EA}' = (1 - \mathrm{R}) \tan \mathrm{EA}', \tag{X}$$

and hence

$$\mathrm{EA} = \mathrm{EA}' + (1 - \mathrm{R}) \tan \mathrm{EA}', \tag{XI}$$

which gives the required quantity as the sum of a function of $\phi$ alone and a correction which is that function multiplied by a small factor which is a function of MC alone. The factor may be tabulated by computation from equation (V).

We must therefore provide three tables for each planet, and the computation proceeds in six stages; an arrow indicates use of a table:

(1) $\mathrm{MM} - \mathrm{Aux} = \mathrm{MC}$, (2) $\mathrm{MC} \to \mathrm{EC}$, (3) $\mathrm{MC} \to (1 - \mathrm{R})$,

(4) $\mathrm{MA} + \mathrm{EC} = \phi$, (5) $\phi \to \mathrm{EA}'$,

(6) $\mathrm{MM} - \mathrm{EC} + \mathrm{EA}' + (1 - \mathrm{R}) \tan \mathrm{EA}' = \text{True place of planet}$.

If one allows only about 2 min. of work for each process involving sexagesimal multiplication or interpolation, it must take at least 15 min. to compute a planetary position with the assistance of such tables. For a horoscope there are seven such positions to compute in addition to finding the ascendant, and it would therefore be difficult to complete the process in less than about 2 hr.

The Equatorie instrument is designed to relieve the computor of a large part of this procedure and spare him the labour of making side-calculations 'on his slate'. With the instrument he has to use tables only twice for each planet to ascertain the current values of the mean motus and mean argument; all else is performed by setting the instrument, and the time taken should be one-half, or possibly even one-third, of that for similar tabular computation—there is also much less chance of making an arithmetical error.

In its convenience and speed, therefore, the Equatorium instrument may well be considered as the medieval equivalent of the modern tables provided by the *Nautical Almanac*; for special purposes high accuracy must be obtained by individual computation, but for routine use the method is labour-saving, rapid and sufficiently accurate.

# VIII

# HISTORY OF THE PLANETARY EQUATORIUM

## I. EARLY HISTORY

It is difficult to trace the history of equatoria before the eleventh century because, without full details of the device, it is impossible to ascertain whether a so-called 'planetary instrument' is an actual device for computation or merely a demonstration model to show the planets moving (uniformly) around the Earth. Such models were by no means unknown in antiquity, for there is some evidence to suggest that Archimedes may have constructed one,[1] and fragments of what was probably a Greek planetarium containing gear wheels and dating from about A.D. 250 have been recovered from the Mediterranean sea-bed off the island of Antikythera.[2] There are also references to the use of such devices in the Orient, for example, the Uranorama of I-Hsing (b. 683, d. 727) which was moved by water-power (?) and indicated the motions of the heavenly bodies relative to the ecliptic.[3] At later dates these demonstration planetaria are found associated with the elaborate astronomical clocks, often erected in cathedrals in the fourteenth and fifteenth centuries. Still more recently this tradition has persisted in the Orrery and other such astronomical toys, as well as the more massive and impressive optical projection planetarium occupying a special building. Perhaps we should also include in this category many of the small parchment volvelles, frequently found in the body or on the covers of astronomical manuscripts and books. Certainly some of these volvelles were designed to assist approximate calculation, but very frequently they seem to serve as nothing more than a moving diagram to illustrate the text in a qualitative fashion. A common type of volvelle which is really qualitative, but serves a useful practical purpose is the 'Sun and Moon' instrument; the Sun and Moon are set to their correct places in the zodiac, and a little aperture indicates the phase of the Moon.[4]

1 See E. L. Stevenson, *Terrestrial and Celestial Globes* (New Haven, 1921), vol. I, p. 16.

2 See J. N. Svoronos, *Das Athener Nationalmuseum* (Athens, 1908), Textband I, pp. 43–51. Better reproductions are given by E. Zinner, *Entstehung und Ausbreitung der Coppernicanischen Lehre* (Erlangen, 1943), p. 48, plates 12, 13 and R. T. Gunther, *Astrolabes of the World*, Item I.

3 Sarton, *Introduction to the History of Science*, vol. I, p. 514.

4 A typical text describing this volvelle is in Bodleian MS. Can. Misc. 517, f. 1r.–1v., 'Canones horoscopi instrumenti'. *Inc.* De instrumentum factum est pro inveniendo gradum ascendentem....*Expl.* ...in duodecim partes equales sive dies fuerit longa sive brevis. Laus deo.

Turning now to the tradition of geometrical calculating instruments in which we are interested, there seems to be no evidence from classical times of any device for computing the positions of all the planets. A method which is not essentially different is, however, described by Proclus Diadochus (*c.* A.D. 450) in his *Hypotyposis Astronomicarum Positionum*, which deals with a device for finding the position and equation of centre of the Sun.[1] Clearly an equatorium *could* have been designed, but the stimulus does not seem to have come until the revival of learning, when the *Almagest* was recovered and made known to the West through the activities of the Hispano-Arabic translators and scholars.

The *Almagest* was translated into Arabic under al-Ma'mūn (*c.* A.D. 830), and it is interesting to see that this work proceeded hand-in-hand with the erection of an observatory, and the calculation of a new set of astronomical tables (the Ma'mūnic tables). Ptolemy's tables were out-of-date by this time, containing the accumulated errors of seven centuries, and hence such correction was highly necessary; furthermore, such correction was obviously very satisfying to the astronomer, for the long interval made it possible to obtain very accurate values for the annual motions of the celestial bodies. This first burst of activity seems to have led to a great rise in instrument-making, in particular to the popularization of the astrolabe, but there does not appear to be any record of equatoria from this period.

## 2. ELEVENTH CENTURY

The first, rather crude planetary equatorium makes its appearance about two centuries later in a text by Abulcacim Abnacahm (Abūlqāsim ibn Moḥammad ibn as-Samḥ *or* Abū-l-Qāsim Aṣbagh ibn Muḥammed Ibn al-Samḥ) of Granada. His main work seems to have been the compilation of a set of tables according to the Siddhānta method, *c.* 1025, and it is probably because of this stimulus that he was led to invent his instrument. Its description has come down to us through the archaic Castilian of the *Libros del Saber* of Alfonso the Wise,[2] but it is probably incomplete, since it contains only the directions for laying out his *Láminas de las VII Planetas*, and does not continue with any account of how the instrument is to be used in practice. Fortunately, the directions for construction are sufficiently complete for there to be no doubt as to the method of use; there is to be a separate plate for each of the seven planets, and on this plate is drawn an eccentric circle (of unit radius) to represent the deferent of the planet, the Earth being taken as the centre of each

[1] Very important for the use of such geometrical methods in calculation is the *Planispherium* of Ptolemy. Cf. the excellent paper by P. Luckey, *Astronomische Nachrichten*, No. 230 (1927), col. 18–46.

[2] It is printed in the edition of Rico y Sinobas, vol. III, pp. 241–71.

plate. A smaller circle is drawn about the equant of the planet, and this circle is divided uniformly in degrees so that it may be used as a sort of protractor for finding a point on the deferent corresponding to the mean motus at any given time. For the epicycles a common plate is constructed, having the radius of the largest planetary epicycle, that of Venus (43/60 of the unit radius taken for the deferents). On this plate are marked also the epicycles of Mars, Mercury, Jupiter and Saturn in decreasing size. In use the epicycle plate is placed so that its centre is correctly positioned on a deferent circle, and its aux line points towards the equant; the planetary place may then be read by means of a thread attached to the point at the centre of the deferent plate representing the Earth. Special additional plates must be constructed for the theories of the Moon, the Sun and Mercury.

The greatest fault in this instrument is the number of separate plates which have to be made; indeed, the only concession to simplicity is the use of a single plate for five of the epicycles. This involves the use of eighteen graduated circles instead of the maximum of twenty-two which might be demanded by Ptolemaic theory of planetary longitudes, and one could not claim that this form of instrument displayed any great ingenuity. Nevertheless, the first step had been taken towards the design of such planetary instruments, and we may regard Abulcacim Abnacahm as the father of this prototype device—any earlier instrument could hardly have been more simple than that which he describes.

Later in the eleventh century the first Hispano-Muslim burst of astronomical activity culminated in the work of Azarquiel (Ibrāhīm ibn Yaḥyā abu Isḥāq ibn az-Zarqālī, b. *c.* 1029, d. *c.* 1087), who designed instruments, made observations of considerable accuracy, and edited the Toledo Tables. These tables became known to European astronomers through the Latin translation by Gerard of Cremona (before 1187) and held the field until they were eventually displaced by the Alfonsine Tables and its modifications late in the thirteenth century.[1] Amongst the instruments invented by Azarquiel is a modification of the *láminás* of Abulcacim Abnacahm, known to us in the Arabic original as well as the Castilian recension which is found in the Alfonsine *Libros del Saber*.[2] The two texts are by no means identical, since the Arabic deals more with the practice of the instrument while the Castilian is almost exclusively concerned

[1] An excellent account of the Toledo Tables with a list of fifty-two manuscript copies and an analysis of the constants used in the tables is given by E. Zinner, *Osiris*, vol. I (Bruges, 1936), pp. 747–73.

[2] A transcription and a Spanish translation of the Arabic text from British Museum MS. Arabic 426[10] is given by José Mª Millás Vallicrosa, *Estudios sobre Azarquiel* (Madrid, 1950), pp. 460–79. The Alfonsine text has been edited by Rico y Sinobas, *Libros del Saber*, vol. III, pp. 241–71, and the instrument itself has been very fully discussed by A. Wegener, 'Die astronomischen Werke Alfons X', *Bibiotheca Mathematica* (1905), pp. 129–89.

with the construction of the plates. Moreover, the Arabic text describes a more primitive version of the instrument in which the procedure for the Moon requires the use of a special separate lamina; the instrument described in the Alfonsine text incorporates this procedure into the general pattern and therefore makes operation more simple and the instrument itself more convenient.

The improvement made by Azarquiel is an extension of the feature of the original instrument which used only one plate for the five planetary epicycles. This feature is now applied to the deferent and equant circles which are thereby arranged on only two plates (or both sides of a single plate) instead of on seven separate plates. The process is effected by taking the planetary deferents of different sizes so that they nest, one inside another, the nesting being as close as the various eccentricities will allow. Thus on one side of the deferent plate are inscribed the deferents of Venus, Mars, Jupiter and Saturn, followed by the four equant circles which nest in the same fashion. On the other side of this plate are found the special circles for Mercury, the Moon and the Sun; a special and noteworthy feature of the instrument described in the Alfonsine text is the use of an elliptical deferent for Mercury, corresponding to the oval which is a consequence of the Ptolemaic theory (cf. Fig. 12, p. 102).

The two plates (three faces) of Azarquiel's instrument contain in all nineteen circles, each of which has to be graduated in degrees, just as in the original primitive form of the instrument. It is certainly more convenient to have only two laminas instead of the original ten, but the improvement has been paid for by the confusing appearance of all these circles in such a small space. A practical instrument of this design would have to show each circle clearly labelled in order to avoid any error in deciding which scale to use for any particular calculation.

Azarquiel's instrument does not seem to have met with any great popularity, possibly because of its very complicated appearance and because of the great labour in constructing so many accurately divided circles. Even Azarquiel himself seems to have seen that the instrument was not the perfect solution for planetary calculations, for he seems to have expended tremendous labour in the compilation of a set of perpetual ephemerides which enable the approximate places of the planets to be found with very little calculation. The tables are based on the fact that it is possible to find some integral number of years in which the planet makes almost exactly some whole number of revolutions; for example Saturn, in a period of 59 years, makes 2 revolutions $+ 1^\circ\ 55'$. The tables give the true place of Saturn at intervals of every 5 or 10 days for a period of 59 years from some given radix date, and at any future date one has only to subtract any number of complete cycles of 59 years, read the planet's place for the remaining date, and apply a small correction. The periods and corrections for a full cycle are as follows:

| | | | |
|---|---|---|---|
| Saturn | 59 years | 2 revolutions | $-1° 55'$ |
| Jupiter | 83 years | 7 revolutions | $+2° 0'$ |
| Mars | 79 years | 42 revolutions | $-1° 0'$ |
| Venus | 8 years | 8 revolutions (5 anomalies) | $-1° 30'$ |
| Mercury | 46 years | 46 revolutions (145 anomalies) | $-2° 45'$ |

Certainly a well-drawn-up set of ephemerides of this type is very convenient for use, rapid in operation, and sufficiently accurate for most purposes during the first few cycles of each planet, provided one does not require greater precision than the tabulated entries which give the place in the zodiac to the nearest degree only. During the remainder of the Middle Ages, and even well into the Renaissance, such perpetual tables must have been regarded as the first choice for astrological purposes and approximate calculations. For more exacting work, the astronomer could still turn to the traditional tables constructed after the method described in the *Almagest*, and the planetary equatorium therefore only served a purpose for those who were not satisfied with the approximate ephemerides but wished to avoid the labour of much sexagesimal calculation.

## 3. TWELFTH AND THIRTEENTH CENTURIES

After the work of Azarquiel there is a blank in the history of equatoria and planetary calculations. The *Almagest* was first translated into Latin (from the Greek) in *c.* 1160 by an anonymous Sicilian author, and was followed in 1175 by the very popular translation from the Arabic by Gerard of Cremona. The stimulus caused by this diffusion of astronomical theory did not have much effect on astronomical calculations and tables until about the middle of the thirteenth century, but for the remainder of that century and the whole of the fourteenth there is a flood of new tables and fresh forms of equatoria. This change must be associated with the introduction of the Arabic forms of numerals which first appear in astronomical tables *c.* 1260 and become standard practice by about 1320, so facilitating astronomical computation and leading to a greater interest in instruments and methods designed to assist this work. The change is perhaps not so great as one might imagine, for the use of the sexagesimal system was conserved until well into the fifteenth century, and in this system there is no need to express any single number greater than 60; if multiplication tables are available it makes little difference whether the numbers up to 60 are expressed in Arabic or in Roman numerals. The greater regularity of the Arabic numerals makes it easier to check errors in the tables, and perhaps also lends itself more readily to the good lay-out of long tables.

The Alfonsine translations of the first two texts dealing with the equatorium have already been mentioned; they were made in 1276–7, and since the Alfonsine Tables of 1272 did not reach Paris until *c.* 1292, it is likely that the descriptions of the instruments were not available in the two great centres of astronomy, Paris and Oxford, before a similar date. Slightly before this date there is mention of an instrument said to have been constructed by John Campanus (Giovanni Campano da Novara, *fl. c.* 1261–92), but it is difficult to disentangle any evidence on this score from the references to the very popular *Theoretica Planetarum* (*Theoretica Campani*) written by this author. John of Linières, writing in *c.* 1320, mentions the *equatorium* of Campanus as an instrument inferior to his own, and a similar situation seems to occur with most of the other manuscript texts purporting to contain a description of the instrument devised by Campanus.[1] If he did actually invent such an equatorium, it does not seem to have become well known except in a form modified and improved by later astronomers. Similar situations are to be found in instruments said to have been devised by Gerard of Sabbioneta, William of St Cloud, Profacius and Peter of Dacia; they were all famous as constructors of tables and as writers on the *theory* of the planets, and from time to time they are mentioned as the inventor of some type of planetary instrument. It is highly desirable that such texts should be studied in detail, and the features of the instruments described and analysed; until this is done the situation must remain confused.

Contemporary with the Alfonsine school of astronomy in Toledo, a similar movement of the greatest significance was taking place in Islam under the guidance of Nāṣir al-dīn al-Ṭūsī (b. 1201, d. 1274). There is the same pattern of translations and revivals of earlier astronomical texts, the calculation of new tables, the design of instruments, and we have also the foundation of the great Marāgha observatory *c.* 1259. The chief deviser and constructor of instruments in this school seems to have been al-'Urḍī (Mu'ayyad al-dīn al-'Urḍī al-Dimishqī), but amongst his list of instruments there does not appear any explicit mention of an equatorium. In spite of this it is reasonable to conjecture that such an instrument may have been developed, for al-'Urḍī certainly had a mechanico-graphical device for computing sines and versed sines, and later Islamic developments such as the *Plate of Zones* of al-Kāshī (p. 131) seem to be closely related to the Marāgha instruments.

[1] A text *Equatoria de veris motibus septem planetarum*, attributed to Campanus, which opens 'Sol habet unum circulum super cuius', appears in MS. Vatican Palatine 1416, f. 198 r. (fifteenth century).

## 4. JOHN OF LINIÈRES

Early in the fourteenth century two forms of equatoria were invented, and, if one might judge from the number of manuscript copies of various recensions and modifications of the text, these two instruments enjoyed a greater popularity and more lasting fame than any other version of the planetary device. The instruments concerned are the equatorium of John of Linières (Joannes de Lineriis, of Lignières in the diocese of Amiens, Picardy, *fl.* Paris, *c.* 1320–50) and the Albion of Richard of Wallingford, Abbot of St Albans (1292?–1336), who was one of the most ingenious of the Oxford school of astronomers associated with Merton College.[1]

The greatest work of John of Linières was his adaptation of the Alfonsine Tables to the meridian of Paris, and his compilation of two sets of canons to the tables (1320 and 1322) which explain their use, give instruction in the art of calculation by means of sexagesimal fractions, and describe the equatorium instrument which may be used with the tables. The printed version of the Alfonsine Tables (Venice, 1483, etc.) represents a minor modification of this version made by John and his pupil and collaborator John of Saxony. Part of the canons to the tables includes a self-contained treatise (or set of texts) on the construction and use of the equatorium of John of Linières. This treatise was frequently copied separately and appears under various titles: *Abbreviatio instrumenti Campani sive aequatorium*, *Instrumentum saphee* (this is either a misnomer, or an error for the Saphea astrolabe projection of Azarquiel), *Tractatus de utilitatibus equatorii planetarum.*[2] As with the rest of John's works the same material appears under different titles, and in this case we have two forms, possibly a third as well. The main title seems to be *Abbreviatio instrumentum Campani sive equatorium* (*incipit*, *Quia nobilissima scientia astronomie non potest scire*),[3] but a shorter version containing the practice, but not the construction, of the instrument appears as *Tractatus de utilitatibus equatorii* (*incipit*, *Descripciones que sunt in equatorio planetarum notificare*);[4] another tract which begins *Recipe tabulam*

[1] An interesting portrait of Wallingford is reproduced in colour by R. T. Gunther, *Early Science in Oxford* (Oxford, 1922), vol. I, frontispiece to Part 2. The abbot is apparently using a pair of dividers to mark out a brass disc; perhaps this disc is intended to represent his Albion.

[2] For an account of several manuscript sources see L. Thorndike, *History of Magic and Experimental Science* (New York, 1934), vol. III, pp. 261–2.

[3] This form appears in the following manuscripts: Bodleian MS. Digby 57, ff. 130r.–132v.; Bodleian MS. Digby 168, ff. 64v.–66r.; Vatican MS. Palatine 1375, ff. 8v.$^{1}$–10v.$^{1}$; Brussels, Royal Library MS. 10124, ff. 142v.–146v. A transcription of the text from the Brussels manuscript is given in Appendix II.

[4] The shortened version occurs in Bodleian MS. Digby 228, ff. 53v.–54r. and Vatican MS. Urban 1399, ff. 16r.$^{1}$–21r.$^{2}$.

*ligneam pergamena* (Vatican MS. Palatine 1416, ff. 168r.–169v.) may be a fragment of adaptation of the main work, or it might be a version of the tract on the *directorium* (see below, this page).

The instrument described by John of Linières is a slight adaptation of that invented by Azarquiel. The deferent circles of all the planets are drawn on one plate, and the epicycles are all indicated by marks on a pointer which may revolve about the centre of a second plate. Linières improves on the older form by drawing all his deferents on one side of the plate, instead of both, but he goes back to the full Ptolemaic construction for Mercury, so ignoring the ingenious construction of an elliptical or oval deferent for that planet. He also reduces the confusion of Azarquiel's equatorium by using the graduated limbs of the plates for all his angular measurements, so avoiding the necessity of marking graduations on all the other circles, and dispensing with the need for separate 'equant circles'. The consequent improvement in simplicity removes the need for any careful nesting of the deferent circles, and Linières is enabled to use a single radius ($\frac{5}{8}$ radius of main plate) for the Moon, Mercury, Jupiter and Saturn; a different radius ($\frac{1}{2}$ radius of main plate) for the two planets having large epicycles, Mars and Venus; and the maximum possible radius ($\frac{24}{25}$ radius of main plate) for the deferent or eccentric circle of the Sun. Two small additional plates are used for the rotating centres of the Moon and Mercury, and a special marked pointer and stretched threads are used for setting the instrument and reading the final result on the limb of the main plate.

With only one exception, the constants used in the construction of the Linières equatorium are in complete accord with the values recorded in the *Almagest* (see Table 4), the exception being the eccentricity of Mars, which is set at 13° instead of the 12° used by Ptolemy and followed fairly closely by other medieval astronomers. This characteristic is especially interesting in connection with our text of the *Equatorie*, where a similarly high value seems to be implied (see p. 69, note on E/58).

There is some evidence that Linières improved his instrument at a later date by the addition of a set of constructions for determining the ecliptic latitudes of the planets. Simon de Phares says that he composed 'ung directoire' with the *incipit*, *Accipe tabulam planam rotundam cuius*..., and John of Saxony, pupil and collaborator of Linières, refers to an instrument called *directorium* which must be made large and graduated in minutes of arc so that it can be used for the determination of planetary latitudes.[1] Since John of Saxony does not associate the instrument with the name of Linières we cannot

[1] See L. Thorndike, *History of Magic and Experimental Science*, vol. III, p. 260 and footnotes 17–19.

be certain in this ascription of the *directorium*, but it seems a very reasonable supposition. An anonymous tract on an instrument which opens *Accipe tabulum planam...* is found in many manuscripts,[1] but this treatise does not describe a planetary instrument at all in the copy which I have seen (Digby 48).

*Constants of the Linières Equatorium*

| | Distance of equant =*dk* | Distance of centre deferent=*dc* | Radius of deferent =*ca* | Radius of epicycle =*ae* | Remainder =*et* | Total parts =*at* |
|---|---|---|---|---|---|---|
| Venus | 1; 15 | ½ (1; 15) | 30; 0 | 21; 35 | 7; 0+* | 60 |
| Mars | 6; 30† | ½ (6; 30) | 30; 0 | 19; 45 | 3; 45* | 60 |
| Jupiter | 5; 30 | ½ (5; 30) | 60; 0 | 11; 30 | 22; 0− | 96 |
| Saturn | 6; 50 | ½ (6; 50) | 60; 0 | 6; 30 | 26; 0+ | 96 |
| Mercury | 1; 00 | Variable 1; 0 to 3; 0 | 20; 0 | 7; 30 | 1; 30 | 32 |
| Sun | No equant | 1; 00 | 24; 0 | No epicycle | 0; 0 | 25 |
| Moon | Equant and centre deferent move on circle of radius 12; 28 | | 60; 0 | 6; 15‡ | 17; 12‡ | 96 |

* The author seems to have miscalculated these two values, taking *dk* instead of *dc* from the total of 60 parts. It would not affect the working of the instrument.

† This is the only deviation from the usual Ptolemaic constants; the customary eccentricity of Mars is 12° 0′ and the distance *dk* should therefore be 6; 0; the value given corresponds to an unusually high eccentricity of 13° 0′.

‡ Either the remainder has been miscalculated (it should be 17; 17) or, more probably, the copyist has misread the epicycle radius as *6 et 4ª* instead of *6 et 3ª*; the latter value is in agreement with Ptolemy's epicyclic radius of *c.* 6; 20.

## 5. RICHARD OF WALLINGFORD

Richard of Wallingford appears to have composed his treatise on the Albion (Albyon=All by one) in 1326, in which same year he also described the Rectangulus, an ingenious skeleton version of the Torquetum.[2] At this time too he may have been occupied with his construction of some sort of complicated astronomical clock for the Abbey of St Albans,[3] but it is possible that the 'clock' was nothing more than an enlarged version of the Albion, perhaps

[1] E.g. Bodleian MS. Digby 48, ff. 91 v.–94 r.; Vatican MS. Palatine 5004, f. 22; and eight manuscripts cited by E. Zinner, *Verzeichnis der astronomischen Handschriften des deutschen Kulturgebietes* (Munich, 1925), nos. 3105–3112.

[2] The torquetum was designed for the calculation or direct observation of ecliptic latitudes and longitudes, so obviating the need for using spherical trigonometry to convert from altazimuth or equatorial co-ordinates. See H. Michel, 'Le Rectangulus de Wallingford', *Ciel et Terre* (1944), nos. 11–12.

[3] Such evidence as there is regarding the existence and form of this complicated machine is collected by R. T. Gunther, *Early Science in Oxford*, vol. II, p. 49.

moved by gearing. The *Tractatus* (*Canones*) *Albionis* has been discussed by Gunther (*op. cit.* p. 31), who has also published a transcription of the second of the four parts of the work found in two Oxford manuscripts (*op. cit.* pp. 349–70); there are many manuscripts of this work in German libraries, some of the copies containing an elaboration called the *Solemne* by Johann of Gmunden (*c.* 1380–1442). The text printed by Gunther has a note after the *explicit* referring to alterations made by Simon Tunsted (d. 1369), 'both in the book and in the instrument'; clearly the Albion seems to have had a considerable popularity comparable with that of the equatorium of Linières.

From the printed text it is evident that the Albion closely resembles the primitive equatorium described by Abulcacim Abnacahm, but is adapted to a construction similar in principle to that used for the astrolabe, where a number of separate plates fit inside a hollowed-out mater. The numerical constants are all taken unchanged from the Toledo Tables, the Alfonsine collection being not yet available in Oxford. One cannot help wondering whether Richard of Wallingford may have borrowed not only the constants, but details of the construction as well from some recension of the Abnacahm tract included in the canons of the Toledo Tables.

There are two features of the Albion which differ materially from the earlier instrument of Abnacahm. First, the main plate (the mater) is also fitted as an observing instrument by the addition of a suspension ring and a pair of pinnules; secondly, the treatment of the orbit of Mercury uses an approximation to the flattened circle demanded by Ptolemaic theory for the shape of the deferent. The approximation is made by means of four circular arcs resulting in a pseudo-ellipse[1] resembling that used by Azarquiel in his equatorium; again it seems as if Richard has borrowed from the Toledo texts in some way.

It is rather surprising that this calculating instrument should be fitted with pinnules for observational use; such a modification is not found in any other text that has been examined, and it is reasonable to suppose that this characteristic may be due to Richard himself. This point is especially interesting to us because the only medieval equatorium still extant[2] incorporates this feature

[1] The author says 'iste circulus non habet errorem sensibilem in instrumento cuius diameter est 60 cubitorum'—is he hinting at his experience with the great clock?

[2] Besides the Merton Equatorium, I know of only one other metal instrument of this type, and this is much later in date. The instrument, or rather the fragment of an instrument, is preserved in the Royal Museum of Brussels, where it is catalogued as no. 1911 together with a Malcote double rete. Both rete and planetary instrument are associated with an astrolabe which is signed 'Iacobus Valerius rectificata Anno Xp̄i 1558', but the planetary instrument may be an early seventeenth-century addition to the astrolabe. (For the identity of the maker see H. Michel, *Traité de l'Astrolabe* (Brussels, 1947),

exactly as described in the *Tractatus Albionis*. The instrument is engraved on the dorsum of one of the astrolabes of Merton College (see Frontispiece); possibly it is the *astrolabium majus* left to the College by Simon Bredon in 1372. It must have been made *c.* 1350, for this date is taken as the radix for a short table showing the amount of precession (1° per 100 years) for a century following that date. The design of the Merton instrument[1] is strikingly similar to that given by Azarquiel; a set of nesting graduated circles are drawn from the equant of each planet. The original Ptolemaic form of the theory of Mercury is used instead of the pseudo-elliptical deferent, but a new note is struck by the use of a circle of little holes for the movable centre deferent of that planet. Unfortunately the 'Epicycle plate' which should go with the Merton instrument has been lost; a little lug at the bottom of the plate seems intended for a string which may have kept the two parts together. Since the plate is marked with all the planetary centres but only the equant circles, it follows that the deferent circles as well as the epicycles must have been represented by the lost portion of the device. The most obvious form would have been a single circle to represent the deferents of all the planets, and an epicycle plate rigidly attached to the circumference of this circle; more convenient would have been a device similar to the Epicycle portion of the Equatorie instrument described in the Peterhouse manuscript.

Considering the Merton instrument and the Albion together there can be little doubt that both are derived, in the same way, from the two types of instrument first described by the astronomers of the Toledo Tables; the manner of derivation is quite different from that used by John of Linières, and Wallingford seems to be responsible for the introduction of sighting pinnules and two major improvements in the pattern of the device. One improvement, the use of a circle of little holes for the centre deferent of Mercury, may be considered fairly obvious. The other and more important change is the omission of the deferent circles and their replacement by some mechanical device which may also be used for the epicycles. Both of these improvements are found again in the text of the *Equatorie of the Planetis*, and we may now consider that text in relation to the historical development of the equatorium. Before doing this we must note that by the end of the fourteenth century such instruments seem to have become more common. Amongst the books and instruments of John Erghome in the library of the Austin Friars of York, not much later

p. 178, *sub* Valerius.) The instrument consists of a brass open-work ring on which a smaller epicycle plate is mounted. The radius of the deferent circle is the same for all the planets, and no provision is made for the eccentricity or equant.

[1] For a full description see R. T. Gunther, *Early Science in Oxford*, vol. II, pp. 208–10.

than 1372, there is an equatorium, and a treatise on the instrument.[1] A little later there are explicit mentions of two instruments belonging to Merton College in one of the periodical distributions (on loan) of instruments and books.[2]

## 6. THE 'EQUATORIE OF THE PLANETIS'

The instrument described in the *Equatorie of the Planetis* shows many features which are found in the Merton instrument but not elsewhere. A circle of holes is used for the centre deferent of Mercury and for the two centres of the Moon; of more fundamental importance, the deferent circles are not drawn out on the plate but are constructed mechanically by the way in which the epicycle plate is attached to the main lamina. The process has been taken one stage further than the Merton instrument, for the Equatorie dispenses with the graduated equant circles, so following the practice of John of Linières in this respect. John's precepts have also been adopted in a conversion throughout the text to the complete sexagesimal system of *signa physica* instead of the *signa communia* of the Albion. Clearly the inventor of the Equatorie was using the Alfonsine rather than the Toledo Tables—this would be apparent even if the translator or adapter responsible for the English text had not included in his volume a set of tables modified from the Linières version of the Alfonsine collection.

By combining the best features of the Wallingford and the Linières instruments, and adding very little to the mixture, the author (or rather the deviser of the Equatorie form) arrives at a system which could hardly be improved without radical change of principles. The Equatorie construction calls for the dividing of only two graduated circles, and the abolition of both deferent and equant circles leads to a minimum of complication. It also leads to an increase in accuracy, for the full size of the instrument may be utilized as the scale of each deferent. The only drawback in the operation of the Equatorie is that the two fundamental parameters of each planetary position, the mean motus and mean argument, must still be obtained by calculation from a set of tables drawn up for the purpose.

[1] M. R. James, in *Fasciculus Ioanni Willis Clark dicatus* (Cambridge, 1909), p. 2. Item 370 in the catalogue is 'Equatorium abbreviatum cum canonibus badcomb' (= William Batecombe?), and no. 378 is 'Item equatorium quondam Magistri Ric*hard* thorpe'.

[2] F. M. Powicke, *The Medieval Books of Merton College* (Oxford, 1931), p. 68. In the Electio of *c.* 1410 no. 37 is an 'Equatorium ligneum cum epiciclo' and no. 50 is an 'Equatorium eneum cum epiciclo'. Perhaps it is one of these which turns up again in the Electio of 1452 as no. 136, 'Equatorium eneum cum epiciclo in dorso cum volvellis solis et lune'. Perhaps this last object is connected in some way with the Merton College quadrant which has on the back a Sun and Moon volvelle containing a circle of holes already noted as a rare and characteristic feature of the Merton equatorium; could it be part of the lost epicycle portion? The volvelle is illustrated by R. T. Gunther, *Early Science in Oxford*, vol. II, facing p. 241.

It must be admitted that much of the simplicity of the Equatorie system would have been lost if the functions of the device had not been restricted to the most simple planetary problems, the determination of the ecliptic longitudes of all planets, and the latitude of the Moon only. The computations of the other planetary latitudes, and the complete determination of lunar and solar eclipses, remain to be tackled—and these are formidable problems on any geometrico-mechanical device; we turn now to an equatorium which is adapted for all these calculations.

## 7. AL-KĀSHĪ

A remarkable type of equatorium is described in a Persian manuscript by the Iranian astronomer and mathematician, Jamshīd Ghiāth ed-Dīn al-Kāshī (died 1429), who was the first director of the Samarkand observatory of Ulugh Beg.[1] The instrument is of the same type as that of John of Linières in so far as the plate contains only the ungraduated deferent circles of the planets, and these are designed to nest so as to avoid confusion. The graduated limb of the instrument is used for all angular measurements, and the plate on which the deferents and centres are engraved is arranged so that it may be rotated within the annulus of the limb. This rotation may be used to allow for the effect of precession—but it seems an uneconomical manner of treating such a small correction. The most novel feature of al-Kāshī's *Plate of Zones* is the provision of a central alidade and a graduated rule which can be placed parallel to it; this enables the epicycle construction to be set out at the centre of the main plate (indeed, the sole plate), using the graduated limb for these angles too. The deferent of Mercury is constructed once more in approximate form, this time as a combination of only two circular arcs; it is inferior in accuracy to the four-arc approximation used by Azarquiel, but quite accurate enough for use in such an instrument.

In addition to the usual mechanism of deferents and equants, al-Kāshī marks points on the instrument which may be used for the geometrical computation of planetary and lunar latitudes, following completely the Ptolemaic theory. A set of graduations on the ruler also enable the instrument to be used for the calculation of lunar eclipses. Besides all this al-Kāshī enumerates a number of other possible designs for laying out the system of deferents on the plate of the instrument; for example, the centre of the plate may be taken

[1] The equatorium is described in detail in four important papers by E. S. Kennedy, *Isis*, vol. XLI (1950), p. 180, and vol. XLIII (1952), p. 42; *J. Amer. Orient. Soc.* vol. LXXI (1951), p. 13; *Scripta Mathematica*, vol. XVII (1951), p. 91. Professor Kennedy has kindly allowed me to use his yet unpublished translation of al-Kāshī's texts.

as the centre of each deferent instead of the centre of the universe, or the Earth may be retained at the centre but the deferents allowed to intersect each other so that they may each be made as large as the whole plate will permit; or thirdly, as a sort of combination of these two variations, the edge of the plate may be used as a common deferent for all the planets—there will then again be one eccentric 'Earth' for each of the planets severally. He enumerates some fifteen possibilities—it is unlikely that these were ever tried in a practical instrument.

It seems very likely that al-Kāshī's *Plate of Zones* is an independent development of the equatorium principle—independent, that is, of the European line which we have just been following, but probably a quite natural product of the Islamic line which has been mentioned above in connection with the Marāgha observatory of Nāṣir al-dīn al-Ṭūsī about a century previously.

## 8. LATER DEVELOPMENTS

We have now taken the history of the equatorium to the end of its medieval development; with the advent of the printed book there appeared numerous texts on planetary instruments—almost every astronomer interested in ingenious devices brought out some tract on a planetary instrument.[1] The process culminated with the appearance of Peter Apian's *Astronomicum Caesarum* (Ingolstadt, 1540)—the most sumptuous scientific book ever printed. The volume contains a large number of elaborate and beautifully decorated paper volvelles, each of them being an equatorium specially designed for one particular calculation, e.g. the longitude of Mars. The beauty of Apian's scheme is that he makes his volvelle function without the need for any additional planetary tables—the mean motus of a planet, for example, is set on the instrument by means of one circle graduated in years, a second in months, a third in days.[2] The style and size of the book unfortunately made it a very expensive instrument, and the complicated paper volvelles are so fragile that they could not long survive any constant use. Nevertheless, this book, by its appearance, closes the whole medieval development of the Ptolemaic planetary instrument.

In conclusion, there is one later renaissance of the equatorium principle which deserves special mention as an illustration of the persistence and

[1] To mention but three: Weidler Camillo Leonardi, *Liber desideratus canonum aequatorii coelestium motuum sine calculo* (1496); Franciscus Sarzosius Cellanus of Aragon, *Novus Commentarius in Aequatorem Planetarum. Prior Fabricam Aequatoris Complectitur, posterior, usum atque utilitatem, hoc est verus motus, ac passiones in zodiaki decursu contingentes* (Paris, 1526, 1535 and 1591); Johann Schöner, *Aequatorii astronomici omnium ferme uranicorum theorematum explanatorum canones* (Nuremberg, 1524 and 1534).

[2] See S. A. Ionides, 'Caesar's astronomy', *Osiris*, vol. 1 (1936), p. 356.

attractiveness of the instrument. It is described by one of the seventeenth-century 'hack-writing' mathematical practitioners, John Palmer; but the equatorium seems to have been made by Walter Hayes, a popular instrument-maker of the period, in the year 1672.[1] What makes this design so remarkable is that the Copernican system is used, the Sun being at the centre of the plate, and the planetary orbits being drawn to scale round it. Each orbit is approximated by an eccentric circle—a very good approximation as has already been shown, and each circle is graduated in degrees marked out from the equant, so that the true place of the planet in its own orbit is known as soon as the mean motus has been calculated. For convenience the orbits are divided into two groups, the inferior and the superior planets; the inferior planets are engraved on one side of the plate, the superior on the other. In each case the outermost orbit is made as large as possible. In the corners of the plate are written planetary radices for 1672 and 1680, and the amount of the mean motus in a day, a year, and in a cycle of four years for each of the planets; with this information it is a simple matter to calculate the mean motus for any date near that of the radices—tables could, of course, easily be provided. The final answer is obtained by measuring the angle of the line drawn from the Earth to the planet concerned, and positions accurate to about 1° may readily be obtained with the pocket-size device. The theory behind this seventeenth-century instrument is modern, but the geometrical principles and the use of circular 'deferents' are but little removed from the first equatoria made in the eleventh century.

The history of the equatorium may be regarded as a valiant but unsuccessful struggle of geometrical calculation against numerical tables. At the beginning of the struggle, the instrument had the advantage because of the distaste for approximation by the neglect of small quantities, and because of the tedious processes of calculation. By the time that the instrument had been developed into a simple and elegant form, tables were becoming more convenient, and astronomers had become more conversant with the techniques of calculation and more sophisticated about the nature of approximation. By the sixteenth century the equatorium seems to have become nothing more than an ingenious toy of little significance to practical astronomers, and inferior to the approximate tables available for routine astrological calculations.

[1] John Palmer, The Planetary Instrument, or the Description and Use of the Theorics of the Planets: drawn in true Proportion, either in one, or two Plates, of eight Inches Diameter; by Walter Hayes, at the Cross-Daggers in Moor-Fields, London, 1672, 6 pp.

# IX

# PALAEOGRAPHY

## I. CONTRACTIONS

THE contractions used in the text of the *Equatorie of the Planetis* are, for the most part, those common in Middle English manuscripts at the close of the fourteenth century.[1] The actual amount of contraction used varies considerably, but on the whole it is employed much more sparingly than in the Latin glosses and notes written by the same hand. The following notes are given merely as examples and are not intended to be at all comprehensive.

(1) A horizontal superior line above a vowel denotes omission of *n* or *m*. The line is usually straight, but in a few places it curves down at the ends; we have, for example, equedistāt [J/21], eccētrik [B/16], ī [K/10], equaciō [I/35], distaūce [C/4], *but* instrumêt [A/2]. This abbreviation mark is also used very frequently in the ending *-ioū*, as in deuisioūs [A/2], but it must be remembered that owing to the confusion caused by the 'two-minim letter' which can be taken equally for *u* and *n*, some doubt may exist as to whether the sign may not be just a scribal habit associated with the ending *-ion*. In rare cases the sign denotes the omission of more than a single letter, as in gēis [=gem*in*is, F/23].

(2) A sign 9 is used at the beginning of a word for *con-* or *com-*, as in 9tene [A/6], 9pas [A/17]. In a few cases the *m* may be written in, for we have comune [D/16, E/22, etc.] but 9mune [D/c, F/2, 24, etc.].

(3) A similar sign, used at the end of a word, signifies *-us*: in some instances the downstroke is finished with a slight curl to the right, as in saturnꝰ [C/2, etc.].

(4) A curl above the line usually denotes the omission of *er* as in man$^{s}$ [A/15], gen$^{s}$al [B/32], but it is also found less frequently for *re* as in p$^{s}$cios [C/33]. In the latter use, one must include a few instances where the *r* has been written in, so that the abbreviation corresponds to final *e* only; e.g. maner$^{s}$ [D/4].

(5) The *p* may be modified into ꝑ=*per* or *par*, and ꝓ=*pro* as in ꝓced [C/2, 9], ꝑties [B/9, 10], ꝓces [A/21]. When such a symbol is followed by another *p* the two letters may be tied, as ꝑpetuel [A/22], ꝓpocioū [C/19].

[1] General discussions are to be found in E. Maunde Thompson, *Introduction to Greek and Latin Palaeography* (Oxford, 1912), pp. 472 ff., and in C. Johnson and H. Jenkinson, *English Court Hand* (Oxford, 1915). A useful study and a list of typical contractions is given by W. W. Greg, *Facsimiles of Twelve Early English Manuscripts in the Library of Trinity College, Cambridge* (Oxford, 1913). Another study which is particularly relevant to the present manuscript is P. Pintelon, *Chaucer's Treatise on the Astrolabe* (Antwerp, 1940), pp. 71–2.

(6) A superior *a*, rather modified in shape (the open *a*), is used to denote the omission of *a*, generally together with other letters, as in m$^{\omega}$rs [H/10], q$^{\omega}$ntite [A/28], lib$^{\omega}$ [N/25]. Throughout the text we find no$^{\omega}$, g$^{\omega}$, mi$^{\omega}$, 2$^{\omega}$ [=no*ta*, *gradus*, mi*nuta*, *secunda*]; the exceptional forms g$^{\omega}$d and m$^{\omega}$ occur in N/1. In our transcription this type of abbreviation has been rendered as mi$^{a}$, g$^{a}$, etc.

(7) The ending *-is* is denoted by two different marks of abbreviation. We have eq$^{\omega}$ntꝫ [E/53], mynutꝫ [E/59], *but* defferētꝫ [E/54], argumentꝫ [G/1–2], equantꝫ [H/17].

(8) A superior *r*, degraded in shape, is used as an occasional abbreviation for *ur* as in sat$^{z}$nꝰ [C/3]. Cf. Greg, *op. cit.* Plate IX.

(9) A crossed or superior *l* denotes the omission of letters before or after, as ƚe [=pl*anete* G/4], ƚre [=l*ett*re I/12], m$^{s}$idi$^{l}$ [=m*e*ridi*onal* K/33], .7.trio$^{l}$ [=*septen*-trio*nal* L/10]; the special use of the numeral 7 in this last example is noteworthy, cf. also .10.*bre* [B/24].

(10) A superior final *t* is used throughout the text in the word-abbreviations þ$^{t}$, w$^{t}$ [=that, with]. These have not been expanded in our transcription because of the uncertainty of the full form which should correspond to them; they are in fact true examples of word-abbreviation. A similar type of abbreviation occurs in cap$^{t}$ [=cap*ut* M/26], but compare cap$^{d}$ [=cap*ud* M/22, 24].

(11) Word-abbreviation also occurs in the tironian symbols & or $\overline{\&}$ for *and*, and in the sign .i. for i*d est*, and in the sign .ſ. for s*cilicet*.

(12) An exceptional set of abbreviations occurs almost at the end of the text [N/37] where the names of six signs of the zodiac are written as 'aries taurꝰ geī.c.l.v$^{z}$'.

(13) Although the double *ll* is sometimes tied, the tie mark does not seem to be used as a contraction for lle as is sometimes the case elsewhere. Cf. Greg, *op. cit.* Plate XII.

(14) What may be a carelessly formed mark of contraction appears at the end of cap$^{s}$corn̉ [D/10], which might be rendered equally well as cap*ri*corn*e* or cap*ri*corn*us*.

(15) A special word-abbreviation is found throughout the text and tables as X$\overline{\text{pi}}$ for *Christi* from the Greek XPICTOC.

(16) Another special abbreviation found in the tables (but not in the text) is R$^{x}$, Ra$^{x}$, Rå, ℞, Rꝫ or Ra for R*adix*. The second of these forms also occurs as R$^{x}$a, probably through a bad positioning of the superior letter. See p. 160.

## 2. PUNCTUATION

As in most Middle English manuscripts the punctuation differs to some extent from that found in modern works in that it has not yet shifted, completely and consistently, from one designed primarily to meet the demands of oral reading by providing breath pauses of different lengths, to a purely grammatical one which is intended to mark off the different parts of the sentence. The position of a pause would, of course, often coincide in the two types of punctuation, and where the usage here differs from that found in modern works it is probably to some extent due to the influences of its physiological function.

The symbols most frequently used are a wedge-shaped sign ⸿, and a solidus /. The first of these, occurring some eighty-three times, marks off short units which are in the main grammatically independent sentences, and would be so punctuated today. It is used more infrequently towards the end of the treatise, and in N approximates more closely to a paragraph sign. It is usually placed at the beginning of the unit, and in five cases is reinforced by a stop, giving .⸿. The solidus, /, is normally used to mark off subsidiary units, such as would today be marked off by a comma or a semi-colon, and only occasionally would it be replaced in modern English by a stop. It occurs more than 200 times, and in seven cases with a stop before it, giving ./. The use of two diagonal lines, //, is comparatively rare (thirteen occurrences), but it becomes a little more frequent after the end of G. So far as can be observed no distinction in usage seems to be made between a solidus and a double diagonal line. In one instance the latter is reinforced by a stop .//. The single stop, also comparatively rare (twenty-nine occurrences), becomes rather more frequent towards the end, and is normally used for the less important pause, or to mark off the subordinate parts of the sentence. Like the single solidus it would normally be replaced by a comma in modern English, but very occasionally by a stop. Single-figure numbers are usually, though not invariably, enclosed by two stops, but two- and three-figure numbers much less frequently. Occasionally single words are treated in the same way, e.g. *.lune.* K/24, *.planete.* J/35. In addition, some particularly important clauses are emphasized by the use of a conventional hand, in two cases alone, in one accompanied by a single diagonal line, and in one by a wedge.

In general the punctuation is comprehensible enough, even by modern principles, but occasionally none appears where it would have been expected, e.g. at A/12, 22, 31, B/32, D/26, E/63, H/9, etc., while on the other hand no punctuation would have been expected at D/36 (second /), G/5, 8, 24 (second /), J/14 (/), L/5 (first /) and M/27.

# X

# LINGUISTIC ANALYSIS

## I. VOCABULARY

OVER a hundred French loan-words occur in the treatise, as well as some fifty which appear to be taken directly from Latin—occasionally from distinctively medieval Latin roots. In addition, there are a further fifteen, mainly scientific terms, which, so far as the form is concerned, could have been borrowed through French or else directly from Latin. The Old Norse loans are naturally of a more popular type, and number ten, or perhaps eleven, in all: *bothe, calle, fro, happ(ith)/(mys)happe, krooke, (ouer)thwart, ren(spyndle), same, scrape, take*, and perhaps also *karte*. From Arabic, probably through medieval Latin, come *alhudda, almenak, aryn, leyk*, while *boistos(ly), boydekyn, she*, are of obscure origin.

Perhaps more interesting is the number of words which, before their appearance in this text, are recorded elsewhere in Middle English only in the works of Chaucer, and more particularly in his *Treatise on the Astrolabe*. The following words occurring in the *Equatorie* are recorded from an earlier date by *OED* only from the works of Chaucer: *ark, boistosly, boydekyn, degre,*[1] *deuysioun, equacioun, equaly, fix, meridional, moeuable, vernissed.* Even more striking is the number of words which are given as appearing first in the *Astrolabe*, and which also appear in the *Equatorie*: *almenak, cancer, capricorne, compas,*[1] *crois,*[1] *difference,*[1] *direct, distaunt, ecliptik, epicicle, equal, fracciouns, label,*[1] *latitude, membris,*[1] *meridie, mote, nadyr, opposit, pol, precise, riet, septentrional, sklat.*[1] In addition, a further seventeen words are not recorded by *OED* until a later—sometimes a considerably later—date: *bakside* (1489), *closere* (1440), *defferent* (1413), *diametral* (1555), *drawe*, 'to subtract' (1425), *eccentrik* (1561, but once in Trevisa), *equant* (1621), *equedistant* (1593), *geometrical* (1552), *lymbe* (1460), *precisly* (1450), *seccioun* (1559), *semydiametre* (1551), *sowded* (1420). The following do not appear in *OED*: *aux/auges* (but see *auge* 1594), *equatorie, renspyndle.*

[1] In this sense only.

## 2. PHONOLOGY

**OE. *a*** is regularly *a*: *laste* A/14, *make* C/11, etc.

When *a* is followed by a back *g*, the *g* has been vocalized to *u* and a diphthong results: *drawe* N/4, *w*[t]*draw* I/29, etc.

**OE. *a/o*** before a nasal is regularly *a*: *many* G/33, *name* A/1, *stant* F/19, *and* B/4, etc.; but before a lengthening group in a stressed syllable *o* appears: *fond* M/28, *long* D/26, *stonde* E/32, etc.

**OE. *æ*** is regularly *a*: *after* C/12, *blake* G/11, *faste* B/31, etc.

When *æ* is followed by front *g*, the *g* has been vocalized to *i* and a diphthong *ai/ay* results: *day* G/35, *may* A/9, *nail* E/1, etc.

In *seid(e)*, *forseid(e)* A/1, 11, etc., the spelling is due either to the falling together of the *ai/ei* diphthongs, or to the influence of the vowel of the infinitive, OE. *secgan*.

*togidere* D/30, shows weakening to *e* in an unstressed word, and later N. EM. raising of *e* to *i* before a dental.

*thennes* M/15, 23, is probably from a late OE. *þænne/þenne*.

**OE. *e*** is regularly *e*: *breke* D/c, *egge* A/7, *euene* C/18, etc.

When *e* is followed by front *g*, the *g* has been vocalized to *i* and a diphthong *ei/ey* results: *alwey* C/1, *ley* B/28, *wey* I/22, etc.

In *agayn* J/9, etc., *agayns* J/26, etc., *ay* is an inverted spelling due to the falling together of the *ai/ei* diphthongs.

*stide* H/14, 16, is probably from the by-form, OE. *styde*, while the *i* in *this* A/1, 9, etc., is due to the influence of other cases of the pronoun.

In some Latin and French loan-words variation occurs between *e* and *i* in a stressed syllable. Hence *defferent* B/21, 22, etc., but *different* G/5, I/32; *descende* M/5, *descending(e)* M/20, 35, N/15, but *discende* M/15, *discendinge* N/30; *deuysioun(s)* B/12, 17, etc., but *diuisiouns* C/27, 33. Such variation is presumably the reason for the inverted spelling *deffe*[a] = *differentia*, on f. 5v.

An unaccented medial *e* has been lost in the loan-word *remnaunt* I/7, 14, 16 (cf. *remena(u)nt* B/26, H/11, etc.).

**OE. *i*** is regularly *i* or *y*. In early ME. *y* was used only before *m*, *n*, *u*, *w*, but variation soon appears. In this text *i* is invariable initially, whatever the following consonant, *w* is regularly followed by *i* not *y*, and when final *y* is usual, though not invariable. Otherwise *y* is normally used before *n*, *m*, but *i* elsewhere, though complete consistency does not occur. When a particular word containing *i/y* occurs frequently it is clear that one or other of the spellings was that normally used by the writer, the other being much less usual, e.g. *by* (30 ×), *bi* (2 ×); *bytwixe* (18 ×), *bitwixe* (1 ×); *lyne* (51 ×), *line* (3 ×); *thy* (105 ×), *thi* (16 ×), but cf. *thin* (22 ×), *thyn* (15 ×). Similarly with loan-words *aryn* (38 ×), *arin* (1 ×); *ecliptik* (6 ×), *eclyptik* (1 ×); *lymbe* (52 ×), *limbe* (2 ×); *signe(s)* (68 ×), *sygnes* (1 ×); *deuyd-* (13 ×), *deuid-* (8 ×); but *-uysioun* (6 ×), *-uisioun* (6 ×).

In *clepe* A/13, J/7, *cleped* B/9, etc., the *e* is from an OE. form with non-WS. back-mutation (OE. *cleopian*).

In *wole* A/13, 26, etc., the *o* is probably due to the rounding influence of the preceding *w*, *wile* > *wyle* > *wule* > *wole*.

The *e* in *(signefy)eng* C/3 is either a mistake by the writer or else due to lack of stress.

In *equacoun* G/9 (cf. *equacioun* C/34, etc.), apparent loss of *i* is probably due to an error by the writer.

**OE. *o*** is regularly *o*: *body* H/15, K/12, *bord* A/5, *north* L/26, etc.

*oe* in French loans appears as *o* in *proued* A/16, F/6, but otherwise remains, *moeuable* A/24, *moeuyng* M/27, *proeue* G/13, etc.

**OE. *u*** is regularly *u*: *ful* C/19, *put* G/2, *thus* A/22, etc.

Before a following nasal *o* is regular, *com* N/35, *some* H/11, *sonne* B/17, etc.

Initially *u* is represented by *v*: *vnder* G/14, *vnto* B/8, *vp* G/20, etc.

In *shollen* (OE. *sculon*) A/20, etc., *o* may be due to the influence of the pt. sg. (OE. *scolde*); in *torne* C/6, D/29 (cf. *turne* B/36, etc.), to the influence of Fr. *tourner*, and in *thorw* D/13, etc., to lack of accent.

Lengthening has taken place before a lengthening group in *bownde* A/8, *founde* I/23.

In *ferther* A/30, *ferthest* A/25, etc. (OE. *furðor*), *e* is probably due to the influence of *fer* (OE. *feorr*).

In loan-words the ending *-oun* varies with *-on*. Since the ending is usually written *oū* the apparent variation may sometimes be due to the accidental omission of the abbreviation, but that the variation did actually occur is clear from the presence of such forms as *equaciō* M/7, etc., by the side of *equacioū* C/34, etc.

Similarly *-aun-* varies with *-an-*, *distaunt* A/30, etc., but *equedistant* G/12, etc.

Apparent loss of *u* in *instrment* I/36 (cf. *instrument* A/2, etc.) is probably due to a mis-writing.

**OE.** *y* is usually *i*: *firste* B/12, *stirte* I/34, *wirke* N/19, etc., but *o* (for *u*) appears in *moche(l)*, I/2, 7, *workest* I/26, and *e* is invariable in *enche* A/10, etc.

**OE.** *ā* is regularly *o*: *knowe* K/21, *more* I/28, *wot* E/44, etc., but *oo* appears in *hoole* I/9 (cf. *hole* I/15, etc.).

Early shortening has taken place in *a* A/4, *an* B/7, *another* D/22, *any* G/4, *anything* M/2, *nat* A/7, *natheles* E/23, etc.

Early shortening and later development of a glide vowel before a back consonant has given a diphthong in *tawhte* M/13.

Loss of *ā* in an unaccented syllable has occurred in *as* A/23, *wheras* A/26, etc.

**OE.** *ǣ*[1] is regularly *e*: *nedle* D/16, *ther* B/14, *thred* E/12, etc.

*lat* A/11, D/25, F/3, presumably shows shortening from a distinctively WS. form with *ǣ*.

**OE.** *ǣ*[2] is regularly *e*: *brede* A/10, *euere* A/3, *teche* E/25, etc.

In *list* D/b, c, F/2, shortening of early ME. *ē* has taken place, and later N. EM. raising of *e* to *i* before a dental.

Early shortening has taken place in *lasse* K/25, *laste* A/22.

*goth* (OE. *gǣþ*) B/8, C/28, is due to analogy with other parts of the verb.

**OE.** *ē* is regularly *e*: *grene* G/35, *sek* B/27, *the* 'thee' A/9, etc., but *ie* appears in the loan-words *contiene* A/19 (cf. *contene* A/6, 11), *riet* K/8.

*doth* (OE. *dēþ*) D/29, is due to analogy with other parts of the verb.

**OE.** *ī* is regularly *i* or *y*: *strid* C/15, *tyme* A/21, *white* G/12, etc.

**OE.** *ō* is regularly *o*: *drow* M/25, *fote* A/6, *mone* B/21, etc., but *oo* appears in the loan-word *pool* G/14, 16, J/22, 33 (cf. *pol* G/30, etc.).

**OE.** *ū* is usually *ow*: *downward* N/35, *how* E/4, *owt* B/23, etc., but *ou* appears in *aboute* K/6, 7 (cf. *abowte* A/21, etc.).

**OE.** *ȳ* is *i* in *litel* A/28, etc., but *y* in *lytel* I/12.

**OE.** *ea* is regularly *a* before *r* plus cons.: *karf* M/32, *mark* J/10, *narwere* A/29, etc.

Before *l* plus cons. *a* was regular in Anglian, and this has given *o* before a lengthening group in *told* L/22, N/36. Otherwise *a* is invariable, *al* B/22, *calle* A/26, *fallith* N/37, etc.

Before *h a* is found in *say* N/4.

After a preceding palatal *a* is regular, *shape* D/27, *shal* E/25, *sharp* B/30, etc.

*myht* N/4, *midnyht* C/29, etc., may show WS. mutation of *ea* before *h*, or are perhaps due to the influence of the following palatal.

**OE.** *eo* is regularly *e*: *erthe* N/14, *kerue* K/33, *werpe* A/7, etc.

In *shortly* B/6, etc., shift of stress and later absorption of the front element of the diphthong has taken place.

*silk* E/12, is from an OE. form with *io*, later smoothed to *i*.

In *sixte* E/33, *i* is due to the smoothing influence of the following consonants.

**OE.** *ēa* is regularly *e*: *grete* C/22, *red* E/36, *shewe* G/21, etc.

In *ney* D/b, F/1, the development of a glide before a following front consonant has resulted in a diphthong.

The mutation appears as *e* in *g(r)ettest* L/32, M/19, *nexte* E/60.

**OE. *ēo*** is regularly *e*: *dep* C/2, *knew* M/34, *whel* A/8, etc.

In *trowthe* A/4, *o* is due to the influence of the following *w*.

In *file* F/1, *filed* D/b, *i* is from OE. forms in *īo*, later smoothed to *ī*, and in *riht* K/10, from smoothing before *h*.

*Consonants*

***c.*** Used as a spelling for the voiceless back stop initially and medially before a following back vowel, *l*, *r*; hence *calle* A/26, *com* N/35, *clepe* A/13, *crois* D/24, *circumference* A/8, *geometrical* A/17, *ecliptik* L/10, etc.

Medially before a following *e* it is used for the same sound in the L. loan *Cancer* D/9, etc., and finally after *i* in *eccentric* E/35 (cf. *eccentrik* B/16, etc.).

In Fr. loan-words initially and medially before a following front vowel *c* is used as a spelling for the voiceless spirant *s*, *centre* D/27, *circumference* A/8, *composicioun* D/1, *distaunce* B/33, *face* E/9, *precise* E/19, *proces* A/21, etc.

In Fr. and L. loan-words medial *-cc-* is used as a spelling for *-ks-*. *eccentrik* B/16, *fracciouns* A/3, *seccioun* D/23, *successioun* G/36, etc.

***ch.*** Used to represent OE. front *c*, and the same sound in Fr. loans, *enche* A/10, *teche* E/25, *which* A/5; *chef* A/2, *towche* C/15, *perchemyn* A/9, etc. *percemyn* D/34, is presumably an error.

Front *c* has been lost finally in *I* A/13, *euery* D/10, etc., and in the OE. endings *-līc*, *-līce*, *dymly* B/18, *only* G/37, etc.

The back consonant in *kerue* K/33, *karf* M/32, *wirke* N/19, etc., is due to analogy with other parts of the verb.

OE. medial front *-cc-* is represented by *-chch-* in *strechche* K/33, *strechcheth* J/3, but by *-ch-* in *streche* G/19.

***d.*** Final *-d* has been lost in *deuyde* A/29, and medial *-dd-* simplified in *midel* M/10, 32.

Already in OE. final *-dþ* had been assimilated to *-tt* and simplified to *-t* in *stant* F/19, H/18 (but cf. *stondith* A/26, E/21).

***g.*** Initially and medially used as a spelling for the voiced back stop, *goth* B/8, *grene* G/35, *glewed* A/13, *bygynne* I/20, *degre* B/6, *regard* K/16, etc.

OE. *-cg* is represented by *gg* in *egge* A/7, *ligginge* H/1, while the same sound initially in loans is represented by *g* in *general* B/32, *generaly* I/30, L/26, *geometrical* A/17, but by *j* in *justli* F/7.

Infinitives *lye* L/4, *sein* B/5, etc., are reformations on the analogy of other parts of the verbs.

***k.*** Initially, medially, and finally, used as a spelling for the voiceless back stop, *kerue* K/33, *karte* A/8, *knowe* K/21, *krooke* A/7, *blake* G/11, *quykly* L/29, *prikkes* E/52, *lik* D/3, *silk* E/12, etc.

*ecliptil* N/30 (cf. *ecliptik* L/10, etc.), is presumably an error.

Medially the double consonant has been simplified in *thyknesse* D/18 (cf. *thikkenesse* D/6), and the single consonant lost in *mad* M/7.

***l.*** The consonant has been doubled medially in *shollen* A/20, etc.

Simplification of the double consonant has taken place in *al* B/22, etc. (cf. *alle*, pl., A/20, etc.).

Loss of the consonant has taken place finally in *moche* I/7, J/5 (cf. *mochel* I/2, 9), and medially in *euerich* B/4, *euery* D/10, *swich* D/22, *which* A/5, etc.

***m.*** Doubling of the consonant has taken place medially in *commune* F/2, etc. (cf. *comune* A/16, etc.). But it should be noted that the form with double consonant appears only when the *com-* abbreviation is used; when written in full the word is invariably *comune* with a single medial consonant.

***n.*** Lost finally when followed by a word beginning with a consonant in *a* A/4, etc. (cf. *an* B/7, etc.), *no* B/31, etc. (cf. *non* E/5, etc.), *my* M/24, etc. (cf. *myn* A/13, etc.), *thy* A/4, etc. (cf. *thyn* D/1, etc.).

On the loss of *-n* in verbal inflexions, see p. 142.

***p.*** The double consonant has been simplified medially in *vpon* A/12, etc., and finally in *vp* G/20.

***q.*** Used as a spelling for OE. *cw-* in *quykly* L/29, but otherwise appears only in loan-words, where it represents Fr., L. *q*, *quantite* I/5, *quarters* A/15, *equacioun* C/34, *equales* B/9, etc.

***r.*** The double consonant has been simplified medially in *ner* A/4, *verey* K/27 (cf. *verrey* G/21, etc.).

Loss of medial *r* in *gettest* M/19 (cf. *grettest* L/32), *capricone* C/28 (cf. *capricorne* D/9, etc.), is probably an error.

***s.*** The medial consonant has been doubled in *compassed* E/8 (cf. *compased* D/5, 13, E/67, F/13), and is represented by *c* in *compaced* E/14.

***sh.*** Used as a spelling for OE. *sc* in *shal* E/25, *shape* D/27, *sharp* B/30, *shaue* A/5, etc.

In loan-words *sc* remains in *ascende* M/14, *descriuen* H/37, etc., but appears as *sk* in *sklat* G/2.

***t.*** The medial consonant has been doubled in *gretter(e)* D/16, I/38, *g(r)ettest* L/32, M/19, *lattere* I/4, *puttest* G/29, H/15.

Final *-t* has been voiced in *capud* D/8 (cf. *caput* K/27, etc.).

***th.*** This is the usual representation of the sound, whether voiced or unvoiced, *thinges* D/3, *tho* A/2, *other* B/28, *the* A/1, etc. (but *þe* K/33), *this* A/1, etc. (but *þis* K/i, etc.).

In the abbreviation *þ*ᵗ, *þ* is invariable, but when the word is written out in full *th* is regular, e.g. *that* D/25, J/7, 13.

This regular use of *þ* in the abbreviation *þ*ᵗ, but its almost complete absence elsewhere is noteworthy.

Assimilation to *t* has taken place in *hastow* B/3, *maistow* I/29, *shaltow* A/15, *list* G/4, etc., and perhaps also in *w*ᵗ A/8, *w*ᵗ*draw* I/29, *w*ᵗ*in* A/11, etc.

Medially *th* has been lost in *or* A/4, etc.

***u/v.*** Medially as a spelling for the voiced labio-dental in native as also in foreign words *u* is regular, *euene* C/18, *euere* A/3, *kerue* K/33, *deuyde* A/15, *leuel* A/5, etc.

Initially, representing the same sound in Fr. and L. loans, *v* is used, *varieth* H/11, *vernissed* A/9, *verrey* I/17, *visage* C/33, etc.

***w.*** Loss of medial *w* has taken place in *also* E/10, *as* A/23, *nat* A/7, *so* B/14, etc.

In *thorw* D/13, etc., *w* may be a spelling for a vocalized back spirant, or it may represent a glide-vowel *u* developed before the following back *h* which has then been lost.

In *folwynge* B/27, *w* represents the OE. voiced back spirant, which elsewhere after a preceding vowel has been vocalized to *u* and become the second element of a diphthong, e.g. *drawe* N/4, etc.

French medial *v* is represented by *w* in *remew* H/22, *remewed* H/33, I/31.

***wh.*** Used as a spelling for OE. *hw*, *whan* A/5, *whel* A/8, *white* G/12, *whom* G/25, etc.

***y.*** Used initially as a spelling for OE. front *g* in *yer* A/14, B/24, *yif* A/9, *yit* A/31, etc. Noteworthy is the complete absence of yogh in the orthography of this writer, and cf. also the A-text of the *Astrolabe* and the Hengwrt and Ellesmere manuscripts of the *Canterbury Tales*.

The same sound medially was vocalized to *i* which joins with the preceding vowel to give a diphthong, *day* G/35, *nayl* D/30, *wey* I/22, etc.

For its use as an alternative spelling for *i*, see above, p. 138.

## 3. ACCIDENCE

### *Nouns*

Those nouns which end with a consonant in the sg. usually have *-s* in the pl., *fracciouns* A/3, *quarters* A/15, *seccions* L/8, etc.; but *-es* appears in *arkes* C/18, *eclipses* N/38, *markes* H/37, *thredes* G/13, I/26, and *-is* in *argumentis* G/2.

Variation appears in *equantis* E/53, H/17, but *equantes* C/12; *mynutis* E/59, D/4, *minutis* N/5, but *mynutes* A/29.

Those nouns ending with a vowel in the sg usually have *-es* in the pl., *lynes* A/15, *names* B/2, *tables* G/2, *thinges* D/3, etc., but *-is* appears in *membris* E/7.

Variation appears in *centres* C/11, E/55, H/17, but *centris* A/20, B/33, C/4, E/53;

*planetes* A/21, D/17, H/11, 17, 37, but *planetis* B/32, C/12, E/5.

Note that for *hed* D/10, etc., *aux* B/24, etc., the plurals *heuedes* L/29, etc., *auges* A/21, etc., are invariable.

*Adjectives*

Those adjectives ending with a consonant in the sg. have *-e* in the pl., and in some words this variation between sg. and pl. appears to be regularly carried out, e.g. sg. *al*, pl. *alle*; sg. *smal*, pl. *smale*; sg. *other*, pl. *othre*; sg. *swich*, pl. *swiche*.

Similarly, in those adjectives ending with a consonant in the sg. the addition of a final *-e* in some words seems to indicate the use of the old weak declension. So, for example, strong *blak*, weak *blake* (but cf. *whiche blake* H/20, due perhaps to a preceding *blake*, and *thy blak* K/13, where *thy* is interlined). Similar variation appears in *whit/white* (but cf. *whiche white* G14), *red/rede*, *rownd/rownde*. In addition the following could also perhaps be regarded as weak forms of the adjective: *firste* (cf. *first*, adv.), *grete*, *laste* (but cf. B/25, C/33), *leste*.

But quite certainly there are numerous examples of the retention or omission of final *-e* for which no explanation appears possible except the whim of the writer. Similar variation occurs also in nouns, adverbs and prepositions. Hence *which* (16)/*whiche* (6), *gretter/grettere*, *descending* (2)/*descendinge* (2), *end* (1)/*ende* (31), *bytwix* (2)/*bytwixe* (16), *euermo* (4)/*eueremo* (1), *ferther* (2)/*ferthere* (2), *forseid* (4)/*forseide* (8), *maner* (8)/*manere* (4), *mark* (3)/*marke* (2), *mor* (1)/*more* (4), *narwer* (1)/*narwere* (4), *reknyng* (3)/*reknynge* (3), *therfore* (1)/*therefore* (1), *thilk* (1)/*thilke* (7). When there are comparatively few forms of one type as against an overwhelming majority of the other, the minority forms may perhaps be regarded as due to mistakes by the writer, but this can hardly be the explanation when the numbers are approximately equal. In such cases it is difficult to avoid the conclusion that an *-e* more or less meant little to the writer.

*Pronouns*

The following pronominal forms appear in the text:

(1) *I*; *me*; *my/myn*.
(2) *thow*; *the*; *thy*, *thi/thyn*, *thin*.
(3) masc. g.sg. *his*, *hise*.
    fem. n.sg. *she*; g.sg. *hir*.
    neut. n.sg. *it*; g.sg. *his*. reflexive *hymself*.
    Pl. g. *hir*; d. *hem*.

Def. art. sg. *the/þe*; pl. *the*, *tho* (1).
Demonstrative sg. *this/þis*; pl. *thise*.

*Verbs*

The inf. normally ends in *-e*, *-en*. Presumably final *-n* was originally lost before a word beginning with a consonant, but remained before one beginning with a vowel. However, only occasionally is it possible to detect any such regularity in this text. So

*be* + cons. (26 ×), but *be* + vowel (3 ×).
*ben* + vowel (3 ×), but *ben* + cons. (9 ×).
*seye* + cons. (1 ×).
*sein* + vowel (3 ×), but *sein* + cons. (6 ×).

When the correct forms appear in the infinitives of other verbs the number of occurrences is, as a rule, too small to be significant. Hence we have

*-e* + cons. (32 ×), but *-e* + vowel (10 ×).
*-en* + vowel (3 ×), but *-en* + cons. (1 ×).

It would seem that the only conclusions to be drawn are that the earlier system has by this time broken down, that *-e* and *-en* can be used side by side, but that *-en* endings are gradually falling into disuse.

In the 3 sg. pr. of strong and weak verbs there is variation between *-ith/-yth* (40 ×) and *-eth* (4 ×), though the former is by far the more frequent. The *-eth* ending is invariable in *dwelleth* (1 ×) and *varieth* (1 ×), but variation appears in *kerueth* (1 ×)/*-yth*, *-ith* (8 ×), and *turneth* (1 ×)/*turnyth* (1 ×). The only syncopated form is *stant* F/19, H/18 (cf. *stondith* A/26, E/21).

In the strong verbs the following forms occur:

| | | | |
|---|---|---|---|
| I. | *strid* imp. sg. | | |
| | *write* inf. | | *writen* pt. p. |
| III. | *bygynne* inf. | | |
| | *kerue* inf. | *karf* pt. sg. | |
| | | *fond* pt. sg. | *founde* pt. p. |
| | | | *bownde* pt. p. |
| IV. | *come* pr. s. sg. | | |
| V. | *lye(n)* inf. | *lay* pt. sg. | *ligginge* pr. p. |
| | *se* inf. | *say* pt. sg. | |
| | *speken* inf. | | |
| | *breke* pr. s. sg. | | |
| VI. | *drawe* inf. | *drow* pt. sg. | |
| | *w*$^{t}$*draw* inf. | | *w*$^{t}$*drawe* pt. p. |
| | *stonde* inf. | | *stondinge* pr. p. |
| | | | *grauen* pt. p. |
| VII. | *knowe* inf. | *knew* pt. sg. | |
| | *fallith* pr. sg. | | |
| | *lat* imp. sg. | | |

In the Weak Verbs there is variation in the pt. p. between *-ed* (62 ×) and *-id* (2 ×), the former being by far the more frequent, the latter occurring only in *endid* I/33 and *stikyd* I/31.

Note also the pt. p. *deuyde* A/29 (perhaps an error), and the contracted forms *leid*, *set*.

In the 3rd Weak Class the following forms occur:

Inf. *haue*; 1 sg. pr. *haue*, 2 sg. pr. *hast/hastow*, 3 sg. pr. *hath*, 3 pl. pr. *han*; 2 sg. pr. s. *haue*.

Inf. *sein, seyn, seye*; 1 sg. pr. *sey, seye*; 1 sg. pt. *seide*, 3 sg. pt. *seide*; pt. p. *seid*.

In the Preterite Presents and Anomalous Verbs the following forms appear:

*shall*: 1 sg. pr. *shal*, 2 sg. pr. *shalt, shaltow*, 3 sg. pr. *shal*, 3 pl. pr. *shollen, shal*; 3 sg. pt. *shold, sholde*.

*may*: 2 sg. pr. *maistow*, 3 sg. pr. *may*, 3 pl. pr. *may*; 1 sg. pt. *myht*.

*must*: 3 sg. pr. *mot*; 1 sg. pt. *moste*.

*wit*: 1 sg. pr. *wot*.

*do*: 3 sg. pr. *doth*.

*go*: 3 sg. pr. *goth*.

*will*: 1 sg. pr. *wole*, 3 sg. pr. *wole*.

*be*: inf. *be, ben*; 3 sg. pr. *is*, 3 pl. pr. *ben*; 3 sg. pr. s. *be*; 3 sg. pt. *was*.

## 4. SUMMARY

To what extent does the preceding analysis of the language suggest that we are here dealing with a holograph manuscript, and what support is given by the language to any theory of Chaucerian authorship?

So far as the first question is concerned, it is only rarely possible to claim that a Middle English work is written in the actual hand of the author, and even more exceptional for definite evidence on the point to be available. Usually the evidence is entirely circumstantial, and in the ultimate resort depends mainly on the general impression gained by the student while working on that particular text. Only two considerable works in Middle English are generally accepted as autograph—the *Orrmulum*[1] and Dan Michel of Northgate's *Ayenbite*

[1] Ed. R. M. White and R. Holt (Oxford, 1878).

*of Inwyt.*[1] So far as the second of these is concerned, a note on f. 2 of the manuscript tells us that it is by Dan Michel, and that it was 'ywrite an Englis of his oȝene hand'. Such a statement could, of course, have been copied from the original by a later scribe; but its presence does at least suggest the possibility that the manuscript may be holograph, and this possibility is confirmed by other evidence. The text is written throughout in a single hand, which is not a formal scribal hand; its spelling and phonology show a remarkable uniformity and consistency, with no such admixture of forms from different times and places as would be expected had the text passed through the hands of different scribes; nor does it exhibit the particular type of scribal errors to be explained only by misreadings and miscopyings at some point in a complicated textual history. On the whole it would seem that we are justified in assuming this to be a holograph manuscript.

So far as the *Orrmulum* is concerned the general impression gained by the reader is even more convincing. There is, to be sure, no initial statement that the text is in the author's own hand, but the consistency of spelling and phonology is as marked as in the *Ayenbite*. In addition, the orthography is so distinctive that it is difficult to believe that it could have been reproduced so faithfully and consistently by anyone but its inventor; and we know from references in the text that the author was himself the inventor of this particular orthography. The work is written in a single hand throughout, which is not a formal one; and there is a lack of the type of error which we associate with copyists. Even more significant is the appearance of the manuscript itself, written as it is on sheets of varying size, presumably as collected together by the author, and not in any uniform volume such as a scribe might rather be expected to use.

Obviously none of these points by itself is of much significance. A statement that the work is in the hand of the author might well be copied into a later manuscript; many Middle English works survive which were copied by persons who had received no formal scribal training; consistency in spelling and phonology is found in works which are certainly not autograph manuscripts, e.g. in the *Ancrene Wisse* and the Bodley manuscript of the *Katherine Group*; while an exceptionally correct text, free from scribal errors, may be nothing more than the result of a simple textual history or of the employment of careful copyists. Nevertheless, when these various points occur together in a single text, there would seem to be good reason—in the absence of definite evidence to the contrary—for supposing that the text may well be in the author's own hand.

[1] Ed. R. Morris, E.E.T.S. 23.

Most of the reasons which have led to the acceptance of the *Orrmulum* and the *Ayenbite* as holograph manuscripts would seem to be applicable to the *Equatorie*. There is, to be sure, no definite statement that the text is in the hand of the author. But it is written throughout in a single hand, which is not a formal scribal hand, and the orthography and phonology are remarkably consistent, so much so that it is difficult to believe that it can have passed through the hands of many scribes. There is not, of course, absolute consistency; this is not to be expected before the introduction of a fixed spelling. Variation appears in the use of the ending *-oun* by the side of *-on*, *hole/pol* occur by the side of *hoole/pool*, nor is the use of *i* and *y* absolutely regular. But the variations are certainly no greater than would be expected in the personal orthography of any writer before the appearance of a fixed spelling.[1] Indeed, the spelling and phonology of the *Equatorie* are much more consistent than those of the vast majority of Middle English texts, and are comparable with those of the accepted holograph manuscripts. Moreover, the very appearance of the text, with its numerous erasures, interlinings, etc., gives the impression of a text which its author is in process of composing, either from the beginning, or at any rate from rough notes. Erasures, corrections, etc., appear frequently enough in texts which are certainly copies, but it would be surprising to find them quite so often in the work of a mere copyist. In addition, they are usually of a type which one would connect rather with the composition than with the copying of a text. It is true that in at least one place there is what might be regarded as a distinctively copyist's error: the omission of a phrase at M/20 which could represent the omission by the writer of a complete line in the original from which he was copying. But other explanations are equally possible; for it is just as likely to be an error in composition as one in copying, or again it may be an indication that the author was working from rough notes. Certainly the consistency of the language and spelling makes it fairly certain that few copies can have intervened between the original and this particular text; and the general appearance of the manuscript suggests that it is in fact most probably a holograph.

Whether or not this particular author was Chaucer is, of course, a very different problem, and one on which the language cannot throw much light. In the first place it must be admitted that, in spite of the volume of his work, very little indeed is known about Chaucer's own linguistic habits. Even the best Chaucerian manuscripts appear to have passed through the hands of various scribes, and it is obviously very difficult to say which of the different

[1] The variation possible in the orthography of a single scribe is well demonstrated by N. Davis, 'A scribal problem in the Paston Letters', *English and Germanic Studies*, vol. IV, pp. 31 ff.

characteristics are due to Chaucer, and which have been introduced by one or other of his copyists. If we compare the numerous extant manuscripts, we can obtain some general idea of the dialect which must underlie all the variations; but it is hardly possible to point to any individual form and to say with certainty that it represents the normal usage of the poet. Only by a study of the rhymes, and to a lesser extent of the metre, is it possible to obtain some knowledge of the forms actually known to and used by Chaucer himself. In prose or in the middle of a line of verse any particular form can be altered by a later scribe without a modern reader being able to detect the change. But a change in the form of a rhyming word may result in the spoiling of the rhyme; in this case, assuming that a perfect rhyme was present in the original, it may be possible to deduce what that original form must have been. Consequently, in this way it is possible to discover something of the actual forms used by the author himself, though whether or not these were his normal forms is not so certain. If he lived in an area of mixed dialect, such as that of London, he might well be familiar with a variety of forms, any of which he could use when necessary for the sake of rhyme. But we can rarely be certain that the forms used in rhyme would also be the forms which would be used by the author in his prose. For example, as far as Chaucer is concerned it may be significant that *e*-forms for OE. *y* seem to be much less frequent in his prose than in his verse. This may indicate that such forms were comparatively rare in Chaucer's own speech, and that in his verse he used them mainly because of the exigencies of rhyme. But, on the other hand, it may merely emphasize the fact that in prose an author's characteristic forms can be changed by later scribes much more easily and with less chance of detection than is the case with verse.

In the main the forms in the *Equatorie* are those that appear most frequently in the best manuscripts of the *Canterbury Tales*, and no forms appear in the *Equatorie* which are not to be found in some of these manuscripts. This does not, however, carry us very far, since much the same can be said of the works of other London writers of approximately the same date, and manuscripts of the *Tales* are so numerous that a wide variety of forms naturally appears in them. But it may be significant that the language of the *Equatorie* appears on the whole to be closer to those manuscripts of the *Tales* which are generally agreed to be the earliest and the nearest to Chaucer's own texts. More particularly does there appear to be a very close similarity between the language of the *Equatorie* and that of the version of the *Treatise on the Astrolabe* preserved in MS. U.L.C. Dd. 3. 53,[1] which seems to present the text closest to the author's original. The main difference is that in the *Equatorie* the language is a good

[1] Ed. W. W. Skeat, E.E.T.S. E.S. 16.

deal more regular and consistent. Indeed, it is the sort of original dialect and orthography that the language of the *Astrolabe* may well have resembled before passing through the hands of one or more scribes. Perhaps too much weight is not to be placed on this; at the best it might only suggest a similar date and place of composition, not necessarily the same author.

In its vocabulary the *Equatorie* contains a fair number of technical words which *OED* otherwise assigns only to Chaucer at this date. This is not particularly surprising; astronomical works in English are not numerous before the fifteenth century. Chaucer's *Treatise on the Astrolabe* is the first English work on the subject of which we know, and naturally many new technical words will appear there for the first time. Nor, indeed, is *OED* intended to be at all exhaustive or definitive. The first quotation for a given word merely indicates that this is the earliest text read for the dictionary in which that word is to be found, but it can certainly not be taken for granted that it represents the first appearance of the word in English. In fact, some of these words appear also in other contemporary, or nearly contemporary, astronomical treatises which are as yet unprinted. The author of the *Equatorie* pretty certainly knew the *Treatise on the Astrolabe*, as is suggested by his reference to the line which 'is cleped in the tretis of the astrelabie the midnyht line' (C/29). The term is a fairly obvious one, so that its occurrence in both works must not be pressed too far, and by itself it is certainly not evidence of common authorship. Nevertheless, the reference does suggest that the writer of one knew the other, and this being so it need not surprise us that the composer of the later treatise has used some of the technical terms which appeared in the earlier one.

In style there are some similarities between the *Equatorie* and the *Astrolabe*. The didactic manner in both is very similar, and many of the usages noted by Pintelon[1] as characteristic of the prose of the *Astrolabe* appear also in the *Equatorie*. The author normally speaks in the first person, with only a very occasional use of the impersonal. He frequently addresses the reader directly, and always speaks of *thy* conclusions, *thy* cross, *thine* instrument, etc., with frequent short warning clauses introducing the sentence (e.g. B/25, C/29, C/13, E/7, etc.). 'Spoken gestures' are not unusual, whether single or reinforced (A/7, C/7, G/13, J/11, etc.); the use of the pt.p. with the value of a dependent clause is not uncommon (A/16, D/5, G/18, H/8, etc.), as also the employment of the relative adjective followed by a substantive already used (A/18, B/9, C/28, D/17–18, D/31, etc.), and the use of the inverted order of verb + subject (B/14, D/34, E/30, etc.). Some of the points noted are absent, e.g. the omission of the pronoun subject, or the use of doublets, but for all that there is a decided

[1] P. Pintelon, *Chaucer's Treatise on the Astrolabe* (Antwerp, 1940), pp. 76ff.

similarity in prose style between the two, and it is a style which has not so far been found to appear in other contemporary works. Certain favourite phrases appear frequently in both, *I seye* 'I repeat', *þat is to seyn*, *yif so be þat*, *as shewith in*, *wheras* 'where', etc., but there is little about these that can be regarded as at all distinctive. Perhaps even verbal echoes may appear:

| *Astrolabe* | *Equatorie* |
|---|---|
| the membres of thin Astrolabe. 3/46. | thyn equatorie w[t] alle hise membris. E/7. |
| tak this rewle general. 5/6. 5. | this rewle is general for alle planetis. B/32. |
| ben compassed certein cerclis. 13/18. 1. | yif thow haue trewely compased thy cercles. D/13. |
| a label þat is schapen like a rewle. 13/22. 1. | this rewle mot be shape in maner of a label on an astrelabie. D/26–27. |
| tho leide I my rewle vpon. 15/1. 13. | tho leide I þ[t] on ende of my thred. N/9. |
| the laste merydye of December. II/44. 3. | the last meridie of decembre. B/25. |

Similarly, in both treatises we find the tendency to use the French construction with the adjective following the noun. So in the *Astrolabe* appear such phrases as *sterres fixe*, *lyne Meridional*, *lettres capitals*, *howres equals*, *sper solide*, *day natural*, etc., and in the *Equatorie*, *centres defferentes*, *parties equales*, *arkes equales*, *remnant forseid*, *enches large*, *of metal rownd*, etc.

There is certainly nothing in the language or style of the *Equatorie* which is definitely against Chaucerian authorship. On the contrary, there are certain facts which offer some support to such a theory; and if none of them is particularly striking by itself, taken together they may well have some significance, though the exact value of this is perhaps more difficult to determine.

# XI

# ASCRIPTION TO CHAUCER

## I. PREAMBLE

In this chapter evidence will be collected and examined in support of the suggestion that the text and the first set of tables in this Peterhouse manuscript are a Chaucer holograph of about 1392.

Since the matter is controversial and involves many points of detail, pains have been taken to separate the issues involved, so as to distinguish those which are reasonably certain from those which are not conclusive in themselves but may produce a cumulative effect. Again, it must be clearly understood that no *decisive* proof is being offered or even suggested. The nature of historical proof, as distinct from scientific proof, is that one can only work by accumulating many 'pointers' and showing that the indicated result is nowhere in conflict with available evidence. It is an inevitable misfortune that this chapter cannot include further evidence and discussion which may become possible once this edition has been published and is available to scholars. For example, the identification of any other specimen of Chaucer's handwriting would materially affect the issue, and so might the finding of the Latin text from which the *Equatorie* has been derived. One might also hope that the problem of how the manuscript came to Peterhouse may eventually be solved, and perhaps there is some chance that other Middle English scientific texts by the same author may yet be found.

Pending further investigation and discussion, the ascription to Chaucer must remain tentative; needless to say, and in spite of these caveats, it would not have been brought to publication if an alternative explanation could have been found to fit the facts and indications so readily.

## 2. THE MANUSCRIPT IS AN AUTHOR'S HOLOGRAPH

The hand of the text and first set of tables is unusually large, broad and clear, and contains but a small number of the palaeographical contractions current at the end of the fourteenth century. Perhaps the most striking feature of the hand is its informality which contrasts so strongly with the tightness and regularity seen in scribal and secretarial writing of the period.

Scribal manuscripts are usually written on folios which have been ruled with lines spaced by pricking with a small pair of dividers, but although the astronomical tables have, of necessity, been spaced in this fashion, such an aid to tidiness is not employed in the text, but the writing has been done freely and naturally across the rather wide pages.

The writer is unquestionably fluent, and he seems mindful of the needs of scribes who might copy the manuscript afterwards. At the heads of sections, spaces have been left for an illuminated letter, and even though the spaces left are all too small for much decoration, the existence of a little guide letter in the margin seems to indicate that the writer intends some later scribe to make a fair copy and leave adequate space at this point for the work of the illuminator.

Apart from the hand itself, the most noticeable characteristic of the pages of text is the presence of numerous corrections and additions (and occasionally, glosses), all of which are in the same hand as the text. The appearance is unmistakably that of a script which has been subject to minute working-over and revision by its writer. Almost every technique of revision has been used—supralinear and marginal insertion, deletion, obliteration and expunging are all there, and, in addition, there is much evidence that text has been written over places where former words have been erased by rubbing and scraping of the parchment.

From the extent of this process of revision (which includes also the deletion of two whole long passages with the comment 'this canon is false') it must be considered unlikely that the text could have been copied rather carelessly from some original and then checked and improved so painstakingly by the same person. Again, it is unlikely that the text is a fair copy of a bad original which has later been revised by the copyist, for any person who could improve the text so materially would certainly have incorporated many of the more trivial corrections into his first copy—it was customary for a copyist to take this much liberty at least.

There remains the possibility that this is an original composition which has been heavily corrected by the writer himself—although, in saying that it is 'original' we cannot preclude the possibility of it being a translation, an adaptation, or a compilation from a set of rough notes, or indeed any combination of these. The alterations to the text therefore suggest that we are dealing with a work which had not previously been composed in this form, and, furthermore, the text seems to be a draft that has undergone the usual process of revision by the author.

Some of the alterations consist in rectifying words omitted or duplicated in error, and since such errors may occur in both original and copied work we cannot use their presence to distinguish one from the other. There are, however, alterations where it is evident that the writer is concerned with an improvement in style or meaning; one particularly striking example near the beginning of the text is discussed in the note to A/12, p. 63. This latter type of alteration can only occur with 'original' work, and we therefore conclude that the text and its corrections were both written by the author himself, or possibly by an amanuensis working at his dictation.

It is difficult to believe that an amanuensis could be responsible for all the writing. It would be much more easy to write many of the tables and calculations than it would be to attempt to dictate them accurately—and the tables here are remarkably good and show no trace of the appallingly frequent scribal blunders which are always present in those medieval astronomical tables which have not been copied by astronomers.[1] Again, it would be remarkable for an author working by dictation to make no corrections whatsoever in his own hand, and yet no trace of a second hand is found anywhere in the text. The only simple conclusion is that some person composed the text and tables (either *ab initio* or by adaptation from previous work) and wrote his draft, which he then proceeded to correct and polish in the ordinary manner.

The homogeneity of language and orthography supports the conjecture that the whole text with its corrections is the work of one writer; certainly if it had been derived from some other work in English one would expect to find more variation in usage and perhaps also in style.

## 3. THE WORK WAS COMPOSED *c.* 1392

It is rarely that one has such definite and unmistakable evidence for the date of composition of a manuscript as exists in this case. Near the beginning of the text (A/14) the author writes 'myn equatorie þ[t] was *com*powned the yer of crist .1392. *com*plet the laste m*er*idie of dece*m*bre', and whether one interprets this as the actual date of construction or as the date for which the instrument was set, it must mean that the time of composition cannot be long after 31 December 1392. At the end of the text three worked examples are given, all of them drawn from the year 1391; again, it would be reasonable to

[1] A rather good example of this is found in the Middle English translation of the *Exafrenon* of Richard of Wallingford (Bodleian MS. Digby 67, ff. 6r.–12v.; see p. 206 of the present work). The translator says, '...the tables are of myn awne makyng. For the tables of the Abbot of sancte Albones (are) made full of errorues. And þat I trewe bee throwgh the vices of writers & nott of the vowshipfull clerke & p*re*late þat þe trectes made.' Nevertheless, this scribal copy contains no less than twenty-four errors in the first small table of forty-eight entries. Another copy of the translation in Bodleian MS. Digby Roll III omits the table completely. (For a facsimile of part of MS. Digby 67 see R. T. Gunther, *Early Science in Oxford*, vol. II, p. 51.)

take these as calculations made not very long before or after the date of composition. The most powerful evidence, however, is to be found in the set of astronomical tables, most of which give the increase of some astronomical variable for any number of days, months, or years. To find the actual value of this variable at any sought time it is necessary to start with the value at some convenient date which is taken as the standard origin or *radix*, and to save labour it is usual to employ a radix near the dates for which one is likely to need calculations. In this set of tables, the chosen radix is nearly always the same as that in A/14, viz. 31 December 1392 complete. If this had been a round number, such as 1400, or perhaps even 1390, some doubt could have been entertained, but with the date in question repeated in so many places it is evident that the person who wrote tables and text was interested in performing calculation for 1392 and the period immediately following.

It must be emphasized that this certain method of dating by radices cannot justifiably be extended to all types of dates mentioned in astronomical tables. In other cases, a list of eclipses, for example, can be handed down from one scribal copy to another of much later date, or even a series of radices for 1380, 1390, 1400, etc., might be copied into a much later work. But in the present case the use of one radix throughout provides a quite clear indication.

Other radices are given from time to time for *Anno Christi* (i.e. the year 'zero'), and these values are drawn from a standard medieval work, the Alfonsine Tables. Furthermore, it seems that the 1392 radices have been specially calculated from the *Anno Christi* values specifically for the author's purpose. There is, of course, no reason why he should choose this year apart from its significance as his date of writing; the number 1392 is not especially convenient on either decimal or sexagesimal systems.

In addition to these radices there are a few places in which other dates are noted. These dates are 1393, 1395 and 1400. Since they only occur rarely, they do not indicate any extended calculations using the whole tables; they may denote nothing more than the working of one horoscope or a single planetary position for such a date. Further, the 1400 radix is probably only taken as a round number, but the 1395 one seems to be inserted (by the same hand) in a quite different ink, and may well correspond to a special calculation made during that year. It is interesting as a sign that the manuscript might well have been still in the writer's possession three years or so after its composition.

Because of the great labour involved in the production and faithful copying and the consequent value of long sets of astronomical tables such as those contained in this volume, it is common to find much evidence of the use of these tables by later workers. Most sets of tables contain emendations, notes, extra radices and glosses in hands later than that of the original compiler. It is therefore remarkable that the Peterhouse manuscript shows very little evidence of such use. There is nothing to show that the volume was used later than the last radix (of 1400), until the owner's or librarian's mark of 1461, by which time the book may have been at Peterhouse, and in any case by then the tables were outdated by better, subsequent productions. The evidence, slight though it is, would suggest that the manuscript was 'lost' or in some way out of circulation between the time when it was in the author's possession and the later period when it was a rather outdated treatise in the College Library.

## 4. SIMON BREDON DID NOT WRITE THIS TEXT

It is remarkable that nowhere in the original manuscript is there any direct citation of the name of the author. The opening of the text has the vital word erased, and at the best, this seems to throw us back only on the author of the Latin or Arabic work from which this Middle English version is drawn. At the end, the somewhat abrupt cutting short of the subject-matter may be taken as an indication that the writer has not reached the last part of his source; and unfortunately he leaves the section without the usual *explicit* which might have given us the title of the work and its author's name. It is just possible that part of the text might have been lost; the first section, containing instructions for making the Equatorie, fills a quire of four folios; the second section, which describes the operation of the instrument, fills a second quire containing four folios of a finer and slightly smaller size of vellum. Perhaps a third complete quire containing a concluding section and an *explicit* has been lost—if the quires were originally unbound while in the author's possession, one of them could easily have been retained amongst other papers or included in some other composite manuscript.

Similarly, the tables have no formal beginning or end, and this is significant, since in spite of the damage to the front of the volume it appears from the collation that neither the first nor the last folios of the tables have been removed. The absence of a formal opening here seems to support the conjecture that the tables were originally meant to follow the text, but in any case this merely transfers the problem of the lack of a formal title or *explicit*.

Whether or not a contemporary ascription has been lost we cannot say. The only evidence which remains in the volume is the note written on the back fly-leaf, presumably by some College librarian *c.* 1600; unfortunately, it has not been possible to identify the hand with that of any known person. The note describes the *Equatorie* text as 'Simonis Bredon æquat plan', and adds the information that Bredon was from Merton College and flourished in 1380.[1] Fortunately, the librarian acknowledges the source of his information, and allows us to find that it is taken almost verbatim from John Bale, *Scriptorum Illustrium maioris Brytannię* (Basel, 1557).

Bale refers to Bredon as the author of a text on 'Aequationes planetarum', possibly deriving his information from John Leland who examined the Peterhouse manuscripts *c.* 1542 and noted the item 'Tab. aequ. planetarum, autore Simone Bredon'. There is no record of any other similar manuscript in the library of the College, and there is no reference to any other copies elsewhere of such a text by Bredon. Hence, unless the sixteenth-century librarian has confused two manuscripts (which, in this case, is very unlikely) we have a curious circular process; the only surviving evidence for the ascription is copied from the index of Bale, yet Leland and Bale must have derived their information from this very manuscript. Clearly there must have been some older evidence of ascription which was available to Leland and Bale, but has since been lost. The title and author might well have been written on the former vellum wrapper, later used as a front fly-leaf and now missing.

Although it is thus established that the manuscript has been ascribed to Simon Bredon since some date prior to 1542, it is now possible to show that this ascription cannot be more than a half-truth, accepted only because of a curiously widespread confusion of dates for this author. In spite of the assertion by Bale and others that he flourished in 1380, it is now known that he was dead by 1372, when his will was proved.[2] Since there is no reason to suppose the existence of another writer of the same name but slightly later date it is not possible that the mathematical, medical, and astronomical writer, Simon Bredon of Merton College, could have written the Peterhouse text dated 1392.

It has already been shown that the text is at least in part an original composition written in that year, and hence the greatest allowable connection with Bredon would be to accept him as the author of a text from which this is a free adaptation or translation. There is little that can be said against such a con-

[1] The note is transcribed on p. 13.

[2] The will has been published by F. M. Powicke, *The Medieval Books of Merton College* (Oxford, 1923), p. 28. An example of the confusion of dates, together with a good account of Bredon's life and works, is given by R. T. Gunther, *Early Science in Oxford*, vol. II (Oxford, 1923), p. 52. This confusion has also been criticized by L. Thorndike, *History of Magic and Experimental Science*, vol. III (New York, 1934), pp. 521 ff. Thorndike distinguishes the date at which the will was made (1368) and the date of proving (1372); he also cites a fifteenth-century manuscript which gives the date 1380 for this author (Bodleian MS. Digby 160, ff. 102–203). This manuscript is probably the source of the error spread by Bale.

jecture except for the fact that no such work by Bredon has been recorded; since many manuscripts of this type remain unexamined there is still a chance that the source text may yet be identified and may be attributable to Bredon. All his known works are in Latin, and there does not seem to be the strong Arabic influence that is found in the present text. Against this, however, is the evidence that Bredon certainly owned an instrument of the same type as that described in the *Equatorie*, though of a rather different pattern, and this particular instrument is still preserved at Merton College. It is reasonable to suppose that Bredon's experience with his own instrument might have led him to design an improved type of instrument, possibly borrowing details from types of equatoria of Arabic ancestry. A text describing such a (conjectural) equatorium of Simon Bredon might well have served as a basis for this description of the *Equatorie of the Planetis*.

On the other hand, it is equally possible that Bredon is not even the author of the source text, but has had his name connected with the Peterhouse manuscript only through the error of some person who sought to identify the anonymous work. Bredon was well known as the author of a *Theorica Planetarum* of great popularity,[1] and if the only date known for him was the erroneous 1380, his name would seem a likely enough guess. Probably he would have been the only English authority of a suitable date whose name was known in connection with this sort of treatise. One would not at first sight suspect that the *Equatorie* with its many tables and its air of professional astronomy might have any relationship to Chaucer's *Treatise on the Astrolabe* which has a more easily understood Prologue and lacks the tabular sections.

If the Peterhouse manuscript is accepted as an author's holograph, it is possible to show independently of date that ascription to Bredon is impossible. This is possible because Bredon's hand has been preserved in his *Conclusiones quinque de numero quadrato* and in notes on an Oxford almanac for 1341 (Bodleian MS. Digby 178, ff. 11v.–14) which contains the note 'Has conclusiones recommendo ego Simon de Bredon volenti circa quadraturas circuli laborare'; the hand is obviously very different from that of the Peterhouse manuscript.[2]

It seems fairly certain that whether or not the Peterhouse text is a translation

[1] This work (which has also been attributed to Walter Britt) must be carefully distinguished from a work of Gerard of Cremona the Elder (Gerard of Sabbioneta) which has nearly the same incipit. See A. A. Björnbo, *Bibliotheca Mathematica* (1905), p. 112.

[2] A portion of a double folio containing Bredon's notes is reproduced by R. T. Gunther, *Early Science in Oxford*, vol. II (Oxford, 1923), facing p. 52. He also remarks that further specimens of Bredon's hand may be found in the following manuscripts: Digby 179, Merton 250 and Digby 176, ff. 71–72 (ascribed to Rede in the Catalogue).

or adaptation from an earlier work by Simon Bredon, there is now no direct evidence to show who might have been the writer of the Middle English author's draft of 1392. We thus have to turn to the less direct evidence contained in the volume.

## 5. CONNECTION WITH CHAUCER'S 'TREATISE ON THE ASTROLABE'

The best known Middle English scientific work is undoubtedly Chaucer's *Treatise on the Astrolabe*,[1] which claims to be an account of that instrument written for his 10-year-old son, 'litel Lowis', and contains worked examples for the year 1391. It is a masterly piece of scientific writing—perhaps the earliest genuinely scientific work in English, and it is still the best explanation of the astrolabe in our language. It is, however, an incomplete work, for Chaucer writes in the Prologue that he will divide the treatise into five parts (ed. Robinson, p. 642), but the extant manuscripts[2] give only the first two parts, and perhaps a few additional sections which might be from the missing parts or might be spurious additions by later copyists or editors anxious to follow the prescribed table of contents. Much of what is missing would have been in the form of astronomical tables, but there are clear references to textual portions, especially a 'theorike to declare the moevyng of the celestiall bodies with the causes' and the 'generall rewles of theorik in astrologie'.[3]

The Peterhouse volume seemed, at first glance, to contain so many of the ingredients of the missing parts of the *Astrolabe*, and the date 1392 which

[1] It was first edited in Chaucer's works by William Thynne, printed at London by Thomas Godfray in 1532, and has since appeared in the editions by Urry, Skeat and Robinson. Separate editions of the treatise have been published by A. E. Brae (London, 1870), W. W. Skeat (London, E.E.T.S. 1872), and P. Pintelon (Brussels, 1940). A version in modern English is supplied by Gunther, *Early Science in Oxford*, vol. v (Oxford, 1929).

[2] The best survey of extant manuscripts is that given by P. Pintelon, *Chaucer's Treatise on the Astrolabe* (Antwerp, 1940), pp. 14ff. To the twenty-three manuscripts recorded there, the following may now be added:

*21. Phillipps 11955* (location unknown to Pintelon). It is now in the Sylvanus P. Thompson collection in the library of the Institution of Electrical Engineers, London. This MS. T seems to be the direct source of MS. K.

*24. Penrose Collection, Philadelphia.* This MS. Y is related to the second group of manuscripts (L, N, O, R), but contains some important variants.

*25. In possession of Messrs Wm. Dawson, London, 1953.* This MS. Z is a curious conflated version consisting of two parts; $Z_1$ is a descendant of the subgroup containing MS. P and $P_1$, while $Z_2$ (in a different hand) is related to the earliest members of group 2. This manuscript therefore provides a slight link between the two main groups.

I am indebted to the three owners mentioned above for allowing me the most ample facilities for studying these manuscripts; it is hoped that fuller details may be published at a later date in a study of the filiation of all extant versions of this treatise.

[3] Since the words were used interchangeably, it is not quite clear whether Chaucer might have meant 'astronomy' rather than 'astrology' in the modern sense.

occurs so frequently compared so well with the accepted date of that other work, that some connection between the *Equatorie* and the *Astrolabe* was immediately suspected. Further study has, however, shown that there is by no means complete correspondence between the contents of the Peterhouse volume and the five parts announced in the Prologue of the *Astrolabe*. The agreement is rather such as would be expected if Chaucer, having written the first two parts, had looked for some convenient sets of tables to include (probably those of John Somer and Nicholas Lynne) and for some suitable tract on the motion of the planets. The equatorium must have been considered as the companion instrument to the astrolabe (Bredon's instrument at Merton College contains both together), and a treatise on the equatorium, if such had come to hand, might well have been considered much more fitting than a translation of the usual, purely descriptive exposition of a 'Theorica Planetarum'. Such deviation from the stated plan is not uncommon—a similar example may be found in the question of the number of Pilgrims mentioned in the *Tales*—and a problem such as this is not unlikely to arise when an author writes the preface before the main chapters of his book.

The evidence from agreement of content between this volume and the requirements of the missing portions of the *Astrolabe* is admittedly slight, and a similar argument could easily apply to almost any collection of miscellaneous tables and Middle English astronomical text that was sufficiently near in date to 1391.[1]

A more specific connection occurs in our text at C/29, where the author describes the construction of a line on the instrument, 'which lyne is cleped

[1] An extensive search at Oxford, Cambridge and London has yielded only two suitable manuscripts, neither of which seems to exhibit any affinity with the *Astrolabe* or the *Equatorie*. MS. Royal 12. D. vi, f. 81 r. at the British Museum contains a fragment of a Middle English version of a Canon to Tables of Ascensions of Signs by Walter Anglus who is said to have flourished *c.* 1400 and to have been educated at Winchester in youth and later at Oxford. The astronomical tables in the same manuscript have a radix date of 1392 and were drawn up in York by Mag. Richard Thorpe (died 1440), who quotes his own (?) calculations for the longitude of London (19° 26′) and the calculations of Simon Bredon for the longitude of Oxford (17° 56′). Perhaps also connected with Bredon or with his equatorium text is a manuscript, now lost, which is marked as no. 378 in the library of the Austin Friars of York (*Fasciculus Ioanni Willis Clark dicatus* (Cambridge, 1909), p. 58), 'Equatorium quondam Magistri Ric' thorpe'. MS. 1307 (=O. 5. 26) of Trinity College, Cambridge, contains a very fine set of Middle English treatises and translations of considerable scientific and linguistic interest. The volume seems to be written in a hand somewhat similar to that of the Peterhouse manuscript, and contains *passim* the date 1397 which may be taken as the approximate time of writing. All the other Middle English astronomical manuscripts which have been inspected contain only more popular compilations such as the 'Wise Book of Astronomy and Philosophy' (Egerton 2433, etc.) and the translation of the *Exafrenon* of Richard of Wallingford (Digby 67, Digby Roll III).

It would be tempting to attribute the rather sudden appearance of so much vernacular writing on astronomy to the influence of Chaucer or even to his own pen; but I can see nothing to support or refute such a suggestion.

in the tretis of the astrelabie the midnyht line'. Chaucer's treatise on the astrolabe is the only English version known to have existed at that time, and if reference is being made to an English work, that must be the one in question. Furthermore, the line noted is given the same name (apparently for the first time) in that work (Part I, concl. 4; Part II, concl. 3, line 21, etc.).

It might also be remarked that Chaucer seems to refer to some previous account of *The Solid Sphere* (Part I, line 92; Part I, concl. 17, line 21) which is not now known. We may perhaps regard the evidence as indicating that Chaucer wrote such a treatise which has been lost. A trilogy containing treatises on the instruments concerned with spherical trigonometry, the stars and the planets would be a good and complete work, admirably suited to an elementary exposition for the use of a student. I regard it therefore as possible (but by no means proven) that the work on the solid sphere was completed at an earlier date, and the tract on the astrolabe was left incomplete because its later sections became diverted into the work on the equatorium which we have in draft, but which never attained a form sufficiently finished for 'publication'.

There are many instances in which technical terms and even whole phrases are common to the *Astrolabe* and the *Equatorie* and apparently to no other work of the period. This similarity, striking though it is, is difficult to assess because so few comparable manuscripts in English have been examined, and none other is available in a printed edition.

It is worthy of note that the *Equatorie* and the *Astrolabe* both contain worked examples based on astronomical calculations for stated dates in 1391; it is, in fact, on the basis of these dates that the *Astrolabe* is usually stated to have been written in that year.[1]

The actual dates contained in the two works are as follows:

*Astrolabe*

12 March 1391 (1st point of Aries), Part II, Conclusion 1.
13 December 1391 (1st point of Capricorn), Part II, Conclusion 1.

*Equatorie*

17 December 1391 (M/24).
19 February 1391 (N/1).
23 February 1391 (N/16).

Now it is perhaps not valid to compare these dates as indicating the actual dates of writing with any greater precision than we have already found from

[1] But see S. Moore, *Modern Philology*, vol. X, pp. 203 ff., who argues for 1392. See also Kittredge, *Modern Philology*, vol. XIV, p. 513 and V. Langhans, 'Die Datierung der Prosastücke Chaucers' (*Anglia*, 1929), p. 240.

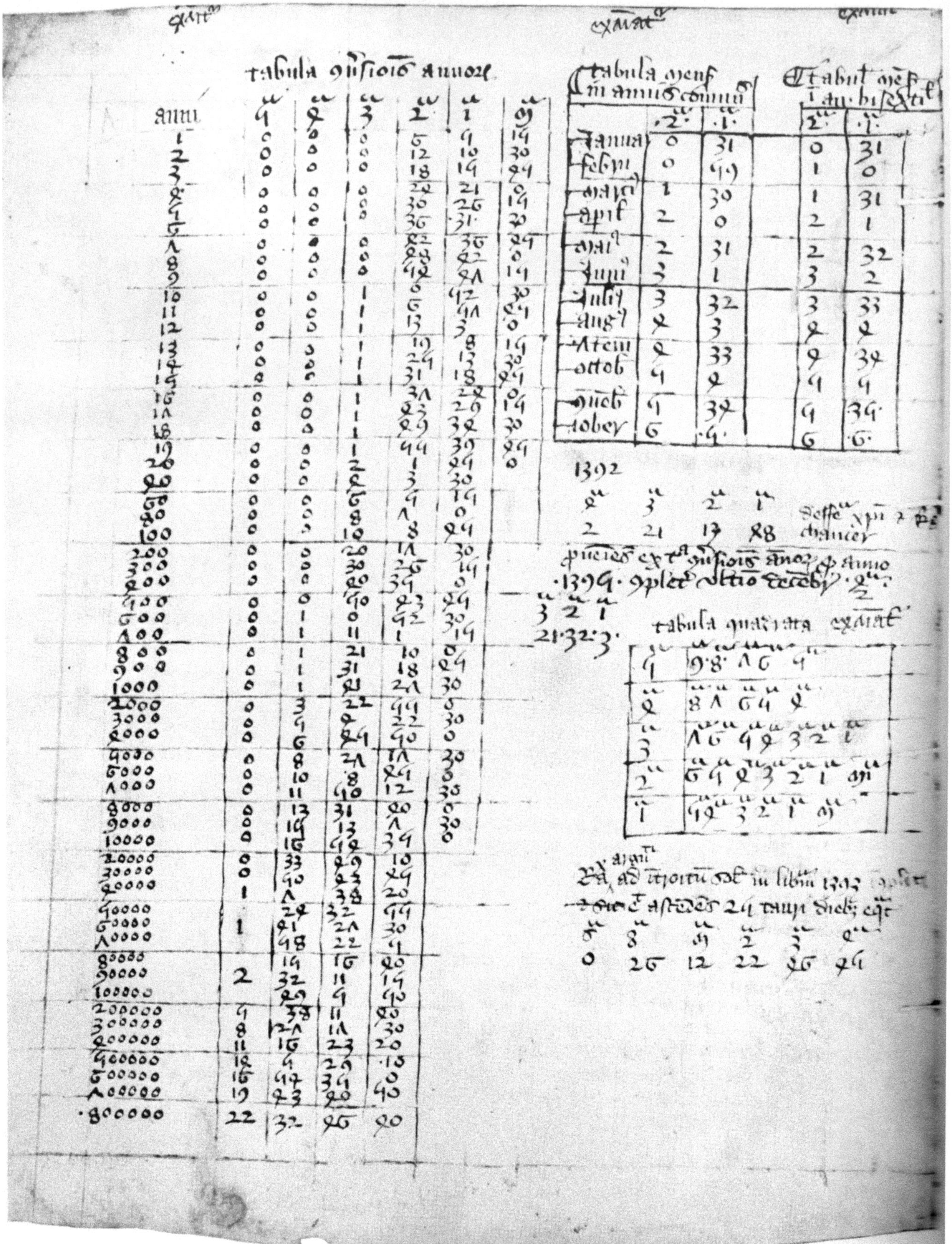

exat' | examinatur | examinatur

tabula conuersionis annorum

| Anni | 5ª | 4ª | 3ª | 2ª | 1ª | [illegible] |
|---|---|---|---|---|---|---|
| 1 | 0 | 0 | 0 | 6 | 5 | 15 |
| 2 | 0 | 0 | 0 | 12 | 10 | 30 |
| 3 | 0 | 0 | 0 | 18 | 15 | 45 |
| 4 | 0 | 0 | 0 | 24 | 21 | 0 |
| 5 | 0 | 0 | 0 | 30 | 26 | 15 |
| 6 | 0 | 0 | 0 | 36 | 31 | 30 |
| 7 | 0 | 0 | 0 | 42 | 36 | 45 |
| 8 | 0 | 0 | 0 | 48 | 42 | 0 |
| 9 | 0 | 0 | 0 | 54 | 47 | 15 |
| 10 | 0 | 0 | 1 | 0 | 52 | 30 |
| 11 | 0 | 0 | 1 | 6 | 57 | 45 |
| 12 | 0 | 0 | 1 | 13 | 3 | 0 |
| 13 | 0 | 0 | 1 | 19 | 8 | 15 |
| 14 | 0 | 0 | 1 | 25 | 13 | 30 |
| 15 | 0 | 0 | 1 | 31 | 18 | 45 |
| 16 | 0 | 0 | 1 | 37 | 24 | 0 |
| 17 | 0 | 0 | 1 | 43 | 29 | 15 |
| 18 | 0 | 0 | 1 | 49 | 34 | 30 |
| 19 | 0 | 0 | 1 | 55 | 39 | 45 |
| 20 | 0 | 0 | 2 | 1 | 45 | 0 |
| 40 | 0 | 0 | 4 | 3 | 30 | |
| 60 | 0 | 0 | 6 | 5 | 15 | |
| 80 | 0 | 0 | 8 | 7 | 0 | |
| 100 | 0 | 0 | 10 | 8 | 45 | |
| 200 | 0 | 0 | 20 | 17 | 30 | |
| 300 | 0 | 0 | 30 | 26 | 15 | |
| 400 | 0 | 0 | 40 | 35 | 0 | |
| 500 | 0 | 0 | 50 | 43 | 45 | |
| 600 | 0 | 1 | 0 | 52 | 30 | |
| 700 | 0 | 1 | 11 | 1 | 15 | |
| 800 | 0 | 1 | 21 | 10 | 0 | |
| 900 | 0 | 1 | 31 | 18 | 45 | |
| 1000 | 0 | 1 | 41 | 27 | 30 | |
| 2000 | 0 | 3 | 22 | 55 | 0 | |
| 3000 | 0 | 5 | 4 | 22 | 30 | |
| 4000 | 0 | 6 | 45 | 50 | 0 | |
| 5000 | 0 | 8 | 27 | 17 | 30 | |
| 6000 | 0 | 10 | 8 | 45 | 0 | |
| 7000 | 0 | 11 | 50 | 12 | 30 | |
| 8000 | 0 | 13 | 31 | 40 | 0 | |
| 9000 | 0 | 15 | 13 | 7 | 30 | |
| 10000 | 0 | 16 | 54 | 35 | 0 | |
| 20000 | 0 | 33 | 49 | 10 | | |
| 30000 | 0 | 50 | 43 | 45 | | |
| 40000 | 1 | 7 | 38 | 20 | | |
| 50000 | | 24 | 32 | 55 | | |
| 60000 | 1 | 41 | 27 | 30 | | |
| 70000 | | 58 | 22 | 5 | | |
| 80000 | | 15 | 16 | 40 | | |
| 90000 | 2 | 32 | 11 | 15 | | |
| 100000 | | 49 | 5 | 50 | | |
| 200000 | 5 | 38 | 11 | 40 | | |
| 300000 | 8 | 27 | 17 | 30 | | |
| 400000 | 11 | 16 | 23 | 20 | | |
| 500000 | 14 | 5 | 29 | 10 | | |
| 600000 | 16 | 54 | 35 | 0 | | |
| 700000 | 19 | 43 | 40 | 50 | | |
| ·800000 | 22 | 32 | 46 | 40 | | |

| Tabula mensium in annis communibus | 2ª | 1ª | Tabula mensium in anno bisextili 2ª | 1ª |
|---|---|---|---|---|
| Januar | 0 | 31 | 0 | 31 |
| Febru | 0 | 59 | 1 | 0 |
| Marc | 1 | 30 | 1 | 31 |
| April | 2 | 0 | 2 | 1 |
| Mai | 2 | 31 | 2 | 32 |
| Juni | 3 | 1 | 3 | 2 |
| Julii | 3 | 32 | 3 | 33 |
| Aug | 4 | 3 | 4 | 4 |
| Septem | 4 | 33 | 4 | 34 |
| Octob | 5 | 4 | 5 | 5 |
| Novemb | 5 | 34 | 5 | 35 |
| Decemb | 6 | ·5· | 6 | ·6· |

1392

| 4ª | 3ª | 2ª | 1ª | |
|---|---|---|---|---|
| 2 | 21 | 13 | 48 | Differentia xpi et chaucer |

procedes ex tabula conuersionis annorum pro anno ·1395· complet ultimo decembris ·4ª·

| 3ª | 2ª | 1ª |
|---|---|---|
| 21 | 32 | 3 |

tabula quadrata examinatur

| | | | | | | | |
|---|---|---|---|---|---|---|---|
| 5ª | 9ª | 8ª | 7ª | 6ª | 5ª | | |
| 4ª | 8ª | 7ª | 6ª | 5ª | 4ª | | |
| 3ª | 7ª | 6ª | 5ª | 4ª | 3ª | 2ª | 1ª |
| 2ª | 6ª | 5ª | 4ª | 3ª | 2ª | 1ª | m |
| 1ª | 5ª | 4ª | 3ª | 2ª | 1ª | m | |

Rx augmenti ad motum oct[illegible] in libro 1392 complet

~~[illegible]~~ [illegible] ascendes 24 tempi diebus est

| s | g | m | 2ª | 3ª | 4ª |
|---|---|---|---|---|---|
| 0 | 26 | 12 | 22 | 46 | 45 |

the radix date of 31 December 1392. The dates from the *Astrolabe* are both 'round numbers' in the sense that they are actually designed for simplicity and correspond to the vernal equinox and the winter solstice respectively. The dates in the *Equatorie* do not seem to be specially chosen in any such fashion.

The *Astrolabe* contains a quite different date in the possibly spurious sections 44 and 45 where a planetary radix of 31 December 1397 occurs, and calculations are also made for 1400. It is very difficult to find any indication for or against the possibility that these passages were part of Chaucer's original text. They do not seem to fit in with the general plan of the treatise—though, strangely enough, they would make an ideal introduction to the planetary tables contained in the Peterhouse manuscript. On the other hand, these two conclusions are equally likely to be the work of one of the later editors who seem to have attempted to augment and rearrange the conclusions so as to bring them into conformity with the stated plan of the five parts. It is just possible that there is some connection between the radix date of 1397 and the same date which occurs *passim* in the Middle English texts in Trinity College (Cambridge) MS. O. 5. 26. It is to be noted that if this radix does in fact conform to the date when these conclusions (44 and 45) were written, then it is an indication that the *Astrolabe* was released for copying before Chaucer's death; there is no evidence that it was known, used or copied by any person other than its writer, until later in the fifteenth century.

Although the *Equatorie* contains no reference to the 'litel Lowis' for whom the *Astrolabe* is written, a similar didactic style is employed, and the exposition is obviously intended for the amateur rather than the professional astronomer.

## 6. THE 'RADIX CHAUCER' NOTE

On folio 5v. there is a table for converting any number of years to the equivalent number of days, at the rate of $365\frac{1}{4}$ days to the year. To the right of this table, and written up to the inner margin, appears a note which mentions Chaucer by name, apparently as the author of the 1392 radix which is used throughout the *Equatorie*.

The note reads:

| | | | | |
|---|---|---|---|---|
| 1392 | | | | |
| $4^{a}$ | $3^{a}$ | $2^{a}$ | $1^{a}$ | deffe$^{a}$ X$\overline{\text{pi}}$ & R$^{x}$a |
| 2 | 21 | 13 | 48 | chaucer |

Now it is easy to verify that the figure 2, 21, 13, 48, does in fact give (in sexagesimal notation) the number of days in 1392 years, and a similar consideration is true of the note which seems to have been added just below at a later date, and gives the corresponding number for the year 1395 and explains 'prouenies ex tabula conuersionis annorum pro anno .1395. completis vltimo decembris $4^a$ $3^a$ $2^a$ $1^a$.'

2 21 32 3

The meaning of the numerals is thus clear, and it remains to interpret the rather contracted phrase containing the word 'chaucer' in some way compatible with this obvious meaning.

Taking the words one by one:

**deffe$^a$.** The omega-like superscript is the usual form of the supralinear letter 'a', compare $4^a$, $3^a$, etc., in the same note, standing for *quarta, tertia*, etc. The most reasonable expansion of this word would be *defferentia* or *defferencia*, where variation has occurred between an *e* and an *i* (see p. 138), for the word *differentia*, the difference.

**Xp̄i.** This is the usual abbreviation for *Christi*, Christ. In this text it is used as well in the special sense of 'Anno Christi', the year zero from which all astronomical calculations are numbered.

**&.** This again is a normal contraction, the ampersand for *and* or *et*.

**R$^x$a.** Unfortunately, one of the slots cut for the nineteenth-century binding has mutilated the top of the superscript 'x'; it is just possible to see the cross of the two arms of the letter, and so it is quite certain that the superscript is not a hook sign, concave downwards, which could be taken as a contraction for 'er'.

Similar examples of R$^x$a and slightly varied forms of it occur at the foot of many of the astronomical tables throughout the volume. The following variations occur: Rad$^x$, Ra$^x$, ℞, R$^x$a, r$^x$a, Rå, Rę, Ra, R$^x$. It is clear from the context of any of these other occurrences that the contracted form stands for 'radix' which is written in full in a precisely similar position at the foot of other tables in the volume.[1] The term 'radix' signifies the astronomical root, or date from which all calculations start in any particular piece of work. The value corresponding to 'Radix Christi' or to 'Radix ultimo decembris anno 1392 completis' is given in this manuscript at the foot of each table of planetary positions.

**chaucer.** The only doubtful point about this word concerns the reading of the 'two minim letter', *u* or *n*. The word can be either *Chaucer* or *chancer*. The second could refer only to the sign of the zodiac, Cancer, and elsewhere in the manuscript (and indeed in all other Middle English texts) this is invariably

[1] If *Radix* had not been written in full elsewhere one might have suggested alternative expansions of some of the abbreviated forms, e.g. R$^x$=Rex, R'a=regula, ℞=Ratio, Recipe, Rubrica, etc., but in any case none of these yields an intelligible meaning for the phrase in its context.

spelt *Cancer*. The reading 'Pri*mo* gra*d*u Chancer' was, at one time, suggested to me but this seems to have been based on an inadequate photograph of the manuscript and is incompatible with the orthography and sense of the passage. The first degree of Cancer has no connection with the number of days in 1392 years. A surname used in this manner is customarily not inflected, hence the slight flourish at the end need not be taken as an abbreviation for 'Chaucer*e*'.

Having examined the words separately, we now turn to the meaning of the whole phrase 'diff*erenti*a Chr*i*st*i* *et* Ra*dix* Chaucer', i.e. 'the difference between Christ and the radix of Chaucer', or better still 'The difference (in number of days) between (the year of) Christ and the (year of the) radix of Chaucer'. Such an interpretation would agree well with the figures giving this number of days, providing we allow the year 1392 to be referred to as the radix of Chaucer. In other words, if this note is to carry the meaning which appears most obvious, it claims that the radix of 1392, which is used throughout the tables and text, must also be the radix of Chaucer. It therefore seems to ascribe the tables and text of the *Equatorie* to Chaucer. In view of the vital character of this explicit note it is essential to examine all the possible objections to this interpretation:

(1) The reading is 'Primo gradu Chancer'; this has already been dealt with.

(2) The Chaucer referred to is not the poet, but some other likely contemporary of the same name. There is no evidence for the existence of such a person, with the possible exception of Lewis Chaucer, for whom the *Astrolabe* was written, and he can have been only 11 years old at the time of writing.

(3) The reading given above cannot be accepted because 'radix' should be inflected. It seems that many such technical words (grada, minuta, introitus) were left in the uninflected state when contracted in any customary form such as we have in R$^x$a.

(4) If the note is in the author's hand he would be unlikely to refer to himself by surname in this fashion. A personal pronoun is more likely, or at the most the given name would be used. Although this would seem a somewhat strong objection, there are other instances in which a medieval writer is cited by surname. For example in the library catalogue of the York Austin Friars, book titles occur in the form: Lectura Holkot, Sermones Waldeby, Conclusiones Herdby, Defensorium Ocham, Questiones Biland de anima, etc. Giraldus Cambrensis (who is believed to have died c. 1220) writes his books in the third person, and when he refers to himself, as he does frequently, it is as Giraldus. An even more striking parallel to the present case is Walter Map (twelfth century, second half), who refers to himself in *De nugis curialium* in

the phrase 'Map said...'. In the present note there might be good reason for not using a pronoun in the possible impiety of juxtaposition with the name of Christ. A pious writer might be unlikely to use any such phrase as 'The difference between (the year of) Christ and my radix'.

(5) The 1392 radix referred to is not that which occurs (coincidentally) in the *Equatorie*, but refers to some other work of Chaucer. This is perfectly valid, but unless the writer is mistakenly referring to the *Astrolabe* (where radix 1392 does *not* occur), it must suppose the existence of some unknown work by Chaucer using that radix, and well known to the writer.

In all, it would seem that the note of 'radix chaucer' implies the ascription of the *Equatorie* to that author. Again the evidence is not certain, for almost all the individual pointers may be explained away separately in one way or another.

## 7. COMPARISON OF HANDWRITING

If the text is an author's holograph, and that author is indeed Chaucer, then it should be possible to compare the handwriting with other known specimens of the poet's hand. Unfortunately, there are no certain holograph or autograph Chaucer documents known, in spite of the very considerable labour that has gone into the assembling of all relevant life records. The search was indeed so barren at one time that Furnivall wrote (*The Athenaeum*, no. 2405 (29 November 1873), p. 698), 'Every single original document drawn up or signed by Chaucer has disappeared from its proper place. Someone who knew the Records thoroughly has systematically picked out—probably scores or hundreds of years ago—all Chaucer's work from every set of Records, and either stolen them or tied them up in some bundle which may be among the unindexed Miscellaneous Records.'

The position has improved in the Record Office since Furnivall's time, but a setback to hopes occurred with the discovery by Professor Manly that the form of oath which Chaucer must have taken on his appointment to the office of Comptroller of the Customs *did* (contrary to previous belief) permit him to appoint a deputy to do all the writing of the counter-roll.

Some six possible documents have been proposed[1] as Chaucer holograph, but none of these has been generally accepted, and they are all in hands considerably more formal than that which occurs in the *Equatorie*. More recently Manly has reported[2] the discovery by Miss L. J. Redstone of a docu-

[1] R. E. G. Kirk, *Life Records of Chaucer*, Chaucer Society, vol. IV, p. 233, p. 251 note 2, p. 278 note 2, p. 325 note 1.

[2] J. M. Manly, *Times Literary Supplement*, 9 June 1927.

ment amongst the Exchequer (K.R.) Bills in the Public Record Office. The document is no. L. 56 in file E. 207/6/2 and reads as follows:

'Geffrey chauce[r] conterollour de le Wolkeye en le port de loundris p*ar* / lauis (^ & assent) du cou*n*seil *nost*re *sire* le Roy a constitut Ric*hard* baret de estre sou*n* lieutenau*n*t / en loffice auant dite de le xvj Jour de maii lan du Roy Ric*hard* le p*ri*mer / Juqz a sa reuenue a loundris'

*Translation.* 'Geoffrey Chaucer, comptroller of the Wool Quay in the Port of London, by the advice and assent of the Council of our Lord, the King, has appointed Richard Baret to be his deputy in the foresaid office, from the 16th May, 1 Richard (II) [i.e. 16 May 1378] until his return to London.'

This is dated only shortly before the known date of Chaucer's second (?) visit to Italy on diplomatic business, and Manly comments 'whether he wrote it or not, it almost certainly passed through his hands'.

Of all the specimens of late fourteenth-century English hand that have been examined in the course of this work it happens that the above-mentioned P.R.O. document bears most resemblance to the hand of the *Equatorie*. Certainly the document stands out very dramatically from all the other bills of similar date which are threaded on the string in that particular *file*. It is quite true that there are many features in common to all informal hands of this period, but in nearly every case dissimilarities are sufficiently widespread to make it quite clear that no other specimens which have been seen are likely to have been written by the same person who wrote Peterhouse MS. 75. 1 or the P.R.O. document.

Since the document is written in Norman French and the *Equatorie* in English (and Latin), it happens that the only word common to both texts is the name of Chaucer which occurs as the second word in the document, and also in the 'radix chaucer' note on folio 5v. of the Peterhouse manuscript. It is unfortunate that in the P.R.O. specimen the last letter ('*r*') of the name has been mutilated and almost lost by the making of a hole for the string which holds the *file* together; this specimen is also rather smaller than the other.[1] There is a striking similarity between the two signatures when these are seen side by side, enlarged to the same size. Some differences may be observed: the loop running from the initial *c* to the top of the *h* is so faint as to be almost invisible on the P.R.O. specimen, the tail on the same *h* is shorter in the Peterhouse specimen; the P.R.O. '*a*' is rather more squat in shape. All these differences are very insignificant, and might readily occur with a quill cut slightly differently in the two cases. Considering that 14 years separates the dates of

[1] Length from first '*c*' to '*e*': P.R.O. 12·6 mm.; Peterhouse 16·6 mm.; ratio 1·32 (*c.* 4/3).

writing, and considering also the very different appearance of many other hands of the period, such agreement is rather dramatic, especially when the eye is assisted by the superimposition of the two specimens (see below).

One must not, of course, accept such calligraphical evidence as any proof that the two writers are the same, and that therefore both specimens are Chaucer holographs. By itself it would be nothing more than a rather weak pointer; but since it has already been shown that there are other reasons for suspecting Chaucer's authorship, it provides some confirmation and makes it even less likely that the *Equatorie* is written by some secretary. It would be improbable that the same secretary or amanuensis should be employed in 1378 and again in 1392 after so many changes in position and fortune.

## 8. ASCRIPTION OF THE PARENT-TEXT

It has already been shown that the text of the *Equatorie* leans heavily on some text of ultimately Arabic origin, and is almost certainly a free adaptation of a Latin version. Since we have not been successful in finding a suitable source amongst existing Latin and Arabic manuscripts describing such equatorium instruments, it may be as well to outline the directions of inquiry which have proved negative, and summarize such information as is at present available.

Disregarding the doxological 'incipit', one may take the opening of the text as 'The largere þt thow makest this instrument...' or possibly 'Tak ther fore a plate of metal...', and use this characteristic as a means of identifying the parent-text. No suitable equatorium text has been found commencing with any reasonable Latin equivalent of the first phrase,[1] and though many instrument tracts commence in a similar fashion to our second alternative, none of them seems to bear any resemblance to the *Equatorie*.

[1] The actual forms of the Latin incipit for which search was made may be of some use: 'Hoc instrumentum quo maius (grandius) facies (feceris) eo maius (grandius) fiunt divisiones (partes principales) (praecipuae)' *or* 'Quo maius facies hoc instrumentum...' *or* 'Quo maius hoc instrumentum facies...', etc.

PLATE XII

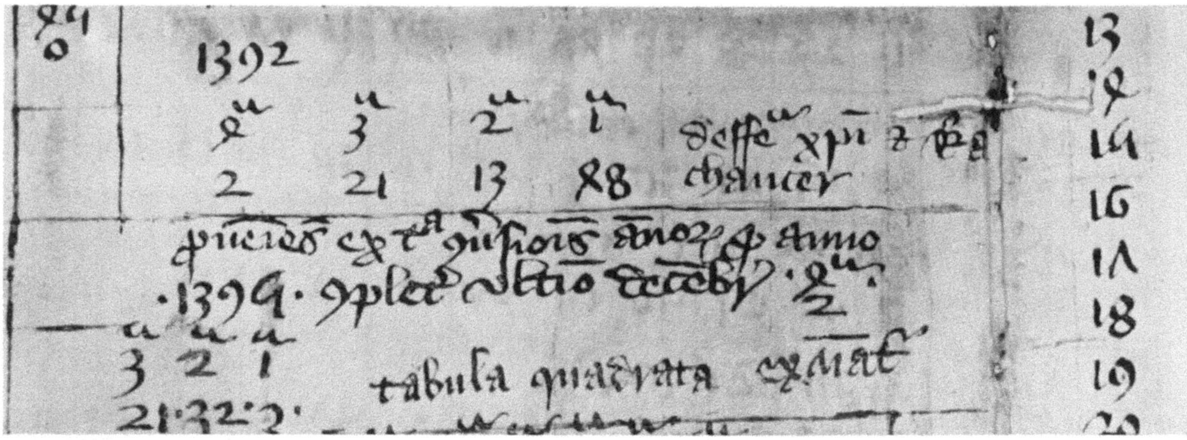

(*a*) PART OF f. 5v. 'RADIX CHAUCER' NOTE

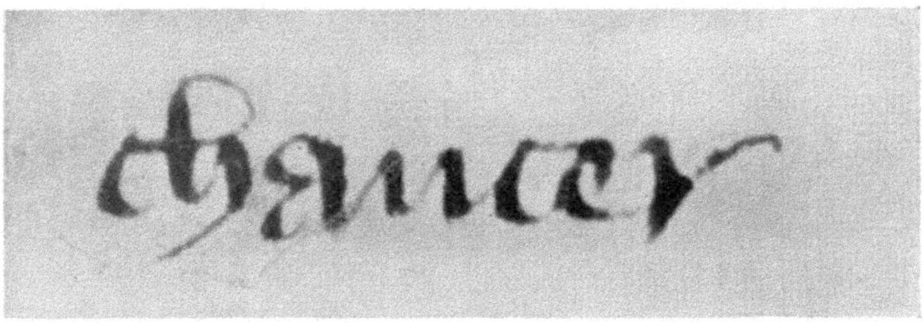

(*b*) NAME ON PETERHOUSE MS. (ENLARGED)

(*c*) NAME ON P.R.O. DOCUMENT (ENLARGED)

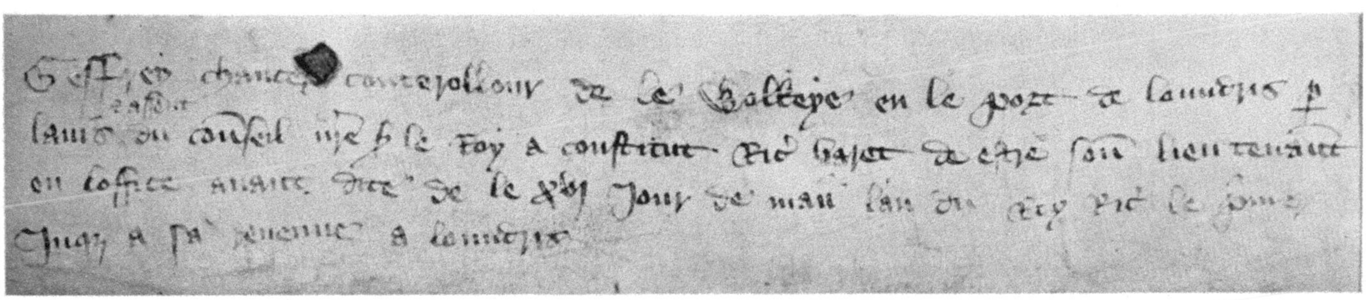

(*d*) P.R.O. DOCUMENT

PLATE XIII

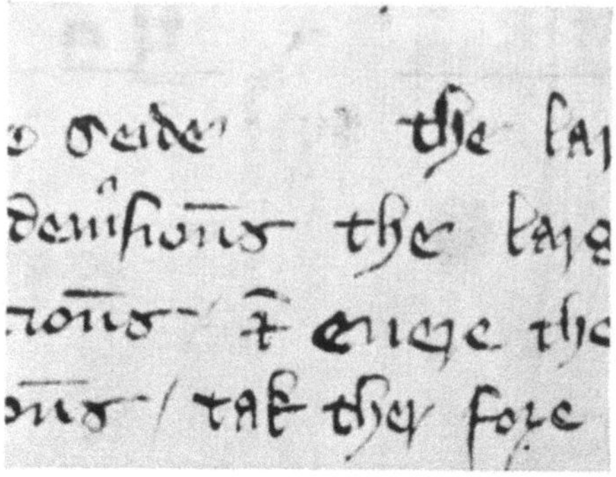

(*a*) FIRST LINE OF TEXT, FOLIO 71 V. WHITE LIGHT

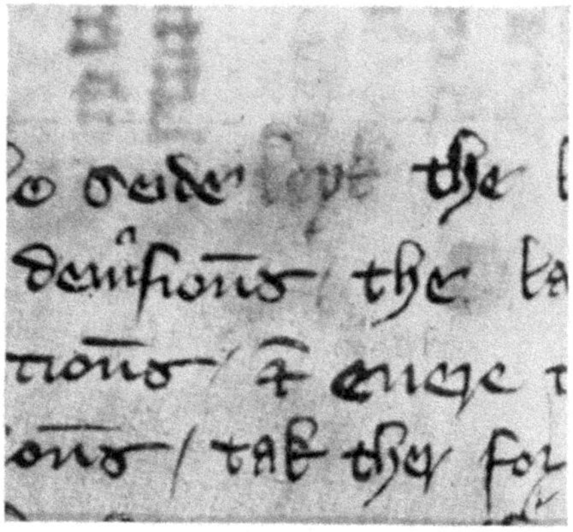

(*b*) THE SAME. ULTRA-VIOLET LIGHT

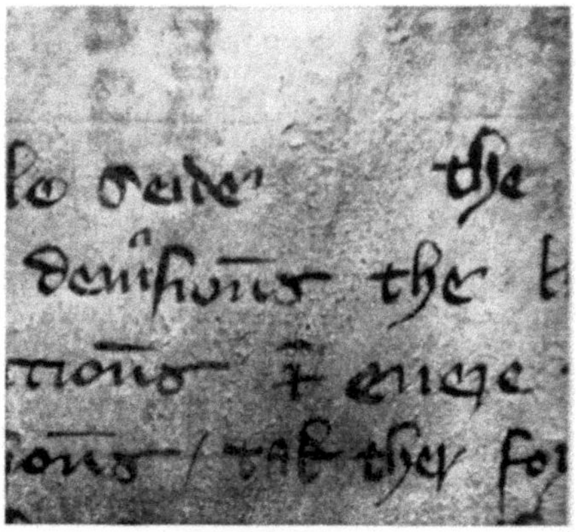

(*c*) THE SAME. INFRA-RED LIGHT

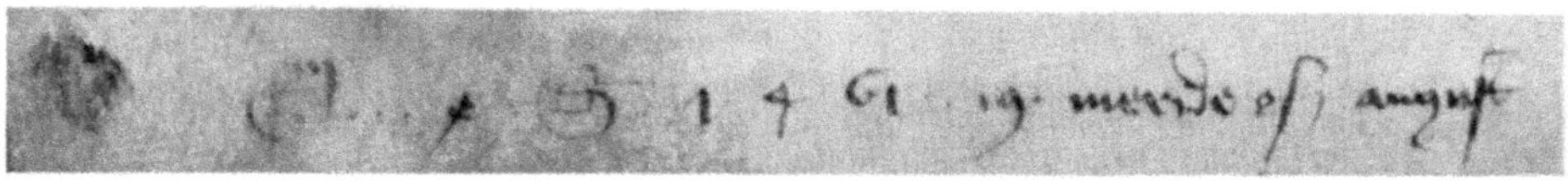

(*d*) OWNER'S MARK FROM 1461, f. 74 v.

*see p.* 8

Since the text itself is untraced, we turn to the instrument described. It seems reasonably certain that the form of instrument, and the constants used in its construction, indicate an origin not earlier than the work of John of Linières, *c.* 1272, and possibly subsequent to that of Richard of Wallingford, 1326. The *Equatorie* is certainly an improvement on the designs of John and Richard, and a very great advance on the original eleventh-century forms of equatorium. This dating would accord very well with the traditional ascription of the text to Simon Bredon (*fl. c.* 1340, d. 1372), and although we know of no extant equatorium text by this author, he may have given the Merton instrument to that College. On the whole this ascription is by far the most simple and satisfactory ascription, and it also explains why the Peterhouse manuscript should be associated with his name.

The matter might rest at this point, and the *Equatorie* be taken as a Middle English adaptation, possibly by Chaucer, of a Latin text by Simon Bredon, if it were not for the fact that the present text contains in its first line a different ascription. This first line contains a single word which has been erased by 'rubbing and scraping' and not replaced by any corrected version. The missing word may be seen very faintly on top of Plate XIII *a*, but this is because the photographic plate sees more than the naked eye; the effect may be considerably enhanced by ultra-violet illumination (Plate XIII *b*) which enables the word to be read without much room for doubt.

We read this word as 'leyk', or possibly as 'Leyk', since the initial letter seems somewhat larger than usual. Only the last letter is at all indistinct, but it agrees completely with the final letter of 'tak' just below in line A/4. At first there was some suspicion that the small circle visible on top of the initial 'L' might be a palaeographical contraction—though this would be most unusual in a proper name cited in this way. Fortunately, it is possible to investigate this by means of a photograph in infra-red light which shows up the surface texture of the parchment but not the remaining traces of ink (Plate XIII *c*); from this one can confidently say that the little circle is a flaw in the parchment and not part of the writing.

Having established the reading as 'seide Leyk' we must endeavour to identify the person so cited, and suggest a reason for erasure of the word. Most likely the word has been erased because of some error, rather than any deliberate attempt to conceal the original authorship, and the likelihood of such an error having been committed makes it very difficult to criticize adequately any interpretation of 'Leyk' which might be suggested.

Amongst Arabic authors, only Costa ben Luca (Qusṭā ibn Lūqā al-Ba'labakkī, died *c.* 912) has a name of which 'Leyk' is a possible form of transliteration;

but although he is known to have written one important work on the solid sphere, he seems much too early in date for a deviser of an equatorium.

Amongst European writers, there seems to be none of this name, but it is quite possible that the word is an error for 'Leyners', i.e. John of Linières—he is cited as 'Lyners' in at least one Oxford manuscript.[1] If it is intended for Linières, it is hard to reconcile this with any ascription to Bredon unless the text is Bredon's version of Linières's canon on an instrument of Arabic origin. This would make our Middle English version at least three stages removed from any original text, and the evidence seems far too tenuous to support or reject any such theory.

It may alternatively be suggested that 'Leyk' is a transliteration of some Arabic word which is not a proper name at all, though taken as such by a translator not completely at home in the language. One such possibility is that the word was originally the subjunctive form *leyaku*, used in a phrase equivalent to 'In the name of God, pitiful and merciful, *it is said that* the larger thou makest...'. Such a word, written without vowel points, might be mistaken for an unfamiliar proper name, but I regard this theory as providing a somewhat improbable explanation.

Professor O. Neugebauer has made the very interesting suggestion that 'leyk' might have originated in the mis-transliteration of an Arabic word read from *left to right* instead of in the correct order for that script. A likely word of the form required, ḳeyl, would be قيل (qīla) meaning 'it is said' [perfect passive form], and such a word might well have been misinterpreted by the translator from Arabic to Latin or by some other person adding a gloss to 'seide' or to 'dixit' at some later date.

[1] Bodleian MS. Digby 168, f. 64v. See L. Thorndike, *History of Magic and Experimental Science*, vol. III, p. 261, n. 20.

# GLOSSARY

ALTHOUGH the glossary is intended to be fairly comprehensive, full references are not given for common words such as *&*, *I*, *the*, etc. Latin words occurring in the text as glosses, etc., are not normally included, but consistency in this has not been aimed at. It is not always possible to be certain that a word is Latin and not Middle English, and where such forms as *zodiac* appear by the side of *zodiacus* it seemed pedantic to gloss one and not the other. Whenever possible the etymology is given, and if it is noted as obscure the student should refer to *OED*. Old English words are normally cited in the WS. form unless it is clear that the form in the text could not be derived from this. Old Norse words are cited in the form found in *A Concise Dictionary of Old Icelandic* (Oxford, 1910), by G. T. Zoëga, but it should be remembered that Middle English loan-words from Scandinavian were usually borrowed from a more archaic form of the language than that represented in Zoëga, and from East, not West Norse. In some cases, so far as the form of the word is concerned, it could have been borrowed from either of two languages, and in such cases the two possible etymologies are both given, separated by a diagonal line, e.g. OFr./L., OE./L., etc. Initial *þ* appears in its regular place under *T*, and following *Te*.

**a**, *ind. art.*, a: A/4, 5, 8, 10, 16, 17, 18, 24, 27, 29, 31; B/1, 16, 21, 27, 29, 30; C/2, 2, 9, 9, 16; D/2, 11, 11, 12, 16, 17, 18, 21, 24, 27, 28, 31, 34, 34, 36; E/2, 11, 12, 25; F/3; G/2, 3, 7, 9, 13; I/24, 24, 37; J/10, 15; K/12, 14 [OE. *ān*].

**abowte**, *adv. and prep.*, round, about: A/21; B/36; D/29; E/46. **aboute**: K/6, 7 [OE. *ābūtan*].

**adde**, *inf.*, to add: I/29; K/2. **added**, *1 sg. pt.*: N/5, 22 [L. *addere*].

**after**, *prep.*, according to: C/12; G/36; H/13; I/1, 4, 9, 11; J/1, 19; K/31; L/3, 19, 32; M/29; N/10. **after þ^t^**, according as: A/21; M/5 [OE. *æfter*].

**agayn**, *adv.*, again, once more: J/18; M/6, 15, 23; N/35 [OE. *ongegn*].

**agayn**, *prep.*, against: J/9; L/17 [OE. *ongegn*].

**agayns**, *prep.*, against: J/26; L/15; N/31. **agains**: N/25 [OE. *ongegn* + *-es*].

**al**, *adj.*, all, the whole of: B/22; I/9, 35; L/1. **alle**, *pl.*: A/20; B/4, 32; C/12; D/3; E/4, 5, 7, 54; F/14. **al hol**, in its entirety: D/26 [OE. *(e)all*].

**alhudda.** See **lyne alhudda.**

**almenak**, *sb.*, astronomical tables: K/22; N/1, 17 [Med.L. *almanac*, from Sp.Arab. *almanākh*].

**also**, *adv. and conj.*, also, in addition: D/a.

**also...as**, as...as: E/10 [OE. *(e)alswā*].

**alwey**, *adv.*, always, all the time: C/1; E/5; I/4 [OE. *(e)alne weg*].

**amidde,** *adv.*, in the middle: C/19 [OE. *onmiddan*].

**amys,** *adv.*, in the wrong place: D/c [OE. *on+miss-an*].

**an,** *ind. art.*, an: B/7; C/3, 10; D/25, 27, 29 [OE. *ān*].

**and,** *conj.*, and: B/4, 10; E/8; G/4, 20; H/7; J/7; K/1; L/14 [OE. *and*].

**&,** *conj.*, and: A/1, 3, 5, 10, etc. [OE. *and*].

**another,** *adj.*, another: D/22; I/19; N/1 [OE. *ān+ōþer*].

**any,** *adj.*, any: G/4; I/32; N/37 [OE. *ānig, ǣnig*].

**anything,** *adv.*, to any extent, at all: M/2; N/33 [OE. *ānig+þing*].

**argument,** *sb.*, an angle measured round the epicycle of a planet: G/26, 34, 36; H/2, 9, 27; J/27; K/11, 13, 18, 20, 29, 30; L/30; M/1, 6, 12, 28; N/8, 23, 33. **argumentis,** *pl.*: G/1 [OFr. *argument*/L. *argumentum*]. See also **equacio(u)n (of the argument)**; **mene (argument)**; **verrey (argument)**.

**Aries,** *sb.*, the Ram, the first sign of the zodiac, from which all reckoning of position begins: D/8, 8, 19, 21; G/10; H/12, 30, 32; I/1, 3, 6, 27; J/1, 19; K/31; L/3, 19, 29, 31; M/2, 9, 25, 30, 32; N/2, 10, 12, 19, 27, 37 [L. *aries*]. See also **hed** (of Aries).

**ark,** *sb.*, a portion of the circumference of a circle, an angle: G/23, 27; H/7, 25, 27, 29, 30, 32; K/17, 19. **arkes,** *pl.*: C/18 [OFr. *arc*/L. *arcus*].

**aryn.** See **centre aryn.**

**as,** *conj.*, as, like: A/23; C/20, 22; D/5, 29, 37; E/6, 11; F/3; H/37; I/16, 18, 37; J/14; K/10, 16, 29; L/21; M/8, 13, 15; N/36. **a**: G/9; while I/26. **as...as,** as...as: D/26; E/20; G/33; J/3, 5; M/10. **as thus,** for example: L/2, 16 [OE. *(e)alswā*]. See also **also, riht, swich, wheras.**

**ascende,** *imp. sg.*, move upwards: M/14. **ascendinge,** *pr. p.*: M/19. **assending**: M/23 [L. *ascendere*]. See also **meridional, .7.trional.**

**astrelabie,** *sb.*, astrolabe, an instrument used to calculate the position of the stars, and for simple observation of altitudes: B/7; C/29; D/27, 29. **astrelabye**: A/16 [OFr. *astrelabe*/L. *astrolabium*].

**astrik,** *adv.*, crosswise: E/2 [OE. *on+strica*].

**at,** *prep.*, at (of place): B/25; H/18; K/13; L/15; N/13. **a**: C/5; (of time): K/24; to M/24; N/36; with K/16. **atte**=at the: C/21 [OE. *æt*]. See also **regard.**

**aux,** *sb.*, apogee, line of apogee and perigee. Position in zodiac of line joining earth, equant, and centre deferent: B/24, 25, 29, 37; C/2, 7, 9, 14, 17; G/22, 27, 28, 28, 30; H/26; I/4, 6, 8, 10, 15, 27, 30, 38; J/24; K/9, 10, 19, 19. **auges,** *pl.*: A/21; B/27, 27; C/13; E/51 [Med.L. *aux, augis*]. See also **lyne (of the aux), mene (aux), table (of auges), verrey (aux).**

**awey,** *adv.*, away, off: B/17; C/4, 24, 27, 32 [OE. *on weg, aweg*].

**bakside,** *sb.*, dorsum (of astrolabe): B/7 [OE. *bæc+sīde*].

**bakward,** *adv.*, in the reverse direction: J/9, 26; L/15 [OE. *bæc+weard*].

**barre,** *sb.*, a strip of metal: D/17, 18, 22, 23, 28, 30 [OFr. *barre*].

**be,** *inf.*, to be, exist: A/8, 9, 12, 18, 21, 30; B/6, 10; C/2, 9, 19, 22, 23, 26, 31, 32; D/4, a, 16, 19, 20, 25, 26, 26, 28, 31; E/8, 12; H/18. **ben**: A/3, 20, 29, 32; B/2, 5; C/27; E/10; H/12, 14, 16; J/34. **ben,** *3 pl. pr.*: A/2, 2, 22; B/4, 9, 23; D/37; E/18, 50, 53, 56; N/38. **be,** *3 sg. pr. s.*: A/5, 10, 16, 17, 27; C/20; D/b, 25, 31; E/16, 19; I/27, 31, 33; K/25; L/12; M/18; whether it be: D/34 34; L/26, 26. **yif so be þ^t^,** if: K/24; M/1 [OE. *bēon*].

**blake,** *adj.*, black: G/11, 12, 18, 21, 31; H/3, 4, 17, 20, 20, 26, 28, 30, 32, 36; J/11, 12, 20, 21, 23, 30, 32, 34; K/8, 15. **blak**: G/3; K/13 [OE. *blæc*].

**body,** *sb.*, body, mass: H/15; K/12 [OE. *bodig*].

**boistosly,** *adv.*, roughly, rudely: E/45 [Obscure].

**bord,** *sb.*, a piece of wood: A/5, 7, 9, 12, 18; E/14 [OE. *bord*].

**bothe,** *adj.*, both: C/15; E/17; G/21; L/27 [ON. *báði-r*].

**bownde,** *pt. p.*, provided with a rim: A/8 [OE. *bindan*: *bunden*].

**boydekyn,** *sb.*, bodkin, large pin: D/11 [Obscure].
**brede,** *sb.*, breadth, width: A/10; D/2, 17, 25 [OE. *brǣdu*].
**breke,** *3 sg. pr. s.*, may break: D/c; F/2 [OE. *brecan*].
**but,** *adv.*, only, but: A/27; B/14; H/29; I/36, 37 [OE. *būtan*].
**but,** *conj.*, but: C/21, 23; E/23, 63; F/16, 26; H/10, except I/20, 21. **but so be þ^t^**, unless: E/16, 19 [OE. *būtan*].
**by,** *prep.*, by, by means of: A/5, 6, 15, 16; C/11, 19, 24, 25, 31; D/21, 32; G/13, 29, 31; H/35; K/21; M/4, 14, 33; N/30. **bi**: N/29, 35; by the side of, near: B/31; C/3, 10; E/3; G/12; H/20, 35, 36; J/21; through: N/12, 27 [OE. *bi*].
**bycause,** *adv.*, because; **bycause þ^t^**, for the reason that: N/19, 25 [OE. *bi*+OFr. *cause*].
**byfore,** *adv.*, before, previously: N/36 [OE. *beforan*].
**byforn,** *adv.*, before, previously: A/23; I/16; L/22; M/16 [OE. *beforan*].
**bygynne,** *inf.*, to begin, start: I/20. **bygynnith,** *3 sg. pr.*: L/7. **bygynne,** *imp. sg.*: L/14 [OE. *beginnan*].
**bytwixe,** *prep.*, between: B/14, 23; C/14, 21; D/32; E/56; G/23, 27; H/7, 26, 27, 30, 32; I/3; K/17, 19. **bytwix**: I/6; M/22. **bitwixe**: B/15 [OE. *betwix*].
**calle,** *inf.*, to name, call: A/26; B/3 [OE. *ceallian*].
**Cancer,** *sb.*, the Crab, the fourth sign of the zodiac: D/9, 9, 12, 15, 23; F/4; H/26; M/3; N/32, 35, 37 [L. *cancer*]. See also **hed** (of Cancer).
**canon,** *sb.*, general rule, formula: I/g; K/i; L/30 [OE. *canon*/L. *canon*].
**Capricorne,** *sb.*, the Goat, the tenth sign of the zodiac: D/9, 10, 23. **capricone**: C/28 [L. *capricornus*]. See also **hed** (of Capricorn).
**caput,** *sb.*, i.e. Caput draconis, the ascending node of the Moon's orbit: K/27; L/12; M/17, 18, 18, 21, 25, 26; N/3, 7, 18, 36, 38. **capud**: D/8; M/22, 24. **caput draconis**: K/23, 26; N/2 [L. *caput*].
**cas,** *sb.*, manner, way: E/25 [OFr. *cas*].
**cauda,** *sb.*, i.e. Cauda draconis, the descending node of the Moon's orbit: L/13; M/13, 17, 22, 22; N/19, 21, 26, 39. **cauda draconis**: M/21 [L. *cauda*].
**centre,** *sb.*, centre, middle point: D/24, 27, 28, 31, 39; E/10, 21, 31, 40, 40; G/19, 25; K/6, 6; N/14. **centres,** *pl.*: D/36; E/55; J/16. **centris**: A/20; B/33; C/4 [OFr. *centre*]. See also **centre aryn, centre defferent, centre equant, centre** (of Epicycle), **comune centre defferent, table (of centris).**
**centre aryn,** *sb.*, a point at the centre of the Equatorie which represents the centre of the earth, see p. 64: A/26, 27; B/4, 8, 12, 16, 18, 20, 28, 30, 34, 36; C/1, 5, 6, 8, 12, 25, 28, 30; D/32; E/19, 32, 39, 46, 57; G/3; H/18, 22; L/6, 9, 24; M/7, 11, 33; N/14, 28, 36. **centre arin**: B/35 [L. *arin*, from Arab. *ozein*, from Ind. *ujjain*].
**centre defferent,** *sb.*, the centre of the main circular orbit of a planet's motion: B/21; C/4, 10, 20, 22; E/26, 47, 57; G/8, 17; H/16; I/17, 25, 35; J/13, 17. **centre D**: C/15, 16. **centre different**: G/5; I/32. **centre deffer**: I/34. **centres, -is, defferentes, -is,** *pl.*: C/11; E/53 [OFr. *centre*+L. *deferent-em*/OF. *deferent*]. See also **comune centre defferent.**
**centre equant,** *sb.*, a point near the centre of a deferent circle about which the angular velocity is uniform, see p. 99: B/33; C/3; G/3; J/14, 15. **centre E**: C/14, 16. **centres equantis,** *pl.*: H/17. **centris equantis**: E/53 [OFr. *centre*+L. *æquant-em*].
**centre** (of Epicycle), *sb.*, middle of the bar of the Epicycle where the label is pivoted: D/24, 31, 39; E/21, 31; G/19; K/6 [OFr. *centre*].
**cercle,** *sb.*, a round piece of wood or metal: A/10, 11, 11, 12, 13; D/2; a circle: A/24, 27, 28, 29, 30, 31, 32; B/1, 2, 3, 8, 16, 17, 21; C/16, 17, 18, 19, 23, 26; D/12; E/34, 36, 45, 48, 64; F/1; I/10, 11, 12, 13, 17, 21; J/12, 24. **cerkle**: I/38. **cercles,** *pl.*: B/3; D/5, 14 [OFr. *cercle*]. See also **cercle defferent, eccentrik (cercle) of the sonne.**
**cercle defferent,** *sb.*, a circle on which moves the centre deferent of a planet: J/12 [OFr. *cercle*].

**chef**, *adj.*, chief, main: A/2 [OFr. *chef*].

**circumference**, *sb.*, circumference: A/8, 12, 25; B/22; C/23; D/2 [OFr. *circonference*/L. *circumferentia*].

**clepe**, *inf.*, to call, name: A/13; *1 sg. pr.*: J/7. **cleped**, *pt. p.*: B/9, 11; C/29; D/16; E/65; G/25; H/9 [OE. *cleopian*].

**closere of the signes**, *sb.*, the innermost of the five circles on the limb and the Epicycle, the names of the signs of the zodiac being written between this and the next larger circle: B/3, 15, 18, 30; C/26, 30; D/b, 12, 14, 19, 20, 33; E/20, 28, 37, 66; F/1, 8, 20. **closere of the sygnes**: B/8. **closeres of the signes**, *pl.*: E/17 [AFr. *closere*].

**com**, *imp. sg.*, come, arrive: N/35. **come**: *2 sg. pr. s.*: M/6; N/35; *3 sg. pr. s.*: M/21, 22, 23 [OE. *cuman*].

**compas**, *sb.*, compasses: A/17, 23, 24, 26; B/11, 13, 20, 35, 36; C/1, 4, 5, 7, 8, 14, 24, 31; D/32, 35, 38, 39; E/2; G/13. **by compas**, as shown by compasses: A/6; by means of compasses: C/25 [OFr. *compas*].

**compas(s)ed**, *pt. p.*, described with compasses: D/5, 13; E/67; figured: E/8; F/13. **compaced**: E/14 [cf. OFr. *compas*].

**complet**, *adj.*, complete: E/7; the year up to its end on 31 December, see p. 63: A/14; B/24 [L. *completus*].

**composicioun**, *sb.*, composition, manufacture: D/1 [OFr. *composition*].

**compowned**, *pt. p.*, constructed: A/14 [OFr. *compondre*].

**comune**, *adj.*, common, ordinary: A/16 [OFr. *comun*].

**comune centre defferent**, *sb.*, a hole in the limb of the epicycle part of the instrument which serves as centre deferent for all of the planets: D/16; E/22; F/6; G/4, 7; I/24. **commune centre defferent**: D/c; F/2, 24; H/21; I/31; J/16. **commune centre different**: G/17; I/33 [OFr. *comun*].

**conclusioun**, *sb.*, calculation, example, problem: A/17. **conclusiouns**, *pl.*: A/4 [OFr. *conclusion*].

**conseile**, *1 sg. pr.*, advise: F/5 [OFr. *conseiller*].

**considere**, *imp. sg.*, calculate, note: I/2, 7; J/6; L/5, 7, 19. **considered**, *pt. p.*: H/8 [OFr. *considerer*/L. *considerare*].

**contene**, *inf.*, to contain, include: A/6, 11. **contiene**: A/19 [OFr. *contenir*/L. *continere*].

**Crist**, *sb.*, Christ: A/14; B/24 [OE. *Crīst*].

**crois**, *sb.*, cross: D/24; E/38. **croys**: A/16 [OFr. *crois*].

**crossith**, *3 sg. pr.*, crosses: H/33 [cf. OFr. *crois*].

**day**, *sb.*, day: G/35. **day by day**, day after day: M/4, 14; N/30. **day bi day**: N/35. [OE. *dæg.*]

**decembre**, *sb.*, December: A/14; B/25. **.10.bre**: B/24 [OFr. *decembre*/L. *december*].

**defferent**, *sb.*, deferent, the circular orbit of the centre of the epicycle in which a planet was conceived to move: B/21, 22; C/5, 10, 10, 15, 16, 20, 22; D/c, 17; E/22, 26, 47, 49, 58; F/2, 7, 25; G/5, 7, 8, 18; H/16, 22, 35; I/18, 23, 24, 25, 32, 36; J/12, 13, 17, 17. **different**: G/5, 17; I/32, 34. **defferentes, -is**, *pl.*: C/11; E/54 [OFr. *deferent*/L. *deferent-em*]. See also **centre defferent**; **(comune) centre defferent.**

**degre**, *sb.*, degree: B/6; M/33, 34; N/28. **degres**, *pl.*: A/31; B/1, 5, 9, 20, 24; D/4, 37; E/60, 62; G/34; J/28; K/29; L/24, 31; M/2, 10; N/4, 5, 5, 6, 10, 13, 15, 17, 18, 31 [OFr. *degre*].

**dep**, *adj.*, deep: C/2, 9; D/31 [OE. *dēop*].

**departith**, *3 sg. pr.*, divides: C/17 [OFr. *depart-ir*].

**descende**, *inf.*, to descend, pass downwards along: M/5. **discende**, *imp. sg.*; M/15. **descending**, *pr. p.*: M/20, 35. **descendinge**: N/15. **discendinge**: N/30 [OFr. *descendre*]. See also **meridional (discendinge)**, **.7.trional (descending).**

**descriuen**, *3 pl. pr.*, describe (a circle): H/37. **descriue**, *imp. sg.*: A/24; B/21. **descryue**: B/16. **descriued**, *pt. p.*: A/27; B/4 [OFr. *descrivre*].

**desirest**, *2 sg. pr.*, wish: G/6. **desired**, *pt. p.*: G/18, 35; I/35 [OFr. *desirer*].

**deuyde**, *inf.*, to divide: A/15; *imp. sg.*: B/7, 18; C/25, 27, 29; D/35. **deuyded**, *pt. p.*: A/30;

B/23; D/4; E/42; F/12. **deuided**: B/5, 6, 6, 10; C/22, 24, 27, 31. **deuyde**: A/29 [L. *dividere*].

**deuysioun**, *sb.*, division: B/12, 15. **deuisioun**: B/13, 14. **deuysiouns**, *pl.*: B/17; C/24, 31, 32. **deuisiouns**: A/2, 2. **diuisiouns**: C/27, 33 [OFr. *devisiun*].

**diametral**, *adj.*, diametral: A/15; radial, see p. 66: B/10 [OFr. *dyametral*].

**diametre**, *sb.*, diameter: A/6, 19; D/3, 26; L/28; M/9; N/11, 26. **dyametre**: A/11; L/1 [OFr. *diametre*]. See also **hole (diametre)**, **semydiametre.**

**difference**, *sb.*, the remainder left after subtracting one quantity from another: J/7, 7; K/28 [OFr. *difference*].

**direct**, *adj.*, direct. **direct w^t^**, at: M/16. **in direct of**, at: D/12, 15, 19, 20: F/3, 8, 21 [OFr. *direct*].

**disposicioun**, *sb.*, position: G/32 [OFr. *disposition*].

**distaunce**, *sb.*, distance: B/33; C/4; D/32, 33, 35. **distance**: M/18. **distaunces**, *pl.*: C/12 [OFr. *distance*].

**distaunt**, *adj.*, distant: A/30, 32; B/1 [OFr. *distant*].

**diuersely**, *adv.*, at different times: I/18 [OFr. *divers* + ME. *-ly*].

**doth**, *3 sg. pr.*, does: D/29 [OE. *dōn*].

**dowble**, *adj.*, double. **dowble signes**, measurements of sixty degrees each: B/26 [OFr. *duble*].

**downward**, *adv.*, downwards: N/35 [OE. *dūn* + *weard*].

**drawe**, *inf.*, to subtract: N/4. **draw**, *imp. sg.*, trace (a line): B/29; D/21, 22; move: J/10. **drow**, *1 sg. pt.*, subtracted: M/25; N/2, 7, 20 [OE. *dragan*].

**dwelleth**, *3 sg. pr.*, remains: B/14. **dwelde**, *3 sg. pt.*: I/14; M/29; N/8, 23 [OE. *dwellan*].

**dymli**, *adv.*, faintly: B/10. **dymly**: B/18 [cf. OE. *dimlīc*].

**eccentrik (cercle) of the sonne**, *sb.*, the off-centre circular orbit of the Sun: B/16; D/a; E/8, 13, 41; F/17. **eccentric**: E/35 [late L. *eccentricus*].

**eclips**, *sb.*, eclipse: N/37, 38. **eclipses**, *pl.*: N/38 [OFr. *eclipse*].

**ecliptik**, *sb.*, the great circle of the celestial sphere which is the apparent orbit of the Sun: L/10, 21; M/12, 35; N/16. **ecliptil**: N/30. **eclyptik**: L/25 [L. *eclipticus*].

**egge**, *sb.*, edge, rim: A/7 [OE. *ecg*].

**ek**, *adv.*, also, in addition: A/20; C/12, 16; G/22; I/2 [OE. *ēac*].

**elles**, *adv.*, else, otherwise: A/4, 6, 9; D/c, 34. **ellis**: A/12, I/20 [OE. *elles*].

**empeiryng**, *sb.*, harm, injury: B/31 [OFr. *empeir-er* + ME. *-ing*].

**enche**, *sb.*, inch: A/10; D/18, 25. **enches**, *pl.*: A/6, 11, 11, 19 [OE. *ynce*]. See also **large (enche)**.

**ende**, *sb.*, end, extremity: B/11, 13, 28, 28, 28; D/18, 20; F/22; G/32; H/1, 2, 18; I/4; J/8, 30; K/13; L/2, 4, 16, 18; M/3, 4, 20, 31, 31; N/9, 11, 24, 26, 34, 34. **end**: K/32. **endes**, *pl.*: L/27 [OE. *ende*].

**endelong**, *adv.*, from end to end, along: D/22; E/1 [OE. *ende* + *-long*].

**endith**, *3 sg. pr.*, ends, finishes: B/29; G/11; I/17; J/10, 28; K/32. **endid**, *pt. p.*: I/33 [OE. *endian*].

**ensample**, *sb.*, example, illustration: C/11; E/3; M/24; N/1, 16 [OFr. *ensample*].

**Epicicle**, *sb.*, part of the equatorie: D/1, a, 6, 7, 13, 19, 21, 24, 26, 31; E/1, 8, 21, 23, 31; F/1, 11; G/5, 8, 14, 15, 16, 19, 20, 22, 23, 26, 30, 31, 31; H/2, 8, 14, 22, 23, 34; J/22, 23, 24, 25, 26, 28, 29, 33; K/7, 9, 11, 12, 13, 20. **Epicle**: D/8; G/28; I/25 [L. *epicyclus*].

**epicicle**, *sb.*, the secondary circle of a planet's motion, see p. 100: D/37; E/5, 6; G/24; H/13; K/6. **epicicles**, *pl.*: E/4; H/37 [L. *epicyclus*].

**equacio(u)n**, *sb.*, the position or angular variation of a planet in the zodiac: F/28; G/4, 6; H/10; I/33, 35. **equacoun**: G/9 [L. *æquation-em*].

**equacio(u)n of (the) argument**, *sb.*, the anomaly caused by motion in the epicycle: G/26; H/9; K/18, 20 [L. *æquation-em*].

**equacion of (the) centre**, *sb.*, the anomaly caused by the eccentricity of the deferent: G/25 [L. *æquation-em*].

**equacioun of the 8^e^ spere**, *sb.*, variation of

position of the stars, precession, see p. 104: C/34 [L. *æquation-em*].

**equacion of latitudes,** *sb.*, variation of latitude of a planet (with respect to the ecliptic): M/7 [L. *æquation-em*].

**equacion of the sonne,** *sb.*, the anomaly caused by the eccentricity of the Sun's orbit: H/28 [L. *æquation-em*].

**equales,** *adj.*, *pl.*, equal: B/9, 19, 22; C/25, 32. **equals**: C/18, 30; D/35 [L. *æqualis*].

**equaly,** *adv.*, uniformly: K/6 [L. *æqualis* + ME. *-ly*].

**equant,** *sb.*, the equant centre, the point about which rotation is uniform: B/33; C/3; E/11, 55; G/3; H/14; I/23, 23; J/15, 15. **equantes,** *pl.*: C/12. **equantis**: E/53; H/17 [L. *æquant-em*]. See also **centre equant.**

**Equatorie,** *sb.*, an instrument for computing geometrically (i.e. equating) the position of the planets: A/13, 20; C/33; D/1, 7, 9, 15; E/7, 9, 30; F/9, 15 [L. *equatorium*].

**equedistant,** *adj.*, equi-distant, parallel: G/12, 13; H/19, 36; J/21; L/28; M/8; N/11, 26 [L. *æquidistant-em*].

**erst,** *adv.*, first: H/35 [OE. *ǣrest*].

**erthe,** *sb.*, earth: N/14 [OE. *eorþe*].

**euene,** *adj.*, exact: C/19 [OE. *efen*].

**euene,** *adv.*, exactly, directly: C/14, 18; D/7, 8; smoothly: A/5 [OE. *efne*].

**euere,** *adv.*, always: A/3 [OE. *ǣfre*].

**euerich,** *adj.*, every: B/4; C/31. **eueryсh**: C/26 [OE. *ǣfre ylc*].

**euermo,** *adv.*, always, ever: H/18; L/7, 26; M/10. **eueremo**: M/8 [OE. *ǣfre* + *mā*].

**euery,** *adj.*, every: D/10, 10, 11; E/6, 10, 11; G/37. **eueri**: B/6 [OE. *ǣfre ylc*].

**face,** *sb.*, face (of the equatorie), that part of the instrument on which are marked the lines of auges: E/9 [OFr. *face*].

**fallith,** *3 sg. pr.*, occurs: N/37 [OE. *f(e)allan*].

**fals,** *adj.*, incorrect: E/23; I/g; K/i [OFr. *fals*].

**faste,** *adv.*, close. **faste by,** near by: B/31; C/3, 10; E/3 [OE. *fæste*].

**fastnyth,** *3 sg. pr.*, fastens: D/30 [OE. *fæstnian*].

**fer,** *adv.*, far, distant: E/20; J/3 [OE. *feorr*].

**ferther,** *comp.*, further: A/30; B/1. **ferthere**: A/31; B/31 [cf. OE. *feorr*, *furðor*].

**ferthest,** *sup.*, furthest: A/25. **forthest**: A/28 [cf. OE. *feorr*, *furðor*].

**ferthe,** *adj.*, fourth: E/62 [OE. *fēorða*].

**figured,** *pt. p.*, sketched: E/44 [cf. OFr. *figure*].

**file,** *imp. sg.*, file, smooth: F/1. **filed,** *pt. p.*: D/b [cf. Ang. *fīl*].

**firste,** *adj. and adv.*, first: B/12; D/5. **first**: K/8; M/8; beginning (of one of the signs of the zodiac): M/3; N/32 [OE. *fyrst*].

**fix,** *imp. sg.*, transfix: G/17. **fixe**: H/21 [cf. L. *fix-us*].

**fix fot,** *sb.*, the fixed arm of the compasses: A/23, 26 [L. *fix-us* + OE. *fōt*].

**fix point,** *sb.*, the fixed point of the compasses: A/17; B/11, 34; C/5, 13; D/39. **fix poynt**: B/13, 15, 19; C/1, 8 [L. *fix-us* + OFr. *point*].

**folwynge,** *pr. p.*, following: B/27 [OE. *folgian*].

**fond,** *1 sg. pt.*, found: M/28; N/1, 12, 16, 27.

**founde,** *pt. p.*: I/23 [OE. *findan*; *fand*; *funden*].

**for,** *prep. and conj.*, for: C/3, 10, 10; D/1, 1; E/3, 63; G/9, 35; H/10; M/7; because of: A/9; B/31; in order to: E/15; as regards: B/32; as: C/34; to: I/37; **for to,** to: I/20, 21; K/5; because: A/20; C/2, 9; G/28; I/19, 33; K/28; M/7; in order that: A/7 [OE. *for*].

**forseid,** *pt. p.*, aforesaid, previously mentioned: A/28; I/6, 17; K/15. **forseide**: A/11; C/7; D/28, 35; G/17; J/11; L/30; N/6 [OE. *fore* + *sægd*].

**forth,** *adv.*, forward: G/20; L/1; M/4 [OE. *forþ*].

**fote,** *sb.*, foot (in measure): A/6, 11 [OE. *fōt*]. See also **fix fot.**

**fracciouns,** *sb.*, *pl.*, fractions, small parts: A/3, 3 [OFr. *fraccion*].

**fro,** *prep.*, from: A/28, 30, 32, etc. [ON. *frá*].

**from,** *prep.*, from: I/12 [OE. *from*].

**ful,** *adj. and adv.*, full: C/19; very, too: D/25 [OE. *full*].

**Geminis,** *sb.*, *pl.*, the Twins, the third sign of the zodiac: F/23; M/4, 15. **Gemini**: N/34, 37 [L. *gemini*].

**general,** *adj.*, invariable: B/32 [OFr. *general*].

**generaly,** *adv.*, as a general rule: I/30; L/26 [OFr. *general* + ME. *-ly*].

**geometrical,** *adj.*, geometrical: A/17 [L. *geometric-us* + *-al*].

**glewed,** *pt. p.*, glued: A/13. **glewed $w^t$ perchemyn,** with parchment glued over it, see p. 62: A/9 [cf. OFr. *glu*].
**God,** *sb.*, God: A/1 [OE. *god*].
**goth,** *3 sg. pr.*, goes: B/8; C/28 [OE. *gān*].
**$g^a$,** *sb., pl.*, gradus, i.e. degrees: B/26, 34, 35, etc. **$g^ad$**: N/1 [L. *gradus*].
**grauen,** *pt. p.*, marked, engraved: G/33; H/4; I/12 [OE. *grafan*; *grafen*].
**grene,** *adj.* (outlined in) green: G/35 [OE. *grēne*].
**grete,** *adj.*, great: C/22; G/11, 22; H/24 [OE. *grēat*].
**gretter,** *comp.*, greater: I/38. **grettere**: D/16 [OE. *grēatra*].
**grettest,** *sup.*, greatest: L/32. **gettest**: M/19 [non-WS. *grētost*].
**happith,** *3 sg. pr.*, happens: I/18; J/14 [cf. ON. *happ*].
**haue,** *inf.*, to have, possess: D/6, 14; G/4, 6, 9; *1 sg. pr.*: A/23; I/16; L/21; M/16; N/36. **hast,** *2 sg. pr.*: F/6; K/11, 30. **hastow**: B/3; C/33; D/35; E/7; I/23; J/16; M/7. **hath,** *3 sg. pr.*: E/5; F/16; H/13; I/36; M/17. **han,** *3 pl. pr.*: H/17. **haue,** *2 sg. pr. s.*: D/13 [OE. *habban*].
**hed,** *sb.*, head (of a nail): E/1 [OE. *hēafod*].
**hed,** *sb.*, beginning (of a sign of the zodiac): D/10, 11; (of Aries): D/8, 21; G/10; H/12, 30, 32; I/1, 3, 6, 27, 28; J/1, 19; K/31; L/3, 18, 31; M/2, 30; N/10; (of Cancer): D/12, 15, 23; F/4; (of Capricorn): C/28; D/23; (of Libra): D/22; L/15, 17; N/25, 31. **heuedes,** *pl.* (of Aries and Libra): L/29; M/9; N/12, 27 [OE. *hēafod*].
**hem,** *pron., pl.*, them: A/3; B/27; C/13; G/2; K/30 [OE. *heom*].
**hir,** *pron.*, her: K/6, 28, 30; M/1, 19 [OE. *hire*].
**hir,** *pron., g. pl.*, their: C/12, 13; G/1; H/17 [OE. *hira*].
**his,** *pron., g. sg.*, its: D/b; E/20; G/25, 26, 26, 27, 27; H/9, 16, 18, 19, 26; I/3, 7, 8, 10, 10, 12, 15, 15, 25, 28, 29, 30, 30, 38; K/18, 19, 19, 20. **hise** (before pl. noun): E/7 [OE. *his*].
**hole,** *sb.*, hole: C/2, 3, 9, 10; D/b, c, 13, 16; I/19; J/13, 14. **holes,** *pl.*: B/22, 23; C/19, 20, 21; I/36, 37 [OE. *hol*].
**hole,** *adj.*, whole, entire: I/15. **hoole**: I/9. **al hol,** altogether: D/26. **hole diametre,** entire diameter, i.e. as compared with semi-diameter or radius: A/6, 10, 19 [OE. *hāl*].
**honestyte,** *sb.*, trueness, i.e. even, level surface: A/10 [OFr. *honestete*].
**how,** *adv.*, in what way, to what extent: E/4; I/2, 7; L/5, 8 [OE. *hū*].
**hymself,** *pron.*, itself: D/6 [OE. *him+self*].
**I,** *pron.*, I: A/13, 23, 26, etc. [OE. *ic*].
**ilike,** *adv.*, alike, similar: E/18 [OE. *gelīce*].
**in,** *prep.*, in, within: A/3, 20, 22, 28, 30, 32; B/2, 25, 27, 33; C/2, 4, 29; D/12, 36, 38; E/11; G/2, 32, 34; H/6, 15, 25, 26; I/10, 12, 17, 18, 19, 23; J/13, 15, 33; K/22; N/1, 16, 34, 35, 37, 38, 39; into: A/15; B/5, 6, 8, 10, 19, 22; C/17, 20, 20, 21, 25, 27, 30, 31; D/4, 4, 4, 5, 5, 35; E/42; on: A/23, 28; B/4, 5, 7, 12, 13, 20, 28, 28, 35, 35; C/1, 2, 5, 6, 8, 8, 13, 14, 24; D/a, 8, 15, 19, 21, 22, 29, 39; E/1, 2, 4, 8, 29, 52; F/19; G/2, 3, 3, 6, 11, 12, 22, 23, 25, 26, 28, 30; H/2, 4, 8, 16, 16, 17, 18, 20, 28, 30, 31, 36, 37; I/1, 2, 5, 25; J/1, 8, 10, 21, 24, 25, 29, 31; K/10, 12, 13, 14, 16, 18, 20, 31; L/3, 5, 19, 23, 31; M/10, 15, 30; N/10; at: B/20, 28, 35; C/6; D/23; H/34, 34; I/4; J/14; L/18; M/19, 32; N/1, 2; as regards: D/2, 3, 3; E/24; F/13; H/11; L/7; M/26; N/15, 24, 25; around: A/24; of: C/34; according to: J/27; comprised in: K/30; by virtue of: A/1; round: I/13, 21; along: M/5, 14. **as in,** as is the case with: I/36, 37. **in direct of,** at: D/11, 15, 19, 20; F/3, 8, 21. **in distance,** distant: M/18. **in maner of,** like: A/8, 15; B/7; D/27; K/14. **in this maner(e),** thus: G/15; N/3, 21. **in the same manere,** similarly: D/4; J/2. **in swich a manere,** so that: D/28. **in proces of tyme,** as time goes on: A/20. **in signefyeng of,** to indicate: C/3. **in some thinges,** to some extent: H/11. **in stide of,** instead of: H/14, 16 [OE. *in*].
**innere,** *comp.*, further in: E/27 [OE. *innor*].
**instrument,** *sb.*, instrument, i.e. the Equatorie: A/2, 22; C/22; D/3; E/14; F/27; K/22. **instrment**: I/36; tool: B/30 [OFr. *instrument*/ L. *instrumentum*].
**into,** *prep.*, into: I/32; M/3; on: M/4 [OE. *intō*].

**is,** *3 sg. pr.*, is: A/5; B/5, 6, 21, 32, 34, etc. [OE. *is*].

**it,** *pron.*, it: A/5, 7, 23; B/37, etc. [OE. *hit*].

**item,** *adv.*, again: K/11 [L. *item*].

**Juppiter,** *sb.*, the planet Jupiter: G/1; H/10; I/26 [L. *Juppiter*].

**justli,** *adv.*, truly: F/7 [OFr. *juste*+ME. *-ly*].

**karte,** *sb.*, cart: A/8 [ON. *kart-r*/OE. *cræt*].

**kerue,** *inf.*, to cut, intersect: K/33. **keruyth,** *3 sg. pr.*: H/5, 24; I/10; J/12, 23, 32; L/28; M/9. **kerueth**: C/18. **kerue,** *3 sg. pr. s.*: G/19. **karf,** *3 sg. pt.*: M/32; N/13, 28 [OE. *ceorfan*].

**knokke,** *imp. sg.*, knock: E/26 [OE. *cnocian*].

**knowe,** *inf.*, to know, determine: K/21; **knew,** *1 sg. pt.*: M/34; N/14, 29 [OE. *cnāwan*].

**krooke,** *inf.*, to become crooked: A/7 [cf. ON. *krók-r*].

**label,** *sb.*, label, a narrow brass rule across the face of an astrolabe or equatorie: D/27, 29, 29, 30; E/1, 3, 4, 6; G/32; H/1, 4, 37, 38; J/30, 31; K/14 [OFr. *label*].

**large,** *adj.*, large: A/17; D/32. **large enche,** Saxon inch, i.e. 1·1 'modern' inches, see p. 62: D/18. **large enches,** *pl.*: A/6. **enches large**: A/19 [OFr. *large*].

**largere,** *comp.*, larger: A/1, 2, 2 [OFr. *large* +ME. *-er*].

**lasse,** *comp.*, less: K/25 [OE. *lǣssa*].

**last,** *sup.*, last: B/25; C/33; **laste**: A/14, 30, 32; B/1, 2. **laste of Geminis,** end of Gemini: M/4, 14 [OE. *latost*].

**laste,** *inf.*, to last, be valid: A/22 [OE. *lǣstan*].

**lat,** *imp. sg.*, cause: A/11; D/25; F/3 [OE. *lǣtan*].

**latitude,** *sb.*, latitude: K/21, 30; L/10, 10, 20, 25; M/1, 11, 17, 19, 34; N/8, 15, 23, 29. **latitudes,** *pl.*: M/7 [L. *latitudo*]. See also **equacion (of latitudes)**.

**latoun,** *adj.*, made of brass: D/24 [OFr. *laton*].

**lattere,** *comp.*, latter. **the lattere ende of Libra,** towards the end of Libra: I/4 [OE. *lætra*].

**lengthe,** *sb.*, length: D/25, 33 [OE. *lengþu*].

**Leo,** *sb.*, the Lion, the fifth sign of the zodiac: N/37 [L. *leo*].

**leste,** *sup.*, least. **atte leste,** at least: C/21 [OE. *lǣst*].

**lettere,** *sb.*, letter (of the alphabet): I/12, 13, 19, 21, 22 [OFr. *lettre*].

**leuel,** *sb.*, a level. **by leuel,** tested by a level: A/5 [OFr. *livel*].

**ley,** *imp. sg.*, lay, place: B/28; D/7; G/11, 12, 14, 19, 32; H/2, 23; J/20, 20, 23, 29, 30; K/32; M/3; N/34. **lei**: H/19. **leide,** *1 sg. pt.*: M/31; N/9, 24. **leid,** *3 sg. pt.*: K/8; *pt. p.*: H/18; K/14. **leyd**: J/34 [OE. *lecgan*].

**leyk**: A/1. See p. 165.

**Libra,** *sb.*, the Scales, the seventh sign of the zodiac: D/9, 9, 20, 22; L/15, 17, 29; M/9; N/12, 25, 27, 31 [L. *libra*]. See also **hed** (of Libra), **lattere (ende of Libra)**.

**lik,** *adj.*, like: D/3; F/15; equal: G/26 [OE. *(ge)līc*].

**likith,** *3 sg. pr.*, pleases: A/9 [OE. *līcian*].

**lippe,** *sb.*, lip, tongue (of metal): F/3 [OE. *lippa*].

**list,** *3 sg. pr.*, pleases: G/4, 9 [OE. *lystan*].

**list,** *conj.*, lest: D/b, c; F/2 [OE. *þy lǣs þe*].

**litel,** *adj.*, small: A/28; C/16, 17, 18, 19, 23; D/13, 16; E/48; F/3; H/29; I/10, 11, 13, 17, 21. **lytel**: I/12 [OE. *lȳtel*].

**loke,** *imp. sg.*, look, find: C/4; G/10; K/22.

**lok,** make certain: A/16 [OE. *lōcian*].

**london,** *sb.*, London: B/24. **londone**: B/25 [L. *Londinium*].

**long,** *adj.*, long: D/26, 34, 34 [OE. *lang*].

**lorn,** *pt. p.*, lost, wasted: I/35 [OE. *loren*].

**lye,** *inf.*, to lie, be placed: L/4, 18. **lyen**: L/27. **lith,** *3 sg. pr.*: F/26; L/6, 8, 23; M/8, 10. **lyen,** *3 pl. pr.*: G/13. **lie,** *3 sg. pr. s.*: D/8. **lay,** *3 sg. pt.*: H/36; M/32; N/11, 26. **laye,** *3 sg. pt. s.*: L/2, 16. **ligginge,** *pr. p.*: H/1 [OE. *licgan*].

**lymbe,** *sb.*, limb, graduated edge of instrument, perimeter of face and epicycle of equatorie: A/13, 15, 16, 25, 28, 29, 31, etc. **limbe**: H/36 [OFr. *limbe*/L. *limbus*].

**lyne,** *sb.*, line: B/8, 10, 11, 12, 14, 17, 18, 20, 29, 31, 35; C/6, 18, 24, 25, 28, 30, 34; D/21, 22, 33, 36, 39; E/42, 43; K/33; L/8, 20, 23; M/2, 5, 11, 14, 33; N/13, 28, 32. **line**: C/28, 29; D/38. **lynes,** *pl.*: A/15; E/50, 52. **lyne of (the) aux,** apse line, line of apogee: B/37; C/2, 7, 9, 14, 17; I/3, 10, 38. **lynes of auges,**

*pl.*: C/13; E/50 [OE. *līne*/OFr. *ligne*]. See also **lyne alhudda, meridional (lyne), midnyht lyne.**

**lyne alhudda**, *sb.*, the line from centre aryn to the head of Cancer on the equatorie: B/11, 12, 17, 18, 20, 35; C/6, 24, 25; D/33, 36, 39. **line alhudda**: D/38 [OE. *līne*/OFr. *ligne* + Arab. *al-ḥaḍiḍ*].

**maistow**, *2 sg. pr.*, can, are able: I/29; K/3. **may**, *3 sg. pr.*: A/9, 21, 22; D/29; *3 pl. pr.*: A/3. **myht**, *1 sg. pt.*: N/4 [OE. *mæg*; *meahte*].

**make**, *inf.*, to make, construct: C/11; D/2, 12. **maken**: E/4. **makest**, *2 sg. pr.*: A/1; determine: F/27. **mak**, *imp. sg.*: A/12, 26, 29, 31; B/1; E/2; J/10; write: C/3, 10. **mad**, *pt. p.* **mad equacion of**, worked out: M/7 [OE. *macian*].

**maner**(e), *sb.*, manner, fashion: D/4, 28; G/15; H/9; J/3; K/14; N/4. **in maner of**, in the same way as: A/8, 15; B/7; D/27 [OFr. *manere*].

**many**, *adj.*, many: G/33; K/29; L/5, 8; M/10, 11 [OE. *manig*].

**mark**, *sb.*, mark: J/10, 11, 31. **marke**: E/2; H/3; K/14. **markes**, *pl.*: H/37 [OE. *mearc*].

**marke**, *imp. sg.*, mark: B/9; C/1, 8; D/34. **marked**, *pt. p.*: H/1; J/29 [OE. *mearcian*].

**Mars**, *sb.*, the planet Mars: E/56; G/1; H/10; I/26 [L. *Mars*].

**may**. See **maistow.**

**me**, *pron.*, for me: N/8, 23 [OE. *me*].

**membris**, *sb.*, *pl.*, parts: E/7 [OFr. *membre*].

**mene**, *adj.*, mean, see p. 107. **mene argument**, mean argument (as distinct from true argument), see p. 100: G/34, 36; H/2; J/27. **mene argumentis**, *pl.*: G/1. **mene aux**, mean aux (cf. true aux), see p. 109: G/27, 28; J/24; K/10, 19. **mene motus**, mean motus (cf. true motus), see p. 100: G/10; H/12, 19, 21, 29, 33; I/2, 7, 8, 14, 15, 28, 29, 30; J/2, 3, 4, 4, 8, 18; K/1, 4, 17. **mene mot**: I/9; J/5, 6; K/2, 3, 4 [OFr. *men*].

**merciable**, *adj.*, merciful: A/1 [OFr. *merciable*].

**Mercurie**, *sb.*, the planet Mercury: C/14, 17; E/49; I/18, 27, 36, 38. **Mercurius**: C/15, 16; I/g, 2 [L. *Mercurius*].

**meridie**, *sb.*, noon: A/14; B/25 [L. *meridies*].

**meridional**, *adj.*, south, southerly (of the ecliptic): L/20. **meridional assending**, south ascending (from the ecliptic); M/23. **meridional discendinge**, south descending (from the ecliptic): N/30. **meridional lyne**, south line, line alhudda divided into 5 degrees: K/33; L/8, 20, 23; M/2, 5, 11, 14, 33; N/13, 28, 32 [OFr. *meridional*/Med.L. *meridionalis*].

**mesure**, *sb.*, measurement: A/7 [OFr. *mesure*].

**metal**, *sb.*, metal: A/4, 10, 19; D/2, 17; E/16 [OFr. *metal*].

**middel**, *sb.*, middle: A/18, 18, 23, 25; K/33; L/6, 8, 23; N/12, 27. **midel**: M/10, 32 [OE. *middel*].

**middes**, *sb.*, middle. **in middes of**, in the centre of: D/30 [cf. OE. *on middan*].

**midnyht lyne**, *sb.*, north line; line from centre aryn to head of Capricorn: C/30, 34; E/43. **midnyht line**: C/29 [OE. *midniht* + OE. *līne*/OFr. *ligne*].

**midwey**, *adv.*, midway: M/22 [OE. *midd* + *weg*].

**minut**, *sb.*, minute, the sixtieth part of a degree: B/29. **minutis**, *pl.*: N/5. **mynutis**: D/4; E/59. **mynutes**: A/29. [OFr. *minute.*]

**mi**[a], *sb.*, *pl.*, minutes: B/6, 20, 26, 34, 35; C/5, 6; D/37, 38; G/34; K/29; L/5, 8, 22, 25; M/10, 11, 24, 25, 26, 28, 29, 32, 33, 34; N/2, 5, 6, 6, 9, 13, 15, 17, 18, 22, 22, 24, 25, 28, 29. **m**[a]: M/31; N/1. **minuta**: N/10 [L. *minuta*].

**mo**, *adv.*, more: A/3, 3 [OE. *mā*].

**moche**, *adv.*, much: I/7; J/5. **mochel**: I/2, 9 [OE. *mycel*].

**moeuable point**, *sb.*, movable point of the compasses: A/24; B/12, 14, 20; C/6, 7. **moeuable poynt**: B/35, 37; C/1, 8, 15; E/2, 2 [OFr. *movable* + *point*].

**moeuyng**, *sb.*, motion: M/27 [AFr. *mov-er* + ME. *-ing*]. See also **verrey (moeuyng).**

**moeuyth**, *3 sg. pr.*, moves: K/6. **moeue**, *imp. sg.*: J/21. **moeued**, *pt. p.*: A/22 [AFr. *mov-er*].

**mone**, *sb.*, the Moon: B/21; C/20, 23; E/48, 58; F/28; G/37; J/2, 5, 6, 12, 13, 17, 19, 31, 33, 35; K/2, 3, 5, 18, 21, 23, 25, 28; L/10, 11, 14, 21, 25, 31; M/1, 12, 16, 24, 27, 28, 34; N/1, 3, 8, 9, 15, 17, 20, 21, 23, 29, 33 [OE. *mōna*].

**more,** *adj. and adv.*, more, greater: I/28; J/6; K/1; L/12. **mor**: N/20 [OE. *māra*].

**mot,** *3 sg. pr.*, must: D/6, 26, 30; E/10, 11; H/14, 16, 18. **moste,** *1 sg. pt.*: N/19 [OE. *mōt*; *mōste*].

**mot,** *sb.*, motion, motus, see p. 109: I/9; J/5, 6; K/2, 3, 4, 25; L/12 [L. *motus*]. See also **mene** (**mot**), **verrey** (**mot**).

**motus,** *sb.*, motus, see p. 109: G/1, 10, 21, 24, 24; H/12, 19, 21, 29, 31, 33, 35; I/2, 7, 8, 14, 15, 28, 29, 30; J/2, 3, 4, 5, 8, 18; K/1, 4, 17, 17, 23, 23, 27, 28; L/11, 13, 14; M/25; N/3, 3, 7, 8, 17, 18, 19, 21, 21 [L. *motus*]. See also **mene** (**motus**), **verrey** (**motus**).

**my,** *pron.*, my: M/24, 31, 32, 34; N/1, 9, 13, 15, 20, 24, 26, 28, 29 [OE. *mīn*].

**myn,** *pron.*, my: A/13; I/36; N/16 [OE. *mīn*].

**myshappe,** *2 sg. pr. s.*, err: E/24 [OE. *mis-* + ON. *happ*].

**nadyr,** *sb.*, point of circle at opposite end of diameter, diametrically opposite point: J/14 [OFr. *nadir*, from Arab *naḍīr*].

**name,** *sb.*, name: A/1; B/32. **names,** *pl.*: B/2; D/5; E/63; F/5 [OE. *nama*].

**narwere,** *comp.*, narrower, smaller: A/29, 31; B/1; E/64. **narwer**: A/27 [OE. *nearu*].

**nat,** *adv.*, not: A/7; C/21, 27, 32; D/a, b, 25; E/8, 15; F/1; H/22; I/21, 31; J/34; N/4 [OE. *nāwiht*].

**natheles,** *adv.*, nevertheless: E/23, 64; furthermore: F/26 [OE. *nā þē lǣs*].

**nayl,** *sb.*, nail: D/30. **nail**: E/1 [OE. *nægl*].

**nayled,** *pt. p.*, nailed, fastened: A/12, 18; D/28 [OE. *næglian*].

**ne,** *adv. and conj.*, not: C/21, 27, 32; D/a, 25; E/5; F/5; I/19, 21, 31; J/34; L/11; nor: A/7; H/13; I/21; N/5 [OE. *ne*].

**nedle,** *sb.*, needle: D/16; E/11; G/7, 17; H/21, 22; I/24, 30; J/18 [OE. *nǣdl*].

**ner,** *comp.*, nearer: A/4 [OE. *nēarra*].

**neuer,** *adv.*, never: L/11 [OE. *nǣfre*].

**newe,** *adj.*, new: D/36 [OE. *nēowe*].

**nexte,** *sup.*, next: E/60 [non-WS. *nēhsta*].

**ney,** *adv.*, near, close to: D/b; F/1 [OE. *nēah*].

**nis,** *3 sg. pr.*, is not. **nis but,** is only: H/29 [OE. *nis*].

**no,** *adj.*, no, none: B/31; D/16; F/5; M/17 [OE. *nān*].

**nombre,** *sb.*, number: L/9. **nombres,** *pl.*: A/32; D/4; E/61 [OFr. *nombre*].

**non,** *adj.*, none, no: E/5; F/16; H/13, 13; J/34 [OE. *nān*].

**north,** *sb.*, north: L/26 [OE. *norþ*].

**nota,** *imp. sg.*, note: A/25; B/2, 10, 24, etc. [L. *nota*].

**nothyng,** *adv.*, nothing: I/19 [OE. *nān* + *þing*].

**now,** *adv.*, now, at present: B/3; C/33; D/1; E/7; H/26; I/22; J/16 [OE. *nū*].

**of,** *prep.*, of, belonging to: A/1, 3, 4, 5, 7, etc.; as regards: B/23; E/18; concerning: E/16; K/5, 16; for: I/19; from: B/23; C/22, 23; I/7, 9, 15, 30; J/4; K/4, 27; L/13; M/26, 27; N/4, 5, 7, 21; in: A/6, 10; L/4; M/24, 25, 32; N/1, 2, 17, 18; made of: A/4, 8, 10, 12, 18; D/2, 17, 24; on: B/20, 24; D/39; possessing: D/2, 2, 17, 18; to: B/31 [OE. *of*].

**on,** *numeral*, one: B/28; D/18; I/18; K/32; L/2, 16; M/3, 31; N/9, 24, 34 [OE. *ān*].

**on,** *prep.*, on, upon: A/18; D/27, 34, 34; E/14, 15; G/9; I/32, 34 [OE. *on*].

**only,** *adv.*, only, solely: G/37; J/35 [OE. *ānlīc*].

**opposit,** *adj.*, opposite: I/13; M/21 [OFr. *opposite*].

**or,** *conj.*, or: A/4, 6, 9, 11, 12; C/20, 21; D/c, 11, 34; E/27; M/17 [OE. *ōþer*].

**other,** *adj.*, other: B/28; D/20; E/39; J/34; L/18; M/4, 31; N/11, 26, 34. **othre,** *pl.*: H/17 [OE. *ōþer*].

**ouer,** *prep.*, above, upon: A/18; J/23; across: M/2; N/32 [OE. *ofer*].

**ouerthwart,** *prep.*, across: D/22; L/1 [OE. *ofer* + ON. *þvert*].

**owt,** *adv. and prep.*, out. **owt of,** from: B/23; C/22, 23; I/6, 8, 15, 30; J/4; K/4, 27; L/13; M/26, 27; N/4, 5, 7, 21 [OE. *ūt*].

**owterest,** *sup.*, outermost: E/59 [OE. *ūtera* + ME. *-est*].

**owtre,** *comp.*, further out: E/27 [OE. *ūtera*].

**part,** *sb.*, part, portion: G/20 [OFr. *part*].

**parties,** *pl.*, parts, divisions: B/9, 9, 10, 10, 19; C/25, 26, 27, 30, 31; D/35 [OFr. *partie*].

**passith,** *3 sg. pr.*, exceeds: L/11; M/13; goes

beyond: M/17. **passith by,** intersects: N/12, 27. **passe,** *3 sg. pr. s.*: M/2. **passed,** *3 sg. pt.*: N/20. **passit**: N/33 [OFr. *passer*].

**perce,** *2 sg. pr. s.*, pierce: D/b; *imp. sg.*: B/22; D/12. **perced,** *pt. p.*: A/20; C/2, 9, 19; G/8 [OFr. *percer*].

**perchemyn,** *sb.*, parchment: A/9, 13. **percemyn**: D/34 [OFr. *parchemin*].

**perpetuel,** *adv.*, perpetually: A/22 [OFr. *perpetuel*].

**pitos,** *adj.*, pitiful: A/1 [OFr. *pitos*].

**place,** *sb.*, position: H/6, 7, 8, 25; J/32; K/15 [OFr. *place*].

**planete,** *sb.*, planet: B/29, 32; E/6; G/4, 6, 9, 10, 18, 23, 24, 35, 37; H/3, 6, 8; I/33, 35; J/35; K/10, 12, 12, 14, 16. **planetes,** *pl.*: A/21; D/17; H/11, 17, 37. **planetis**: B/32; C/12; E/5 [OFr. *planete*/L. *planeta*].

**plate,** *sb.*, a metal disc: A/4, 18, 19, 20, 21, 24, 25; metal strip at the rim of the equatorie: A/8; the face of the equatorie: G/6, 9, 11; H/5, 16; I/25, 32, 34; L/1 [OFr. *plate*].

**pol,** *sb.*, pole of the epicycle, centre where the label is pivoted: G/30; H/14, 23, 34; J/23; K/7, 9. **pool**: G/14, 16; J/22, 33 [L. *polus*].

**polised,** *pt. p.*, polished: A/5 [OFr. *poliss-*].

**possible,** *adj.*, possible: C/20 [OFr. *possible*].

**poynt,** *sb.*, point, sharp end: B/13, 15, 19, 35, 37; C/1, 1, 8, 8, 15; D/14; E/2, 2; prick: E/1. **point**: A/18, 24; B/11, 13, 15, 20, 34; C/5, 6, 7, 13; D/39. **middel poynt,** centre: A/25 [OFr. *pointe, point*]. See also **fix point, moeuable point.**

**precios,** *adj.*, precious: C/33 [OFr. *precios*].

**precise,** *adj.*, exact: E/19, 22, 28, 31 [OFr. *precis*].

**precisly,** *adv.*, exactly: E/18; L/31 [OFr. *precis* + ME. *-ly*].

**prikke,** *sb.*, a small hole: C/2, 2, 9; D/31, 31; F/18. **prikkes,** *pl.*: E/52 [OE. *pricca*].

**prikke,** *imp. sg.*, make holes for: C/13 [OE. *prician*].

**procede,** *inf.*, to go, proceed: I/22; N/30; *imp. sg.*: I/12; L/15 [OFr. *proceder*].

**proces,** *sb.*, process. **in proces of tyme,** as time goes on: A/21 [OFr. *proces*].

**proeue,** *imp. sg.*, test, prove: G/13. **proued,** *pt. p.*: A/16; F/6 [OFr. *prover*].

**proporcioun,** *sb.*, proportion. **by euene proporcioun,** evenly spaced: C/19 [OFr. *proportion*/L. *proportion-em*].

**puttest,** *2 sg. pr.*, put, place: G/29; H/15. **put,** *imp. sg.*: G/2, 4; I/23; J/15 [OE. *putian*].

**quantite,** *sb.*, amount, number: I/5, 14; J/7; L/20, 22, 24; distance: A/28 [OFr. *quantite*].

**quarters,** *sb., pl.*, quarters: A/15; B/5 [OFr. *quarter*].

**quykly,** *adv.*, properly: L/29 [OE. *cwicu* + ME. *-ly*].

**red,** *adj.*, red: E/36, 67. **rede**: F/1 [OE. *rēad*].

**regard,** *sb.*, consideration. **at regard of,** with respect to: K/16 [OFr. *regard*].

**rekne,** *1 sg. pr.*, reckon, calculate: I/37; *imp. sg.*: G/35; I/1, 4, 9, 11, 14; J/1, 2, 8, 18, 25, 27; K/30. **rekned,** *1 sg. pt.*: M/29; *pt. p.*: H/12; K/11 [OE. *(ge)recenian*].

**reknyng,** *sb.*, reckoning, calculation: I/21; K/32; L/7. **reknynge**: J/29; L/14; M/5 [OE. *(ge)recen-ian* + ME. *-ing*].

**remedie,** *sb.*, remedy, cure: E/25 [AFr. *remedie*].

**remenaunt,** *sb.*, remnant, remainder: B/26; H/11; J/8, 10. **remnaunt**: I/7, 14, 16. **remenant**: N/38 [OFr. *remenant*].

**remew,** *imp. sg.*, remove: H/22. **remewed,** *pt. p.*: H/33; I/31 [OFr. *remeuv-*].

**renspyndle,** *sb.*, pivot rod: D/11. **renspindle**: D/14 [ON. *renn-a* + OE. *spinel*].

**requerith,** *3 sg. pr.*, necessitates, requires: M/6 [OFr. *requer-*].

**rewle,** *sb.*, ruler: B/27; D/21, 34; strip (of metal): D/24, 26, 27; procedure, rule: B/32 [OFr. *reule*].

**riet,** *sb.*, probably the open-work epicycle portion of the equatorie, see p. 71: K/8 [L. *rete*].

**riht,** *adv.*, exactly. **riht as,** just as: K/10 [OE. *riht*].

**rownd,** *adj.*, round, circular: A/6, 19. **rownde**: A/7 [OFr. *round-*].

**sadly,** *adv.*, firmly: D/7 [OE. *sæd* + ME. *-ly*].

**same,** *adj.*, same: A/27, 31; B/14, 18; C/23, 25; D/2, 2, 4, 21; G/13, 20, 32; H/4, 31, 34;

I/3, 11, 13, 28; J/2, 13, 31; K/11, 24; L/24; M/30; N/6, 23 [ON. *sam-r*].

**Saturne**, *sb.*, the planet Saturn: B/25, 33, 37; C/7, 9; E/3, 3. **Saturnus**: B/24; C/2, 3, 5, 10, 11; D/37; G/1; H/10; I/26 [L. *Saturnus*].

**saue**, *conj.*, except: D/5; E/55; G/37; J/35 [OFr. *sauf*].

**Scorpio**, *sb.*, the Scorpion, the eighth sign of the zodiac: N/18, 22, 23 [L. *scorpio*].

**scrape**, *imp. sg.*, scrape. **scrape awey**, erase: B/17; C/24. **scraped**, *pt. p.*: C/27, 32 [ON. *skrapa*].

**se**, *inf.*, to see: C/17; D/13. **say**, *1 sg. pt.*, saw, realized: N/4 [OE. *sēon*].

**seccioun**, *sb.*, intersection: D/23; E/38. **seccions**, *pl.*, divisions: L/8 [OFr. *section*/L. *section-em*].

**2ª**, *sb.*, *pl.*, seconds, sixtieth parts of a minute: B/26 [L. *secunda*].

**sein**, *inf.*, to say, repeat: B/5; G/27; K/7; M/27. **seyn**: D/10; M/21, 26; N/9, 32. **seye**: C/18. **sey**, *1 sg. pr.*: G/7; I/14; K/5; L/30. **seye**: B/25; C/29; G/15; I/11; J/27; L/7. **seide**, *1 sg. pt.*: M/8; *3 sg. pt.*: A/1. **seid**, *pt. p.*: A/23; I/16; M/16. **last(e)**, **firste seid**, last, first mentioned: A/30, 32; B/1, 3; C/33; D/5 [OE. *secgan*].

**sek**, *imp. sg.*, seek, look up: B/27, 32; D/36; G/1 [OE. *sēcan*].

**semydiametre**, *sb.*, radius: B/9; D/36; **semydiametres**, *pl.*: E/4 [L. *semi-diameter*]. See also **diametre**, **hole (diametre)**.

**.7.trional**, *adj.*, northern (from the ecliptic): L/10; M/1, 19. **.7.trional ascendinge**, north ascending (from the ecliptic): M/18. **.7.trional descending**, north descending (from the ecliptic): M/20, 34. **descendinge**: N/15 [L. *septentrional-is*].

**seruen**, *inf.*, to serve, be used for: C/34. **seruyth**, *3 sg. pr.*: I/19, 21 [OFr. *servir*].

**sette**, *inf.*, to set, place: C/13. **set**, *imp. sg.*: A/17, 23; B/11, 19, 34; C/5; D/39; E/1; *pt. p.*: F/7 [OE. *settan*].

**shal**, *1 sg. pr.*, shall, must, will: E/25. **shalt**, *2 sg. pr.*: D/1, 12; G/7; I/20, 22. **shaltow**: A/15; C/11, 13, 17; D/13; I/28; K/2; L/13; M/5; N/30. **shal**, *3 sg. pr.*: A/6, 7, 8, 11, 18, 19, 30; B/5, 6, 10; C/2, 9, 19, 23, 26, 31; D/4, a, 14, 16, 19, 20, 25, 28; E/8; G/21; J/34; K/33. **shollen**, *3 pl. pr.*: A/20, 29; B/2; C/34; L/27. **shal**: A/32; C/21, 27, 32. **shold**, *3 sg. pt.*: L/4. **sholde**: L/18 [OE. *sceal*; *sceolde*].

**shape**, *pt. p.*, shaped; D/27 [OE. *scieppan*; *sceapen*].

**sharp**, *adj.*, sharp: B/30 [OE. *scearp*].

**shaue**, *pt. p.*, shaven, planed: A/5 [OE. *sceafan*; *sceafen*].

**she**, *pron.*, she: M/17, 17, 18, 18, 20, 20, 22, 23, 23 [See *O.E.D.*].

**shewe**, *inf.*, to show, indicate: G/21; I/20, 22. **shewith**, *3 sg. pr.*: E/6; F/3; G/34; H/20; K/9, 10, 15, 16. **shewed**, *pt. p.*: G/28, 30; H/35 [OE. *scēawian*].

**shortly**, *adv.*, in brief: B/6; K/5; finally: E/16 [OE. *sceort* + ME. *-ly*].

**signe**, *sb.*, sign (of the zodiac): D/10, 10, 10, 11; I/37. **signes**, *pl.*: B/2, 3, 5, 15, 19, 26, 30; C/26, 31; D/5, b, 12, 15, 19, 20, 33; E/17, 20, 29, 37, 63, 65, 66; F/2, 5, 8, 21; G/33, 37; H/13; I/1, 5, 10, 12, 29; J/1, 9, 19, 27; K/2, 26, 29, 31; L/3, 12, 16, 17, 19, 32; M/8, 13, 18, 20, 26, 27, 30, 30; N/10, 11, 20, 24, 25, 31. **sygnes**: B/8 [OFr. *signe*]. See also **closere of the signes**, **dowble (signes)**, **succession (of signes)**.

**signefyeng**, *adj.*, denoting: C/3 [OFr. *signifi-er* + ME. *-ing*].

**silk**, *adj.*, silk: E/12 [OE. *sioluc*].

**sixte**, *adj.*, sixth: E/33 [OE. *sixta*].

**sklat**, *sb.*, slate: G/2 [OFr. *esclat*].

**smal**, *adj.*, small: C/2, 9; D/31; E/10; slender: D/16. **smale**: A/3, 3; C/19; thin: E/50 [OE. *smæl*].

**smothe**, *adj.*: A/5 [OE. *smōþ*].

**so**, *adv. and conj.*, so, thus: B/14; E/6, 16, 19, 30; K/24; L/1; M/1, 4; N/35; in such a way: D/8; G/19; as: I/9; K/29; M/11 [OE. *swā*].

**softely**, *adv.*, carefully: B/36; C/6; H/23; J/22 [OE. *sōfte* + ME. *-ly*].

**some**, *adj.*, some: H/11 [OE. *sum*].

**somtyme**, *adv.*, sometimes: I/18, 19 [OE. *sum* + *tīma*].

**somwhat**, *adv.*, slightly, little: A/29, 31; B/1 [OE. *sum+hwæt*].

**sonne**, *sb.*, the Sun: B/17; D/a; E/5, 8, 13, 35, 41; F/18; H/12, 13, 15, 21, 25, 27, 29, 31; J/3, 4, 6, 8; K/1, 4 [OE. *sunne*]. See also **eccentrik (cercle) of the sonne, equacion of the sonne.**

**sothly**, *adv.*, in this case: C/21; in fact: G/28 [OE. *sōþ*+ME. *-ly*].

**sowded**, *pt. p.*, soldered: D/19, 20 [OFr. *souder*].

**sowth**, *sb.*, south: L/26 [OE. *sūþ*].

**space**, *sb.*, space, distance: A/28; B/23; D/38; E/59, 60, 61, 62. **spaces**, *pl.*: B/23; C/21 [OFr. *espace*].

**sparing**, *sb.*, saving. **for sparing of**, to save: E/15 [OE. *spar-ian*/ME. *-ing*].

**speken**, *inf.*, to speak: K/5 [OE. *specan*].

**spere**, *sb.*, sphere. **8e spere**, the Firmament, containing the fixed stars, auges, etc.: C/34. **9 spere**, the Crystalline, revolving once in 49,000 years, carrying with it all the others, see p. 105: A/22; B/26; H/6, 15, 25; J/33; K/16 [OE. *espere*].

**stide**, *sb.*, place. **in stide of**, instead of: H/14, 16 [OE. *styde*].

**stike**, *inf.*, to transfix: G/7; *imp. sg.*: I/24; J/16. **stikyd**, *pt. p.*: I/31 [OE. *stician*].

**stille**, *adv.*, firmly: C/1, 8; G/15, 16, 31 [OE. *stille*].

**stirte**, *3 sg. pr. s.*, shift: I/34. **stirt, styrt ouer**, *imp. sg.*, cross, proceed: M/2; N/32 [OE. *styrtan*].

**stonde**, *inf.*, to stand, remain: E/32; F/3. **stant**, *3 sg. pr.*: F/19; H/18. **stondith**: A/26; E/21. **stonde**, *3 sg. pr. s.*: E/19, 27. **stondinge**, *pr. p.*: C/1; G/14, 15, 31. **stonding**: C/7 [OE. *standan*].

**strechche**, *inf.*, to extend, proceed: K/33. **strechcheth**, *3 sg. pr.*: J/3. **streche**, *3 sg. pr. s.*: G/19 [OE. *streccan*].

**strid**, *imp. sg.*, extend: C/15 [OE. *strīdan*].

**strik**, *sb.*, mark: E/3 [OE. *strica*].

**subtil**, *adj.*, small, fine: B/22 [OFr. *subtil*].

**subtili**, *adv.*, skilfully: A/11 [OFr. *subtil*+ ME. *-ly*].

**succession**, *sb.*, order. **after succession of signes**, anticlockwise (round the zodiac): I/1, 9; J/1; L/3, 32. **after successioun of signes**: G/36; H/13; I/5, 11; J/19; K/31; L/19; M/29; N/11. **agayns succession of signes**, clockwise (round the zodiac): N/31. **agains succession of signes**: N/25. **agayns successioun of signes**: J/26; L/15. **agayn successioun of signes**: J/9; L/17 [OFr. *succession*/L. *succession-em*].

**suffisaunt**, *adj.*, enough, sufficient: D/6, 18 [OFr. *sufficient*].

**suppose**, *1 sg. pr.*, suppose, assume: L/2 [OFr. *suppos-er*].

**sustene**, *inf.*, to support: D/6 [OFr. *susten-ir*].

**swich**, *adj.*, such: D/22, 28; G/9. **swiche**: D/37 [OE. *swilc*].

**table**, *sb.*, a tabulated list. **table of auges**, table showing position in zodiac of auges of each of the planets at any given era: B/27. **table of centris**, table showing distances of centre equant and centre deferent from Aryn for each of the planets severally: B/33; C/4. **table of centres**: D/36. **tables**, *pl.*: G/2, 35 [OFr. *table*].

**tak**, *imp. sg.*, take: A/4, 10, 17, 22; B/16, 27; C/4; D/7, 11, 17, 24, 32, 38; G/18; H/5, 19, 24; I/17; J/23, 24, 32; L/9, 32 [ON. *taka*].

**Taurus**, *sb.*, the Bull, the second sign of the zodiac: N/37 [L. *taurus*].

**teche**, *inf.*, to show, teach: E/25. **techith**, *3 sg. pr.*: E/4. **tawhte**, *1 sg. pt.*: M/13 [OE. *tǣcan*].

**than**, *adv.*, then, next: B/36; D/35; G/18, 21; I/28, 29; K/2, 15; L/12, 18; M/7; N/37, 38 [OE. *þan*].

**than**, *conj.*, than: D/16; I/28; J/6; K/1, 25; L/12; N/20 [OE. *þan*].

**thanne**, *adv.*, then, next: A/10, 17, 22, 27, 29; B/7, 11, 17, 19, 22, 27, 32, 34; C/4, 5, 13, 16, 24, 28; D/7, 11, 13, 17, 21, 24, 32, 36, 38; E/30; G/2, 10, 12, 23; H/19, 21; I/5; J/4, 11, 15, 16, 22, 23; K/3, 12; L/5, 19; N/32, 36 [OE. *þanne*].

**þt**, *pron.*, the, that: A/21; B/21, 28, 28; C/2, 5; D/18, 20; E/2, 39; G/35; J/6; K/28, 28, 32; L/2, 4, 16, 18; M/3, 3, 31, 31; N/9, 11, 19, 24, 26, 34, 34. **that**: J/7, 13 [OE. *þæt*].

**þᵗ**, *rel. pron.*, that, which: A/5, 10, 14, 17, 27; B/5, 8; C/26, 28; D/12, 16, 30, 31, 37; E/1, 36; F/1, 19; G/6, 8, 27, 33; H/3, 15, 17, 18, 26; I/4, 10, 12, 14; J/7; K/7, 7; L/22, 28; M/9, 21, 26, 27; N/9, 12, 14, 27, 32. **that**: D/25. **þᵗ of**, of which: G/4 [OE. *þæt*].

**þᵗ**, *conj.*, that, so that: A/1, 2, 10, 25; B/2, 10, 24; C/11, 17, 18, 33; D/5, b, 6, 8, 13, 14, 29; E/5, 5, 8, 10, 13, 16, 17, 19; F/6; G/7, 13, 19; H/37; I/22, 27, 30, 35; J/33; K/1, 6, 24; L/6, 16, 22, 26; M/1, 16, 28, 34; N/4, 14, 25, 27, 37. **after þᵗ**, according as: M/5. **til þᵗ**, until: B/36; C/7; F/6 [OE. *þæt*].

**the**, *def. art.*, the: A/1, 1, 2, 2, 3, 4, 4, 6, etc. **þe**: K/33 [OE. *sē*, *þē*].

**the**, *pron.*, thee, you: A/9; E/25; F/5; G/4, 9; I/20, 22; M/13 [OE. *þū*, *þē*].

**thennes**, *adv.*, thence: M/15, 23 [OE. *þænne* + *-es*].

**theorike**, *sb.*, matter, subject: K/5 [Med.L. *theorica*].

**ther**, *adv.*, there, in that place: B/14, 16, 29; C/16; D/30; G/11; H/5, 24, 25; I/17; J/20, 24, 29, 32; L/9, 32; M/3, 31 [OE. *þǣr*].

**therfore**, *adv.*, therefore, for this reason: A/4. **therfor**: H/14 [OE. *þǣr* + *fore*].

**therin**, *adv.*, in it: C/9 [OE. *þǣr* + *inne*].

**thilke**, *pron.*, that same, the: B/7; C/24, 26; G/6; I/14; L/28; M/20. **thilk**: I/16 [OE. *þē* + *ilce*].

**thinges**, *sb.*, *pl.*, things. **in alle thinges**, in everything: D/3; F/14. **in some thinges**, to some extent: H/11 [OE. *þing*].

**this**, *dem. pron.*, the, this: A/1, 9, 11, 11, 12, 12, etc. **þis**: K/i. **thise**, *pl.*: B/9, 23; C/27, 31, 32, 33; E/54 [OE. *þēs*; *þis*].

**tho**, *pron. pl.*, the, those: A/2 [OE. *þā*].

**tho**, *adv.*, then: M/25, 28, 29, 32; N/2, 5, 7, 8, 9, 12, 19, 20, 23, 24, 27 [OE. *þā*].

**thorw**, *prep.*, through: D/13; I/31; K/8, 14 [OE. *þurh*].

**thow**, *pron.*, thou, you: A/1; D/1, b, 12, 13; E/24; F/6, 27; G/6, 7, 29; H/15; I/20, 22, 26; K/11, 30; M/6; N/35 [OE. *þū*].

**thred**, *sb.*, thread: E/12; G/3, 3, 11, 12, 12, 14, 16, 18, 21, 29, 31, 33, 36; H/3, 5, 15, 17, 17, 19, 20, 20, 23, 24, 27, 28, 31, 33, 34, 35, 36; I/24; J/11, 12, 15, 20, 20, 21, 23, 23, 23, 25, 30, 32, 34; K/8, 10, 13, 15, 32, 33; L/2, 4, 6, 9, 16, 23, 27; M/3, 8, 10, 31, 32; N/10, 13, 24, 26, 28, 34. **thredes**, *pl.*: G/13; I/26 [OE. *þrǣd*].

**thridde**, *adj.*, third: E/61; N/16 [OE. *þridda*].

**thus**, *adv.*, thus, in this way: A/22; B/16; F/26; I/8; L/2, 16; N/30 [OE. *þus*].

**thy**, *pron.*, thy, your: A/4, 16, 23, 25, 26, etc. **thi**: A/2, 24; B/5, 6, 13, 33, etc. [OE. *þīn*].

**thykke**, *adj.*, thick: D/25 [OE. *þicce*].

**thyknesse**, *sb.*, thickness: D/18. **thikkenesse**: D/6 [OE. *þicnes*].

**thyn**, *pron.*, thine, your: D/1, 15; E/7, 22, etc. **thin**: A/22; D/3, 7, 8, etc. [OE. *þīn*].

**til**, *conj.*, until: B/37; C/15; E/27; I/33; M/6, 18, 20, 22, 23; N/35. **til þᵗ**, until: B/36; C/7; F/6 [OE. *til*].

**to**, *prep.*, to, towards: C/28; D/b, 19, 20, 22, 23, 28; G/17; H/22; I/13, 29; J/11; K/3; M/6, 14, 21; N/6, 6, 22, 22. **lik to**, like: D/3; F/15; G/25. (preceding infin.): B/5; C/18; D/6, 10; E/4; G/6, 9, 27; I/20, 22; K/5, 7, 21; M/21, 26, 27; N/9, 32 [OE. *tō*].

**to**, *adv.*, too: D/b; F/1 [OE. *tō*].

**togidere**, *adv.*, together: D/30 [OE. *togǣdre*].

**told**, *pt. p.*, said: L/22; N/36 [Ang. *tellan*; *tald*].

**torne**, *inf.*, to turn, move: D/29. **turnyth**, *3 sg. pr.*: E/6. **turneth**: H/38. **torne**, *imp. sg.*: C/6. **turne**: B/36. **turned**, *pt. p.*: A/21 [OE. *turnian*/OFr. *tourn-er*].

**toward**, *prep.*, towards: I/13 [OE. *tōweard*].

**towche**, *inf.*, to touch: C/15; *3 sg. pr. s.*: B/37; C/7. **towched**, *pt. p.*: D/14 [OFr. *tuch-ier*].

**tretis**, *sb.*, treatise: C/29 [AFr. *tretiz*].

**trewe**, *adj.*, correct: A/16; in good order: A/17 [OE. *trēowe*].

**trewely**, *adv.*, accurately, truly: D/13. **treweli**: F/7 [OE. *trēowlice*].

**trowthe**, *sb.*, truth, accuracy: A/4 [OE. *trēowþ*].

**two**, *adj.*, two: J/16 [OE. *twā*].

**tyme**, *sb.*, time: A/21; K/24 [OE. *tīma*].

**vnder**, *prep. and adv.*, under, underneath: G/14, 29; H/23; J/22, 34. **vndir**: G/16 [OE. *under*].

**vnderstond,** *imp. sg.*, understand: L/29 [OE. *understandan*].

**vnto,** *prep.*, to, as far as: B/8, 18, 30; C/25, 30; G/20; L/1; N/31, 32 [OE. *un*+*tō*].

**vp,** *adv.*, upwards: G/20 [OE. *uppe*].

**vpon,** *prep.*, on, upon: A/12, 27; B/4, 11, 13; D/7, 8, 9, 9, 10, 11; E/28, 32; G/5; H/1, 3; J/30. **stike...vpon,** transfix: G/8; I/25; J/17 [OE. *uppe*+*on*].

**vpperest,** *adv.*, uppermost, furthest: G/20 [OE. *uppe*+ME. *-rest*].

**vpward,** *adv.*, upwards: M/14; N/30 [OE. *ūpweard*].

**varieth,** *3 sg. pr.*, differs: H/11 [OFr. *varier*].

**Venus,** *sb.*, the planet Venus: G/e, 1; H/10; I/27 [L. *Venus*].

**vernissed,** *pt. p.*, varnished: A/9 [OFr. *vernisser*].

**verrey,** *adj.*, true, actual: I/17. **verrey (argument)**: K/29, 30; L/30; M/1, 6, 12, 28; N/8, 23, 33. **verrey (aux)**: G/22, 27, 30; K/9, 19. **verrey (moeuyng)**: M/27. **verrey (motus)**: G/21, 24; H/31, 35; K/17, 22, 23, 27; L/11, 13, 14; M/25; N/2, 3, 7, 7, 17, 18, 20, 21. **(mot)**: L/12. **verre (mot)**: K/25. **verey (motus)**: K/27. **verre (motus)**: N/19. **verrey (place)**: H/6, 7, 8, 25; J/32. **verre (place)**: K/15 [AFr. *verrey*].

**Virgo,** *sb.*, the Virgin, the sixth sign of the zodiac: L/5; M/24; N/1, 6, 7, 37 [L. *virgo*].

**visage,** *sb.*, the face of the equatorie: C/33; D/1, 7 [OFr. *visage*].

**was,** *3 sg. pt.*, was: A/14; B/25, 26; H/35; I/15; M/24, 25, 34; N/15, 17, 18 [OE. *wæs*].

**wel,** *adv.*, well, certainly: E/44; N/4, 14 [OE. *wel*].

**werpe,** *inf.*, to warp: A/7 [OE. *weorpan*].

**wey,** *sb.*, way, fashion: I/22 [OE. *weg*].

**whan,** *adv.*, when: A/5; F/27; H/33, 36; I/8, 14, 31; K/8, 11, 13; L/30; M/12, 16, 17; N/33, 37 [OE. *hwanne*].

**whel,** *sb.*, wheel: A/8 [OE. *hwēol*].

**wher,** *adv.*, where: G/10; I/20 [OE. *hwǣr*].

**wheras,** *adv.*, where: A/26; B/29; H/4, 24; I/16; J/9, 11, 23, 31; K/32; L/6. **whereas**: J/28 [OE. *hwǣr*+ME. *-as*].

**wherfor,** *adv.*, therefore, consequently: I/37; N/14 [OE. *hwǣr*+*for*].

**which,** *adj. and pron.*, which: A/5, 18, 30, 32; B/2, 34; C/28; D/31, 33; G/29; H/29; I/22, 23; L/10; M/33; N/29. **whiche**: A/7, 28; B/9; D/18; G/14; H/20 [OE. *hwylc*].

**white,** *adj.*, white: G/12, 14, 16, 29, 33, 36; H/15, 16, 19, 23, 24, 28, 34, 35; J/20, 23, 23, 25; K/10. **whit**: G/3; I/24; J/15 [OE. *hwīt*].

**whom,** *pron.*, whom, to which: G/25 [OE. *hwām*].

**widnesse,** *sb.*, width: D/2; E/18 [OE. *wīdnes*].

**wirke,** *inf.*, to work, use: N/19; *1 sg. pr.*: N/25. **workest,** *2 sg. pr.*: I/26. **wirk,** *imp. sg.*: I/25; N/36. **wyrk**: M/13 [OE. *wyrcan*].

**w$^{t}$,** *prep.*, with, by means of: A/8, 9, 24; B/30; C/1, 8; D/34, 38; E/2; G/7, 17; H/21; I/24, 25, 26, 26; J/18; M/13; N/19, 25, 36; **with**: E/67; along with: E/7. **direct w$^{t}$,** at: M/17 [OE. *wiþ*].

**w$^{t}$draw,** *inf.*, to subtract: I/29; L/13. **w$^{t}$drawe**: K/3. **w$^{t}$draw,** *imp. sg.*: I/5; J/4; K/26; *pt. p.*: I/15. **w$^{t}$drawe**: I/8 [OE. *wiþ*+*dragan*].

**w$^{t}$in,** *adv.*, within: A/11 [OE. *wiþinnan*].

**wole,** *1 sg. pr.*, will, shall: A/13, 26; B/3; *3 sg. pr.*: C/15; E/30 [OE. *willan*; *wolde*].

**wot,** *1 sg. pr.*, know: E/44 [OE. *witan*; *wāt*].

**write,** *inf.*, to write: F/5. **writ,** *imp. sg.*: B/31; E/3; G/2; K/28. **writen,** *pt. p.*: A/32; B/2 [OE. *wrītan*].

**yer,** *sb.*, year: A/14; B/24 [OE. *gē(a)r*].

**yif,** *conj.*, if: A/9; C/20; D/13; E/24; I/27, 33; K/1, 24; L/11, 16; M/1 [OE. *gif*].

**yit,** *adv. and conj.*, yet, again: A/31; B/1, 18; C/25; I/4; J/18; L/29 [OE. *gīet*].

**yren,** *sb.*, iron: A/8 [OE. *īren*].

**zodiac,** *sb.*, zodiac: K/7. **zodiacus**: G/25 [L. *zodiac-us*].

# APPENDIX I

## CIPHER PASSAGES IN THE MANUSCRIPT

As there are certain instances of the solution of cipher passages where much doubt has been cast on the validity of the methods of solution, a summary analysis is here appended. Perhaps an outline of this method will also be of assistance to others encountering this most frequently used form of medieval secret writing. One may in fact doubt whether any more complicated system was ever used before the invention of special diplomatic methods towards the end of the fifteenth century.

The cipher which is used in five places in the large (Oxford) set of astronomical tables in the Peterhouse manuscript has all the appearances of a simple substitution cipher, i.e. one in which each letter of the message or plain text has been substituted by some cipher symbol which invariably corresponds to it. This is indicated by the 'word lengths' which seem to agree in length with a fairly normal sample of such lengths found in the English or Latin languages; it is also indicated by the range of frequencies of the twenty-three different symbols employed which agrees approximately with the similar range of letter frequencies in either of these languages.

It is readily seen that the most frequent symbol used is 'o', of which there are at least 50 % more than there are of any of the next most frequent symbols (2, b, U). Since the most frequent letter in English or Latin is 'e', it seems reasonable to suggest that this corresponds to the symbol 'o'—of course, pathological examples can easily be constructed with other letters occurring more frequently, but since the total amount of code writing is here quite large, one may assume that there is a fair statistical sample.

Next, it might be observed that a particular three-letter word, 'U6o', occurs more often than any other short and recognizable combination. Assuming that this does actually represent a three-letter word in plain text, we now have to find a common three-letter word ending in 'e'. If the language used is Latin this would be difficult—in that case one would expect to find a common three-letter word with 'e' at the beginning for *est*. With the English language there is no difficulty in suggesting tentatively that the word is *the*, and we thus obtain the symbol U$=T$, 6$=H$, as well as o$=E$.

Writing in these equivalents, we look for a word with only one letter unsolved; such a case is found in 'U6VU' = *TH-T*, from which we are led to suppose that V = *A*. A similar trial with 'U6V3' and 'U6V330' indicates that the unknown letter (which cannot be *T*, which has already been found) must be *N*, so 3 = *N*.

Next we take the words '23' and 'U623' (i.e. *-N*, *TH-N*) and find 2 = *I*. In the same way '21', 'U621' and 'U6210' (i.e. *I-*, *THI-* and *THI-E*) give 1 = *S*, and 'Ub' and 'b3' (i.e. *T-*, *-N*) yield b = *O*. Taking note of the words 'b8' and '8bθ', and seeing that 8 cannot represent *N* which has already occurred, we derive next the pair of equivalents 8 = *F* and θ = *R*.

The solution is now reasonably well advanced, and if no errors have been made each additional symbol may be indicated by a number of words that lack equivalents for that symbol only. Taking the symbol k as an example we note 10θk2U6, k2U6, U6bk, 6bkθ01, 6bkθ21 (i.e. *SER-ITH*, *-ITH*, *THO-*, *HO-RES*, *HO-RIS*), which shows that 'k' is the symbol used uniformly for the modern letters *U*, *V*, *W*. The solution continues in this fashion and yields the following equivalents taken in one of the possible orders in which they may occur: R = *D*, W = *L*, 9 = *G*, Λ = *C*, Э = *Q*, H = *M*, y = *P*, 7 = *K*, α = *B*, И = *X*, T = *Y* (or perhaps 3), and ⊙ = &.

This gives a complete solution of all the symbols used as follows:

| | | | | | | | | | | | | | | | | | | | | | |
|---|---|---|---|---|---|---|---|---|---|---|---|---|---|---|---|---|---|---|---|---|---|
| Plain text: | *A* | *B* | *C* | *D* | *E* | *F* | *G* | *H* | *I* | *K* | *L* | *M* | *N* | *O* | *P* | *Q* | *R* | *S* | *T* | *U* | *X* | *Y* | *&* |
| Cipher: | V | α | Λ | R | o | 8 | 9 | 6 | 2 | 7 | W | H | 3 | b | y | Э | θ | 1 | U | k | И | T | ⊙ |

In actual practice this alphabet was obtained by using only four of the five cipher passages, and later testing the agreement by decoding the remaining passage—it was, for purpose of record, that on folio 14r. The omission of this passage was by accident rather than design, but the decipherment of it as intelligible Middle English effectively removed doubt as to the possibility of any arbitrary factor in the solution of the other four passages.

The greatest difficulty experienced in the above solution, as indeed with the decipherment of any passage in a medieval document, is the actual identification and recording of the symbols used. An example of an actual error will make the point clear. In the note on folio 62v. the seventh and eighth words were first read as 'oИΛoHk1 9θ1RkkH', which gave after solution '*EXCEMUS GRSDUUM*'. It is not difficult to see that in the first word '11' has been read as 'H' and in the second 'V' with a faint 'right-hand' stroke has been read as '1'—the correct reading is *EXCESSUS GRADUUM*. In all the above analysis the forms of the cipher symbols have been only approximately rendered by the letter or numeral nearest to them in shape. As can be seen from any of the reproductions, this agreement is by no means perfect, and in fact many of the symbols do not agree with any reasonable contemporary letter or numeral or Greek character, while some are ambiguous and might be represented in two or more different ways.

Another factor which has been omitted in the above discussion is the existence amongst the code symbols of certain words which are left uncoded and written in ordinary (contracted) script—e.g. in the passage on folio 62v. we have at the end 'H23ko U6V330 equacō dierę'. There seems to be no readily discernible reason for the writer having left these few words in plain text—unless we take it as being due to a certain laziness or tiring with the task of encipherment, and allow that those few words would not help to disclose the secret message to any person not in possession of the key.

## POSSIBILITY OF A KEY-MNEMONIC

It will be remarked that the alphabet of cipher equivalents given above seems to be merely a random arrangement of special symbols. Neither the symbols nor their order shows any sign of a planned system, and if this is actually the case it would follow that any person to whom the writer intended his message to be transmitted would be obliged either to have a copy of the key as written above, or to have memorized the entire set of symbols. Such a demand on the memory, or such a committing to writing of the secret of the process, would be a rather unusual course to adopt. It is surely more reasonable to suppose that the writer had some system which could be transmitted or memorized in a less obvious and more permanent fashion.

Many examples of medieval cipher occur, most of them constructed according to such a system rather than to any random identification with symbols. A frequently observed system in scribal signatures was to substitute for the vowels alone, sometimes using numbers $\frac{a\ e\ i\ o\ u}{1\ 2\ 3\ 4\ 5}$ (or to prevent confusion of the 'i' with '1' perhaps $\frac{a\ e\ i\ o\ u}{2\ 3\ 4\ 5\ 6}$) or using the next adjacent letter $\frac{a\ e\ i\ o\ u}{b\ f\ k\ p\ x}$. An example of the numeral system occurs in MS. Royal 17. A. iii (folios 123r. and 124v.), and an example of the latter method occurs in MS. Royal 8. C. ix (folio 157v.).

An excellent example of a complete substitution cipher[1] comparable in many respects with that of the Peterhouse manuscript is to be found in one of the famous Roger Bacon ciphers. This particular cipher seems to have been

[1] I.e. a cipher in which *all* the letters of the plain text are given cipher equivalents.

first noted by Newbold[1] in MS. Sloane 1754 which contains the *Tractatus Trium Verborum* or *Epistolae Tres ad Johannem Parisiensem* which is discussed by A. G. Little, *Roger Bacon Essays* (Oxford, 1914), p. 398. The Sloane MS. 1754 might be thought to contain these three code sentences at the end of the sections of the *Tractatus* merely as a scribal addition, but similar sentences are to be found in the earlier (late thirteenth/early fourteenth century) manuscript Cott. Jul. D. v, folios 152–8, 160–4. The absence of corruption in these code sentences seems to indicate either a very careful scribe or the common knowledge of this particular cipher device.

Using the same technique as that employed for the Peterhouse manuscript code, the key alphabet (which uses ordinary letters throughout) is readily found to be:

| | | | | | | | | | | | | | | | | | |
|---|---|---|---|---|---|---|---|---|---|---|---|---|---|---|---|---|---|
| Plain text | A | B | C | D | E | F | G | H | I | L | M | N | O | P | R | S | T | U | Z |
| Cipher | B | Z | E | C | D | F | H | I | G | O | M | N | P | L | R | U | S | T | A |

and in this case it is clear from the pattern of the lower line that a definite pattern of arrangement has been used. It is impossible to assert the employment of any particular system, since so many plausible devices might be constructed to yield the same set of substitutions. Perhaps the most simple method would be to write the alphabet in rows of three letters in zigzag fashion starting at the middle of a block, so:

L O P
U T S
Z A B
E D C
G H I

where the letters which are not coded (F, M, N, R) and those which are not used (J, K, Q, W, X, Y) are omitted altogether from the block.

Either by writing the block of letters round a little cylinder (like a pencil) or by repeating the right-hand column on the left, we obtain the pattern

P L O P
S U T S
B Z A B
C E D C
I G H I

[1] W. R. Newbold [ed. R. G. Kent], *The Cipher of Roger Bacon* (Philadelphia, 1928), Plate IX, facing p. 68, facsimile of part of MS. Sloane 1754. We are not here concerned with the discussion of the Voynich manuscript which Newbold ascribes to Roger Bacon and decodes in a somewhat tortuous and complex manner. Both the ascription and the decoding have been subject to much criticism, and J. M. Feely, *Roger Bacon's Cipher* (Rochester, N.Y. 1943), suggests that the manuscript is written in a more simple substitution cipher of the common type. (Newbold gives a different solution for the Sloane 1754 cipher.)

and this may now be used for the cipher by taking the letter on the left of that intended when one wishes to code the message, and by taking the letter on the right of that written when one wishes to read the message so encoded.[1] The solution in this case is complicated considerably by the fact that the words are not preserved in units, but the code symbols are run together in such a way that recognition of likely or familiar words demands successful substitution of many letters previously.

Returning to the cipher in the Peterhouse manuscript, it will be noted that there is no obvious indication of an underlying regular scheme as has been found with the Roger Bacon cipher. Part of the difficulty in looking for such a scheme is caused by the use of a completely different set of symbols instead of the ordinary letters of the alphabet, since it is hard to determine whether these symbols are drawn from any one set of symbols (e.g. an outlandish alphabet) or a combination of such sets.

On the other hand, some sort of indication of a plan is pointed at by the use of symbols more artificial than the rest for the rarer letters K, Q, X, & which are usually omitted from a cipher system, especially if it has been constructed originally for messages in Latin rather than in English.

The remaining symbols seem to be a mixture of letters and numerals, but the symbols $\alpha$, $\theta$, and a W which is nearer in form to *omega* suggest that these and perhaps others are drawn from some medieval concept of the Greek alphabet.[2]

It is most unfortunate that this difficulty in determining the nature of the set of symbols prevents us from setting the plain text equivalents in any order other than their natural alphabetical sequence. There is good reason to expect that if this could be done, some simple scheme of formation for the code might become apparent. The scheme could be of the geometrical variety exhibited in the key to the Bacon cipher, or more probably it could be based

[1] For completeness we transcribe and decode the not very significant sentences as written in Sloane 1754:

(1) Explicit mzinsm & orhmsm mcnezdhsm rlicrh azdsn ze hlgznnc̄ de
magnum primum mendacium rogeri bacun ad iohannē
ozrhd' mendacium...
paric'

(2) Explicit vcrdhsm mcnezdhsm rlicrh azdsn ad $\overline{\text{frem}}$ hlgznnc̄ de ozrht Alk.
tercium mendacium rogeri bacun iohannē paris

(3) Explicit mzhst mcnezdhsm eisdem ut supra
maius mendacium

[2] Such alphabets, containing highly corrupted and sometimes 'invented' letters, are to be seen in the alphabets contained in manuscripts of Mandeville's *Travels* (cf. ed. M. Letts, Hakluyt Society, London, 1953). A puzzling note is introduced by a consideration of the letters and numerals which are *not* employed as symbols—there are no symbols which might be taken for any reasonable form of C, D, E, F, or G; similarly, there is nothing that looks like a number 5.

on some key-word or key-sentence. It was not uncommon to use a proper name or a family motto for such purposes, and hence the key, if it could be found, might throw light on the identity of the writer of these code passages. Since there is a good chance that the writer of the code is the same as the writer of the text and tables of the *Equatorie*, there may be a clue here as to the ascription of the *Equatorie*.

The type of key to be expected may be exhibited in the following partial attempts (which are by no means suggested as anything more than mere coincidence and artificial construction):

Cipher symbols: 9 8 7 6
Plain text: G F C H=*GeF*fray *CH*aucer

Cipher symbols: H o Λ b R o k i=(John Holbroke)
Plain text: M E C O D E U S=‘my code’

# APPENDIX II

# COMPOSITIO EQUATORII SECUNDUM JOHANNEM DE LINERIIS

(MS. Royal Library, Brussels 10124, f. 142v., following)

QUIA nobilissima scientia astronomie non potest sciri complete sine instrumentis debitis propter quod necessarium fuit componere instrumenta in ea. Composuerunt quidam ex eis antiqui multa et diversa, ut sunt saphea et astrolabium cum quibus sciuntur plura ex motibus firmamenti, et ut armille et trebetum et regule cum quibus verificantur loca stellarum fixarum, et chillindrum et quarta pars circuli quibus utebantur antiqui in accipiendo horas et umbras et solis altitudinem et hiis similia. Sed nullus philosophorum antiquorum composuit aliquid instrumentum ad verificandum loca planetarum sive stellarum erraticarum nisi nunc ultimo quidam vir bonus et deo benedictus Campanus nomine, qui composuit quoddam instrumentum valde necessarium per quod sciuntur loca planetarum vera et eorum stationes et retrogradationes. Sed eius compositio est magis tediosa propter multitudinem tabularum in eodem instrumento contentarum cum earum concavitatibus diversis. Et eciam propter magnitudinem hujus instrumenti eo quo de levi non potest deferri de loco ad locum sive de regione ad regionem. Quam propter ego Johannis de lineriis virtute divina et eius auxilio volo abbreviare eius opus ita quod in una superficie unius tabule tantum, et satis parue quantitatis possent omnes planete satis veraciter equari. Ita quod non indigemus alio instrumento neque alia equacione.

Nunc autem ad componendum huius instrumenti cum effectu redeamus. Accipe ergo in nomine domini nostri Jesu Christi tabulam unam auricalceam ad quantitatem unius pedis et si plus vel minus non est vis, que sit valde plana et polita ex utraque parte vel saltem ex una parte tantum, et ex utraque parte eleventur limbi aliquantulum ad modum membris astrolabii. In cuius medio faciam punctum d loco centri orbis signorum. Super quem describam circulum maiorem quam potero in limbo elevato, at sub eo faciam tres circulos sibi correspondentes ita quod primus tantum distat a 2° quod inter eos possint scribi nomina signorum 12, scilicet Aries Taurus etc. Et $3^{us}$ a 2° tantum quod inter eos possint scribi numerum graduum distinctorum per $5^{tam}$ et $5^{tam}$. Et $4^{tus}$ a 3° quod inter possint scribi divisiones graduum. Et totum istud faciam in limbo.

Sequitur istius tabule elevatio.

Deinde quadrabo istos circulos super punctum d orthognoaliter et dividam quamlibet $4^{tam}$ in $3^{es}$ partes equales pro inscriptione 12 signorum que sunt in universo circulo 12 spacia; et ulterius dividam quodlibet spacium in $3^{es}$ partes vel 6 ubi scribam numerum graduum sic 5 10 15 etc. Et ulterius partes dividam ita quod sint in unoquaque duodena partes 30 ex quarum ductu in 12 resultant 360.

Deinde in exteriori parte limbi scribam nomina signorum. In interiori vero divisiones graduum. In medio vero numerum graduum signorum. Et ista inscriptio erit loco circuli signorum. Deinde ab auge unius cuiusque planete ad oppositionem augis traham lineam per centrum d in instrumento isto ad ostendendum auges planetarum ut melius possim.

Preterea in hiis lineis invenies centrum equantis et deferentis et similia. f. 143 r. Et sunt 5 linee, una pro Saturno alia pro Jove $3^a$ pro Marte $4^a$ pro Sole et Venere quia unam habent augem et $5^a$ pro Mercurio. Luna vero non habet augem fixam sicut ceteri planete. Semper enim cum fuerit in coniunctione cum Sole et in oppositione a Sole est in auge, et quadris circuli a Sole in oppositione augis sui ecentrici.

Nunc igitur perveniam ad situationem centrorum ecentricorum et equantium. Accipe quandam tabulam planam valde et ligneam et in ea traham quemdam circulum ad quantitatem circuli predicti cum aliis circulis sibi concentricis ad similitudinem orbium signorum cum divisionibus et inscriptionibus signorum ut dictum est supra in materia auricalcea. Et sit centrum d ut supra. Deinde quadrabo istum circulum duabus dyametris sese secundum d centrum secantibus et protraham unam dyametrum extra orbem signorum ad quantitatem quam voluero et sit ista extremitas punctus t. Deinde dividam istam dyametrum inter d et t in 96 partes equales, quod faciliter facere possum primo in 24 partes et quamlibet istarum in $4^m$ quia quater 24 sunt 96.

De istis autem 96 partibus a puncto d versus t computa 12 partes et 28 $mi^{ta}$ pro quibus ponam dimidium unius et ubi veratur ponam notam c. Eritque centrum deferentis lune. Et centrum illius ecentrici sive deferentis move in circumferentia orbis signorum super d. Describam igitur circulum kc super centrum d qui erit circulus parvus quem describit centrum deferentis lune circa centrum orbis signorum ut predixi. Item a puncto c versus t accipiam de 96 partibus 60 partes et ubi terminabitur pone notam a. Eritque a aux deferentis lune et linea ac semidyametrum deferentis. Prorsus a puncto a versus t de predictis 96 partibus accipiam 6 partes et quartam[1] unius partis et ubi

[1] *For* tertiam?

terminatur ponam notam e. Eritque punctus e aux epicycli lune quando centrum epycicli est in auge ecentrici.

Remanebit igitur de predictis 96 partibus 17 partes et 12 mi[ta] inter e et t de quibus nil est curandum quantum ad compositionem istius instrumenti nec etiam ad instrumentum campani. Et sic perveniam ad centrum equantis et centrum deferentis et ad semidyametrum circuli parui in cuius circumferentia movetur centrum ecentrici circa centrum terre super punctum d et erit semidyametrum ecentrici et epicycli. Et hec est figura:

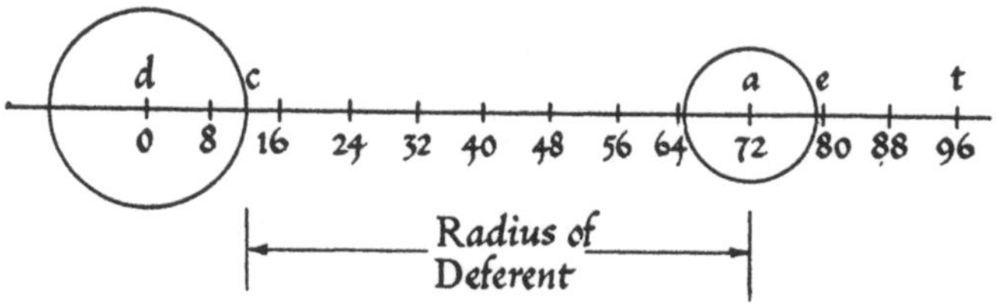


In mercurio vero dividam videlicet inter d et t in 32 partes et quamlibet iterum intelligam esse divisam in alias 3 partes et erunt in universo 96 partes ut in lunam. Accipiam ab d versus t unam partem ex 32 partibus et ibi pone k. Erit igitur k centrum equantis. Deinde a puncto k versus t sumam lineam kh que sit equalis linee dk. Eritque h centrum parui circuli quam describit centrum ecentrici motu suo versus occidentem uniformiter. Super ipsam autem h [f. 143 v.] describam circulum c k qui erit circulus paruus predictus. Sitque punctus c in dyametro predictam erit igitur c in centrum deferentis in horam in qua centrum epycycli fuerit in auge ecentrici. Rursus a puncto c versus t accipiam de dyametro predictam ac que sit 20 partes de illis 32 partibus. Eritque punctus a aux ecentrici et centrum epycicli et linea c a semidiametrum ecentrici. Deinde a puncto a versus t accipiam de predictam dyametro de 32 partibus 7 partes et medietatem unius partis et sit ibi e. Erit igitur e punctus augis epycicli et eius semidyametrum linea ae. Remanet igitur de 32 partibus predictis pars una et medietas partis, de quibus nil ad presens quantum ad nostram materiam. Sequitur figura:

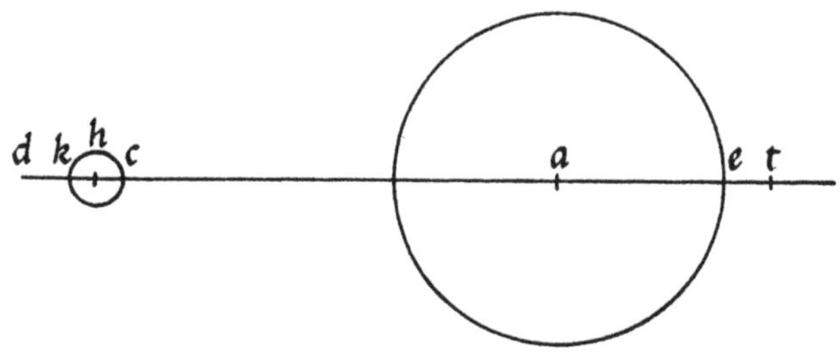

In venere autem et marte dividam lineam d t in 60 partes et sumam ex eis in venere a puncto d versus t unam partem et quartam partis et ubi terminabitur linea pone notam k. Eritque k centrum equantis. Dividam etiam lineam d k per medium sive puncto c, eritque c centrum deferentis. Sumam igitur ex ea a puncto c versus t 30 partes de 60 partibus et ubi terminabitur linea pone notam a. Eritque a aux deferentis et centrum epycicli veneris. Deinde a puncto versus t accipiam de predictis 60 partibus 21 partes et etiam $3^{am}$ et $4^{tam}$ unius partis. Et ubi terminabitur pone notam e. Eritque a e semidyametrum epycicli.

In marte autem sumam ex predictis 60 partibus a puncto d versus t 6 partes et medietatem partis que sit linea dk quam dividam per medium in puncto c. Eritque c centrum deferentis et k centrum equantis. Deinde a puncto c versus t accipiam 30 partes de hiis 60 partibus et ibi pone notam a. Eritque a centrum epycycli et aux deferentis sive ecentrici. Et ab a versus t sumam 19 partes et $3^{es}$ quartas unius partis de predictis 60 partibus sitque nota e. Eritque e aux epycicli. Et linea c a semidyametrum ecentrici. Et linea a e semydyametrum epycicli. Remanebuntque in venere de predictis 60 partibus 7 partes fere et in marte 4 partes minus $4^{ta}$ parte unius partis de quibus nulla est vis quantum ad compositionem huius instrumenti. Hec est figura omnium superiorum planetarum.

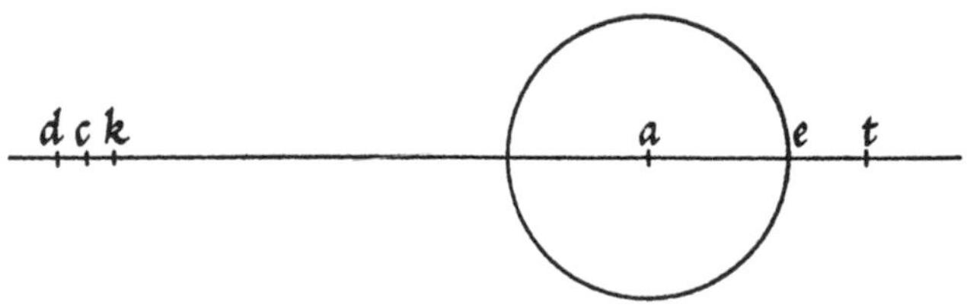


In Jove autem et Saturno dividam predictam lineam dt in 96 partes equales f. 144 r.
ut in luna et mercurio predictum est. Et pone lineam dk in Jove 5 partes et medietatem partis quam divide in medio in puncto c. Et lineam a c ponam 60 partes et lineam ae undecim partes et medietatem unius partis.

In Saturno vero pone lineam dk 6 partes et 5 sextas unius partis quam dividam per medium in puncto c. Et lineam c a ponam 60 partes et lineam a e 6 partes et medietatem partis. Erit igitur c a semydyametrum ecentrici eorum. Et linea a e semidyametrum epycicli istorum 4 planetarum videlicet Saturni Jovis Martis et Veneris. Remanebunt igitur in Jove 22 partes et in Saturno 26 partes fere. Exemplum istorum duorum patet etiam in figura precedenti.

In Sole vero dividam antedictam lineam dt in 25 partes de quibus sumam unam partem a parte d versus circumferenciam et ubi terminabitur pone notam k. Eritque nota k centrum deferentis solis. Et linea kt erit semidyametrum deferentis. Sed hic attende quod linea ista solis non excedit materiem sed extendit se a puncto d usque ad circulum signorum in materie cum puncto t. Et sic habes semidyametrum ecentricorum equantium et epicyclorum 6 planetarum. Sol vero non habet epycyclum sed ecentricum tantum in quo movetur uniformiter. Situabo igitur centrum equantium et deferentium planetarum in materie auricalcea predicta in lineis eorum directis a centro orbis signorum quod est d ad augem ipsorum in circulo signorum secundum composicionem quam inveni in tabula lignea. Centrum vero equantis et deferentis lune non ponam in materie sicuti in aliis planetis quia centrum deferentis lune et equantis movetur circulariter circa centrum d secundum quantitatem quam predixi.

Quam propter oportet quod inveniam aliud ingenium per quod centra predicta possint moveri et describam circulum quendam circa centrum orbis signorum quod est d et est tale—accipe unam tabulam auricalceam vel regulam cuius semidyametrum sit plus quam linea dk. In linea lune ita quod d sit centrum illius linee super quod faciam circulum ad quantitatem dk et sit circulus ille k c ut predixi loco suo. k vero est centrum equantis et c deferentis lune. Tunc a puncto k transeat lingua una ad calculandum quantum aux ecentrici lune transit omni die versus occidentem. Et in puncto c figatur unus clavus sine capite. Et hec est figura:

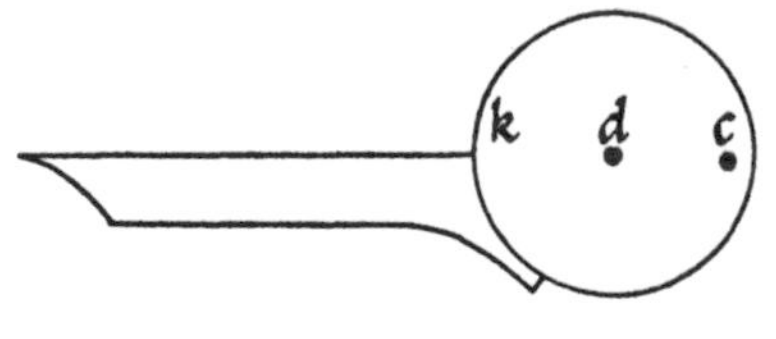


Accipiam tunc aliam laminam minoris quantitatis pro centro deferentium videlicet ad quantitatem k in linea mercurii predicta que sic intitulatur circulus parvus mercurii quia centrum ipsius ecentrici movetur circulariter supra centrum h et non supra centrum orbis signorum [f. 144 v.] sicut in luna sed extra ad quantitatem ad quam in mensuratione linee mercurii predicte....Et etiam eodemmodo erigam duos stilos in illa lamina sicut in lamina deferentis centrum ecentrici lune videlicet in puncto k unum et alium in puncto c ut patet in figura subscripta: Ista figura non debet habere linguam vel lineam fiducie sed filum ligatum.

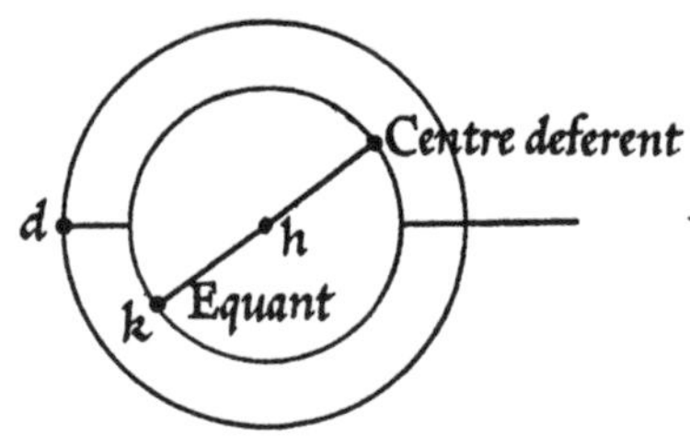


Accipiam postea tabulam unam auricalceam planam valde et politam quam dyametrabo orthognoaliter in centro a et erit sic ista lamina tante quantitatis ut possim bene dividere limbum ipsius in 360 partes ut predixi in compositione minoris. Ducam ergo tria paria circulorum iuxta limbum illius dividamque illam in 12 partes equales et quamlibet duodenam dividam in 30 partes. In exteriori vero numerum graduum cum eorumdem distinctionibus. In interiori vero scribam ista nomina signum primum signum $2^m$ $3^m$ $4^m$ $5^m$ usque ad duodenam. Incipiendo a dextris versus sinistram et iterum in eisdem spaciis eundem numerum incipiendo a sinistris versus dexteram. Videlicet ponam in loco ubi 12 situantur $p^r$ et ubii 11 $2^m$ et ubi 10 $3^m$ et ita de ceteris quod in primo pertingat $12^m$.

Deinde evacuabo totam superficiem tabule infra concavitatem circulorum dimittendo tantam partem tabule ex utraque parte dyametrorum que posset dyametros sustinere a limbo usque ad centrum ad modum crucis. Et sic habes quemdam circulum paratum cuius est epycyclus communis omnibus planetis. Et hec est figura:

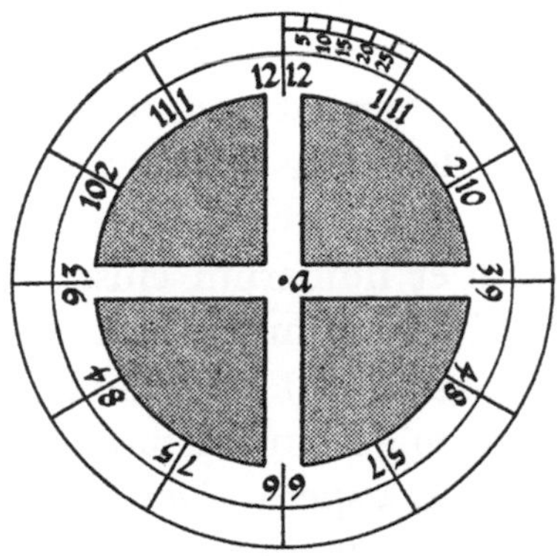


Eodem modo accipiam tabulam que sit maioris quantitatis. Parabo quod eam eodemmodo sicut istam pro epicyclo paravi excepto quod dyametrum debet totaliter evacuari ita quod nil maneat de tota lamina nisi limbus tantum cum divisionibus et inscriptionibus et ista sicut antea dividi debet et inscribi nisi (?) quod iste numerus qui incipit a sinistris versus dexteram non debet hic scribi in ista lamina. Et sic est iste circulus paratus omnibus planetis pro equante. Et in isto circulo debemus invenire centra planetarum cum quibus intramus tabulas ad accipiendum equationem centri et minuta proportionalia. Et incipit eius computatio semper ab

auge ecentrici ipsius cuius fuerit. Bene propter eius levitatem non oportet f. 145 r.
ponere figuram sed figura communis epicycli erit etiam in hoc loco excepto quod predictum est de situatione numerorum. Deinde accipiam regulam quandam auricalceam longam ad quantitatem semidyametro remotissimi a centro proprio ad eius augem perforando quod eam secundum longitudinem iuxta quantitatem remotissimum et propinquissimum augis omnium planetarum et in capite regule faciam foramen rotundum. Et sic est hec regula communis omnibus planetis loco deferentis. Et hec est figura:

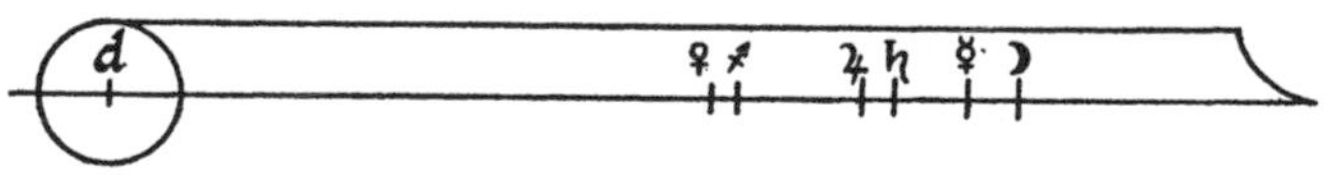


Et signabo loco augium ecentricorum secundum quod ipsam inveni in lineis eorum predictis in tabula lignea probationis. Deinde accipiam regulam auricalceam et protraham in ea lineam rectam a summo usque deorsum. Et in linea illa signabo semydyametros epycyclorum planetarum omnium prout inveni in lineis eorum predictis. Et hec linea erit loco epicyclorum circulationis omnium planetarum. Deinde lineabo regulam illam ex utraque parte usque in lineam rectam in medio nisi in illo loco ubi planetarum corpora signavi dimittendo tamen in uno capite tantum spacium in quanto possum rotunditatem facere ad designandum centrum epycycli. Et hac lingua debet figi volubilis super a centrum epicycli et erit tante longitudinis quod describat gradus limbi ipsius. Hec est figura.

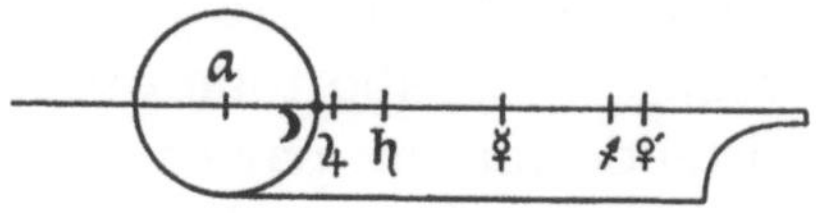


Cum vero volueris planetas per istud instrumentum equare a sole quidem incipiendum est cum inter omnes planetas dignior existit a quibusdam reputatur. Et motus aliorum a motu ipsius quodammodo regulantur. Cum igitur locum solis circum investigare volueris sume argumentum eius ad tempus datum et numerum eiusdem argumenti quere in ecentrico equante solis et fac ibi notam mentalem. In centro eius (spere corporis solaris, i.e. in centro orbis signorum) super quam extende filum ligatum ad punctum ubi ad orbem signorum. Et ubi istud filum ceciderit in orbe signorum ibi est verus locus solis in 8$^{va}$ spera. Debes eciam figere equantem communem omnibus super minorem instrumenti cum duobus clavibus ita quod centrum eius sit super centrum ecentrici solis videlicet secundum punctum c in linea que terminat augem solis. Et quod aux equantis ipsius sit in auge solis et oppositum in opposito.

Cum volueris equare lunam pone centrum tabule in nota centri deferentis lune super centrum d in minore instrumenti et fige ibi clavum unum
f. 145 v. Deinde ad punctum c liga unum filum. Postea quere medium motum lune et eius argumentum et eciam medium motum solis ex tabulis prout mos est. Deinde quere medium motum lune in facie instrumenti incipiendo ab ariete et fac ibi notam mentalem. Et quere filum in medium motum solis et ubi finierit fac ibi notam aliam. Deinde quere locum qui tantum antecedit locum solis medium quantum locus solis medius antecedit locum lune medium, ita quod sol sit in medio utriusque ibique fac notam 3$^{am}$. Post move tabulam motus centri deferentis lune quousque lingua eius ceciderit super notam ultimam i.e. augem deferentis etiam situatur aux deferentis lune prout in celo. Deinde pone filum ligatum ad punctum d super primam notam que terminat medium motum lune et manente tabula motus centri ecentrici fixa secundum situm in quo posuisa augem ecentrici. Tunc volue regulam motus centri epicycli communis quousque centrum epicycli ceciderit sub isto filo terminante medium motum lune eciam tunc est centrum epycycli. Dispositum in isto mirabili instrumento quodadmodum est in celo. Et punctus ipsius augis deferentis sub eodem puncto orbis signorum sub quo debet poni sol quoque secundum medium

eius motum cadit in medio eorum. Post hoc quere argumentum lune in epicyclo communi incipiendo ab auge epicycli versus dexteram videlicet versus occidentem et ubi numerus argumenti terminabitur fac notam. Tunc accipe filum ligatum ad punctum k in punctis extensis fac transire per medium epicycli usque ad extremam circumferentiam ipsius c a motus epycycli communis quousque aux ceciderit sub filo directe. Deinde pone virgulam vere circulationis epycicli ad notam prius figuratam que terminat argumentum lune Et tunc extende filum ligatum ad punctum d ultra corpus lune vere circulationis. Et supra quem gradum ceciderit filum in orbe signorum ibi lunam invenies penes eius verum locum in $8^{va}$ spera.

Cum volueris equare Mercurium pone centrum tabule motus eccentrici deferentis mercurii super punctum h in minore et fige eam cum clavo ibi. Deinde ad punctum c liga unum filum et ad punctum k aliud filum. Deinde fige equantem communem omnibus cum duobus clavis in facie instrumenti super punctum k et super centrum eius. Et in isto equante debes numerare medium motum mercurii. Et super centrum h quod est centrum parui circuli alium equantem quoniam in eodem debes numerare motus centri versus dexteram videlicet versus occidentem. Deinde accipe medium motum solis ad tempus quod volueris et ex motu eius medio augem mercurii subtrahe.
Et quod relinquitur erit centrum quia medius motus solis et mercurii idem f. 146 r.
sunt. Deinde quere argumentum eius ex tabulis. Est autem argumentum medium mercurii distancia corporis eius ab auge epycicli. Et centrum eius medium est distancia centri epycycli ab auge equantis de partibus equantis. Invento igitur centro medio et argumento quere numerum medii centri in circulo equantis et in circulo motus centri deferentis similiter et facies utrobique notam unam. Numerum quoque medii argumenti quere in epycyclo communi et fac ibi notam. Post accipe filum ligatum ad punctum h et extende ipsum super notam factam in circulo motus centri deferentis et move ca motus centri deferentis versus occidentem quousque punctus c ceciderit directe sub illo filo. Deinde extende filum ligatum ad punctum k super notam factam in equante que terminat numerum medii centri et manente hac parua tabula fixa circumvolue regulam motus centri epycycli communis continentem epycyclum quousque epycyclus ceciderit directe sub isto filo et eris circumvolue tabulam epycycli communis quousque aux epycycli ceciderit directe sub isto eodem filo. Eritque hec dispositio huius circuli deferentis et centri eiusdem et epycycli et augis ipsius medie ad tempus datum prout est in celo. Deinde circumvolue regulam siue lingulam fixam ad centrum epycycli communis continentis corpora planetarum quousque eius linea fiducie ceciderit super notam factam in epycyclo qui terminat numerum argumenti. Post extende

filum ligatum ad punctum d usque ad corpus mercurii in linea epycycli vere circulationis et super quem gradum ceciderit filum in orbe signorum ibi est verus locus mercurii in $8^{va}$ spera.

Cum volueris equare venerem martem jovem et saturnum sume medium centrum cuiuslibet eorum et medium argumentum similiter ad quod tempus volueris et numerum medii centri quere in equante. Ipsa prius disposita in minorem ut supra dixi in mercurio. Numerum vero argumenti quere in epicyclo communi et fac notam in termino utriusque scilicet in equante et in epycyclo etiam extende filum ligatum ad punctum k. Similiter notam equantis et circumvolue regulam motus centri epycycli communis quousque centrum epycycli ceciderit sub isto filo. Circumvolue etiam epycyclum communem quousque aux epicycli ceciderit sub eodem filo. Deinde circumvolue linguam deferentis corpora planetarum super notam terminantem medium argumentum in epycyclo communi. Post extende filum ligatum ad punctum d per corpus planete in epycyclo vere circulationis. Et super quem gradum ceciderit filum in orbe signorum vide quoniam ibi est verus locus planete. Medium autem centrum veneris est medium argumentum solis. Ad cetera sunt tabule.

*Laus omnipotenti deo.*

Explicit equatorium planetarum secundum Joh. de lineriis.

# APPENDIX III

# SPECIMENS OF MIDDLE ENGLISH SCIENTIFIC TEXTS

SINCE so few Middle English scientific texts are available in printed editions,[1] we make no apology for giving here a selection of extracts from such works having some affinity with the *Equatorie of the Planetis*. The first extract is taken from the *Newe Theorik of Planetis* which is one of twelve highly important texts in the Trinity College (Cambridge) MS. O. 5. 26. The other eleven include the astrological works of Albumasar, Zael, Ptolemy and Messahalla, the *Exafrenon* of Richard of Wallingford, and directions for making a type of portable sundial called the *Shippe of Venyse* (Navicula); it is the best scientific corpus in Middle English which I have seen.

The second extract is a fragment of a canon by Walter Anglus, included because the manuscript from which it is taken uses the same radix date (1392) as the Peterhouse manuscript (see p. 157*n.*). Its prose style is remarkably good.

The last extract is from one of the most popular Middle English texts on astronomy, a translation of the *Exafrenon* of Richard of Wallingford. It occurs almost as frequently as a more astrological text 'The Wyse boke of phylosophi & astronomye compiled out of Grew in to Ynglish'.[2]

### *Trinity (Cambridge) MS. O. 5. 26* (c. *1375*)

[f. 125 (118)] Here bigynneþ þe newe theorik of planetis þat schal eu*er*more dure w$^{t}$outen end aft*er* þe almagest of ptholome:~ (F)or in þe theorik of planet*is* þe meuy*n*ge of þe su*n*ne is necessarie forto fynde þe meuynges of oþ*er*e planetes firste we schul discreue þe meuynges of hit & þo þat beþ necessarie about þe knowleche & enserchinge and meuy*n*ge of hit // It is to knowe soþly þ$^{t}$ sol haþ oon centre excentrik. þe centre of which diu*er*seþ fro*m* þe centre of þe erþe twey p*ar*ties of þilke p*ar*ties of whiche wiþout aft*er* almagest i*n* þe .3. boke i*n* þe .4. t°. þe .2. p*ar*tie .m$^{a}$.29./.2$^{a}$.30. of þe half diametr*e* excentrik /. it is diuided into .60. p*ar*ties in þe circu*m*ference of which excentrik / c*er*cle. þe centre of þe body of þe su*n*ne is born in ech natural day fro*m* þe west to

[1] The most useful texts are R. Steele (ed.), *The Earliest Arithmetics in England*, E.E.T.S. Extra Series, no. CXVIII (1922), and J. O. Halliwell, *Rara Mathematica* (London, 1839).

[2] See, for example, British Museum MSS.; Sloane 965, f. 143b; Additional 12195; Egerton 827; Royal 17. A. xxxii; and Cambridge, Caius College, 457 (395).

the est to þe successioun of signes .m$^{a}$.59.2$^{a}$.8. of þe parties of þe forseide excentrik // Þerfore ymagine we a lyne led. from þe centre of þe forseide excentrik til to þe centre of the body of the sunne. which bereþ þe centre of þe body of þe sunne vppon þe circumference of þe forseide excentrik passinge vppon þe centre of þe forseide excentrik in euene tymes euen angles & of þe circumference of þe forseide circle excentrik euen bowes i.e. arcus // Se we what is aux solis & what is þe opposicioun of augis & what aux in þe secunde significacioun. what þe argument & what þe mene mote. what þe verrey place & what þe equacioun. // Aux soþly is þilke place in þe excentrik which diuerseþ more from þe centre of þe erþe // Opposicioun augis is þilke place in þe excentrik which is neiȝer to þe centre of þe erþe // Aux in þe secunde significacioun is arcus zodiaci þ$^{t}$ is þe bowe of þe zodiac. from þe begynnynge of arietis til to þe lyne led from þe centre of þe erþe by þe centre of þe excentrik & by augem til to þe zodiak. ȝif alyne be led. from þe centre of þe erþe passinge by þe centre & by augem of þe excentrik til to þe zodiak it schal termyne. in 17.g$^{a}$.50.mi$^{a}$. of geminorum & for from þe bigynnynge of arietis til to .þe.17. g$^{a}$.50.mi$^{a}$. of geminorum. beþ .2. signes .19.G$^{a}$.50.mi$^{a}$. // Argument soþly & partye & proporcioun beþ sinono$^{a}$ þ$^{t}$ is signifiynge þe same // Þe argument of solis is þe bowe of þe zodiak from auge til to þe lyne of þe mene mote. // Aux solis of tabularijs is cleped .2. signes .17.G$^{a}$.50. minutis. // Þe mene mote is þe bowe of the zodiak fallinge bitwixe þe bigynnynge of arietis & þe lyne led from þe centre of þe erþe til to þe zodiak. Þe lyne I saye euen diuersinge or distaunt. to þe lyne led from þe centre of þe excentrik to þe centre of þe sunne // Þe verrey place is þe arke of þe zodiak. fallinge bitwixe þe bigynnynge of arietis & þe lyne led. from þe centre of þe erþe by þe centre of þe body of þe sunne. til. [f. 125v.] to þe zodiak // Equacioun is distaunce in þe zodiak bitwixe þe lyne of þe mene mote & þe lyne of þe verrey place // Þerfore ymagine we alyne outgoynge from þe .17 g$^{a}$.50.mi$^{a}$. of sagittarij. passinge by þe centre of þe erþe & by þe centre of þe excentrik & by augem. til to þe .17.g$^{a}$. þe .50.mi$^{a}$. of Gemini which is þe lyne from .h. to .k. & which is þe diametre of þe zodiak & of þe excentrik of þe sunne which euermore remayneþ fixe // ymagine we also alyne outgoynge from þe centre of þe excentrik which bereþ þe centre of þe body of þe sunne in þe circumference of þe excentrik which is þe lyne from .p. to .l. // ymagine we also alyne outgoynge from þe centre of þe erþe ended in þe zodiak whiche goþ ech natural day of þe parties of þe forseide zodiak .59.mi$^{a}$.8.2$^{a}$. as þe lyne beringe þe centre of þe sunne. goþ of þe parties of þe excentrik which is þe lyne from .d. to .m. and it is cleped þe lyne of þe mene mote // ymagine we also alyne outgoynge from þe centre of þe erþe ouerpassinge by þe centre of þe sunne. which is þe lyne from d to q & it is cleped þe lyne of þe verrey place. // ymagine we þerfore þ$^{t}$ þese þre lynes þ$^{t}$ is. from p. to l. & fro .d. to .m. & fro .d. to q beþ meuable & alle also beþ to gidre. vppon þe lyne from .h. to .k. which is þe diametre aboueseide. // And saye we þ$^{t}$ þe lyne of .p. & .l. bereþ þe centre of þe body of þe sunne vppon þe excentrik .G$^{a}$.42.mi$^{a}$.10. it is necessarie þ$^{t}$ þese twey lynes be euermore euen distaunt in þe .2. place. Putt we ensaumple of geometere. þ$^{t}$ þer beþ twey circles excentrik in oon ouer partie. & oon is more þan þ$^{t}$ oþer or euene. & from þe centre of euerech of hem goþ out alyne til to þe circumference of his cercle which is meued from auge togidre & euerich of hem goþ by oon degre or moo. from þe parties of his circle. it is necessarie þ$^{t}$ þese lynes euermore euenly be distaunt in þ$^{t}$ þat þai makeþ vppon þe diametre of necessite euen angles. which may be proued þus. ȝif anoþer lyne falleþ vppon þe lynes euenly distaunt it bihoueþ þ$^{t}$ þe angles opposite. as wel excentrik as internal or w$^{t}$yneforþe beþ euene. And whan it were in auge or in opposito augis þei beþ coniuncte // Saye we þerfore þat þe lyne from .p. to .l. bereþ þe centre of þe sunne into þe eccentrik .42.g$^{a}$.m$^{a}$.10. & þe centre of þe sunne euermore bereþ wiþ hym þe lyne. from .d. to .q. ended in þe zodiak. which is cleped þe lyne of þe verrey place. Þe distaunce soþly þat is bitwixe þe lyne of þe mene mote & of þe verrey place is cleped equacioun. And for from auge til to þe opposite of þe partie of þe est. þe lyne of þe mene mote. goþ bifore þe lyne of þe verrey place founden by þe tables in þe mene mote of þe sunne. þe forseide equacioun it behoueþ be mynnschide from þe [f. 126] mene mote of þe sunne. & hit þat remayneþ schal be þe verrey place. And for aȝenward. from þe opposite of augis til to augem. of þe partie of þe west Þe lyne of þe verrey place. goþ bifore þe lyne of þe mene mote þe forseide equacioun. it bihoueþ be added. to þe mene mote. and hit þat were gadered togidre of þe mene mote & equacioun. schal be þe verrey place. And it is to knowe. þ$^{t}$ þe forseide þre lynes in auge. & in þe opposite of augis euermore beþ coniunte. & þerfore þere is noon equacioun for þer is no distaunce bitwixe hem. // It is to knowe also þ$^{t}$ þe zodiak in þe lengþe of hit in þe circumference is diuided into .360. partyes which is cleped gradus or degrees & alle þe .360. degrees beþ diuided into .12. euen parties whiche beþ cleped signes. And euerech of þese signes. conteyneth .30. degrees. whiche þe zodiak haþ in latitude or brede .12. degrees // Þerfore ymagine we in þe forseide zodiak a circulere lyne which diuideþ þe zodiak in to twey euen partyes. þ$^{t}$ is. þ$^{t}$. 6. degrees in latitude. as from þe forseide lyne toward þe norþe. & .6. degrees toward þe souþe. which is cleped lynea ecliptica vnder which lyne & in þe ouer partie of hit is þe excentrik of þe sunne & þerfore it is necessarie þ$^{t}$ þe sunne euermore passe. vnder þe forseide ecliptik lyne. diuidinge or departinge þe zodiak by þe myddel. // Oþere planetis soþly. goþ not so. but þei beþ oþerwhile of þe Norþe partie. oþerwhile of þe souþe from þe ouer partie of þe forseide ecliptik lyne. & oþerwhile þei beþ in þe ouerpartye of hit as we schul saye in oþere planetis. // It is to knowe also þ[t] þe sunne haþ twey meuynges. þ[t] is. þe meuynge from þe west into þe est .59.mi[a].8.2[a]. as we hau seide aboue. // Also he haþ anoþer meuynge which is cleped þe meuynge of þe .8. spere which is in an hundred ȝeres .1. degre after ptholome in almagest li°.7.c°.2°. And after thebit it is meuynge of accessioun & recessioun. as in þe meuynge of þe .8. spere it schal be expowned:— // It is to knowe also þ[t] aux of þe excentrik of þe sunne is fixe after ptholome but after auctores þat beþ nowe it is meued oon degre in .66. ȝeres & .8. moneþes. Philosophres þat maden tables. for þei myȝt not fynde þe verrey place of planetis. but ȝif þei fonde raþer þe mene mote firste þei enforced. to knowe þe mene mote. wiþ þe which þei fonde þe verrey place. in þe maner wiþynne writen. // Þei fonde soþly by instrumentes at a tyme where þei wolde bigynne her tables þe place of þe sunne. but of þe centre of þe sunne & þei putt in þe bigynnynge of þe table of þe sunne in saturno. ioue marte luna And whan þe mene mote was founden þei souȝt þe centre of þe epicicles which ȝaf þe mene mote of hem // In venere soþly & mercurio þei wrote not þe mene mote. but only þe argument for þe mene mote of veneris & mercurij. is þe mene mote of þe Sunne for þ[t] þe lyne of þe mene [f. 126v.] mote of þe sunne & þe lyne of þe mene mote of boþe .ſ. of veneris & mercurij euermore beþ oon & þe same in þat þ[t] þei beþ meued euenly .ſ. þe centres of þe epicicles of veneris & mercurij & þe lyne of þe mene mote of þe sunne beþ oon & þe same // Soþly þe place of þe Sunne bifore writen in þe bigynnynge of þe table where þe sunne was at þat tyme bifore schewide. þei cleped þat place þe rote of þis table. After þese þei fonde þe cours of it in oon ȝere. And þan in many ȝeres // fforewhy sommen maden tables from .20. into twenty ȝeres and sommen from .30. into .30. þ[t] is. as þei þ[t] made þe tables of tolete. And sommen þ[t] made from .28. into .28. ȝeres. // In þe ende soþly of þe forseide .30. ȝeres þei sawe where was þe sunne. and þei wrote vnder þe rote. & so þei dide sewingly. from .30. into .30. ȝeres. & þei wrote to hit þe title of anni collecti. or ȝeres gadered. Afterward [f. 127] soþly þei loked how mych þe sunne went in oon ȝere. þei wrote hit in partie & how mych in tweyne & þei wrote vnder þe firste & so sewingly til into .20. ȝeres which is þe space bitwixe þe ȝeres collecte & þe ȝeres collecte. to which þei wrote þe title of anni expansi or of þe ȝere strauȝtout. or of þe residue ȝere. þei beþ cleped expansi. for. þ[t] þei beþ strauȝt out in þe space bitwixe þe ȝeres collecte & þe ȝeres collecte And þan þei loked how mych þe sunne went in oon moneþe & how mych in tweyne & so folewingly how mych in .12. moneþes which is þe space bitwixe þe ȝeres collecte & þe ȝeres expanse. // After þis þei loked how mych þe sunne went in oon daye & how mych in tweyne & so folewingly into .20. // But þei þat made tables to þe ȝeres of *cri*st til into .31. which is þe space bitwixe amoneþe & amoneþe And þan þei loked how mych þe su*n*ne went in oon hour*e* & how myche in tweyne & so folewingly. til i*n*to 24. houres which is þe space bitwixe aday & aday // And þan þei loked how mych þe su*n*ne went in twey mynut*is* & how myche in four*e* & so folewingly. til into .30. which is þe space bitwixe an hour*e* & an hour*e*. // Also þe makeres of tables enserchide by art of geomet*er*e

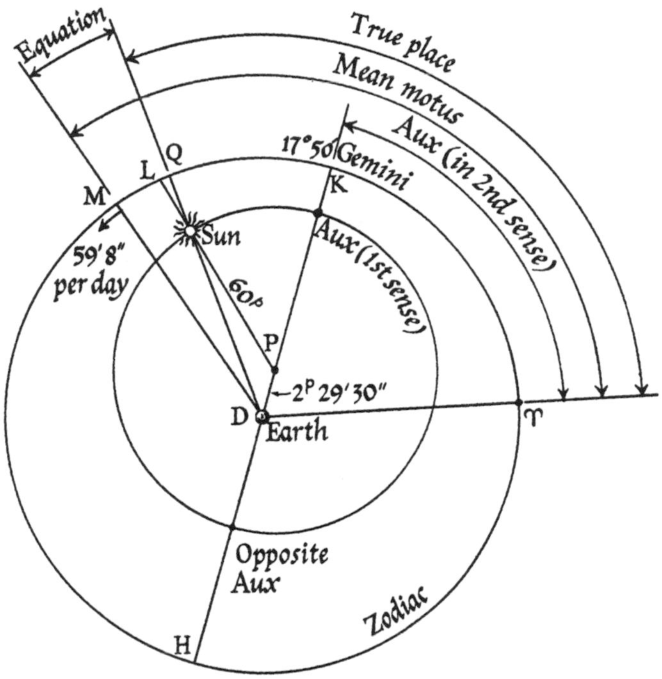


Fig. 21. Theory of the Sun in Trinity MS. O. 5. 26.

& arismetik. þe su*n*ne beynge i*n* auge of his excentrik & goynge oon degr*e* of þe p*ar*ties of þe forseide excentrik how mych was þe distau*n*ce bitwixe þe lyne of þe mene mote & þe lyne of þe v*er*rey place. & so þei made in Gª.2.3. & 4. & so folewingly. to. Gª .180. of þe p*ar*tie of þe est. which is in opposiciou*n* of aug*is* And also þei dide of þe p*ar*tie of þe west // ffor þilke same distau*n*ce in degrees. which is bitwixe þe lyne of þe mene mote & of þe v*er*rey place of þe p*ar*tye of þe est. it is in þe 1.Gª. of þe p*ar*tie of þe west. And also þilke distau*n*ce which is in .Gª.2. or .3. or .4. &cª of þe p*ar*tie of þe est. it is in Gª.2. or .3. or .4. of þe p*ar*tie of þe west & so folewingly. til to opposiciou*n* of aug*is* whiche distau*n*ces founden bitwixe þe lyne of þe mene mote & of þe v*er*rey place. þei cleped equaciou*n* of þe su*n*ne // Þei diuided also þe forseide c*er*cle excentrik of þe defferent into .12. signes. bigy*n*ninge fro*m* auge & þei fonde þ$^{t}$ oon degre of þe raþer signe .scilicet. þe distau*n*ce of hit fro*m* auge is like to þe distau*n*ce which is þe .11. signe & 29. g$^{ra}$d / fro*m* auge. & also twey g$^{ra}$d / of þe firste signe. beþ distau*n*t so mych. fro*m* auge. of þe p*ar*tie of þe est. as .11. signes & 28. degrees. of þe p*ar*tie of þe west goynge ret*ro*gr*a*de or bacward & so by ordre dressinge of þe p*ar*tie of þe est & goynge ret*rogra*de of þe p*ar*tie of þe west & alle þe degrees of oon longitude or of þe distau*n*ce fro*m* auge beþ like in equaciou*n*. & þei wrote in signes of .o. (i.*e.* aries) & i*n* þe firste degree of hit & in directe of hit in signes .11. i*n* þe .29. degr*e* And after þei wrote in þe signes of .o. i*n* þe .2. g$^{ra}$d / [f. 127v.] & in þe directe of hit i*n* signes .11. & in þe .28. degre & so þei passide dressinge as we hau seide bifore. til þei come to .6. signes & .6 signes on eu*er* eiþ*er* p*ar*tye which is in opposiciou*n* of aug*is* which desc*ri*pciou*n* þei cleped. lineas nu*meri* or þe lynes of nou*m*bre // In þe directe soþly of þe signes. of o. þe .1.g$^{ra}$d/& of þe signes .11. & 29.g$^{ra}$d / þei wrote how mych was þe distau*n*ce bitwixe þe lyne of þe mene mote & þe lyne of þe v*er*rey place & also þei dide i*n* þe directe of þe signes of .o.2. g$^{a}$d/& .11. signes .28. g$^{a}$d/& þus þei passed folewingly. til to .6. signes & .6. signes which desc*ri*pciou*n* þei cleped þe equaciou*n* of þe su*n*ne Whan soþly þei wolde knowe þe v*er*rey place of þe su*n*ne þei souȝt raþer þe mene mote of þe su*n*ne of which soþly þei w$^{t}$drowe auge*m* solis in þe secunde significaciou*n*. & hit þat remayned þei cleped þe argument // ȝif soþly. aux. solis which is þe secunde signe. þe .17. degr*e* þe .50. minute were founden more þan þe mene mote. þan þei added to þe mene mote .12. signes of which addiciou*n* þei wiþdrowe auge*m* solis & þe residue þei cleped þe argument // Þe argument soþly is to be souȝt i*n* þe lynes. of nou*m*bre. & in þe directe of þis þe equaciou*n* of þe su*n*ne. whiche equaciou*n* soþly is to be mynnschide of þe mene mote of þe su*n*ne. ȝif þe argument is lasse þan 6. signes. or to be added to þe mene mote. ȝif þe argument is more þan .6. signes // Þe resou*n* soþly why þe equaciou*n* is added or mynnschide is þis. Whan þe argument is lasse þan .6. signes þe lyne of þe mene mote. goþ bifore þe lyne of þe v*er*rey place. & þ*er*fore þe space or þe equaciou*n*. which is bitwixe hem. it behoueþ be mynnschide. And whan þe argument is more þa*n*ne .6. signes Þe Lyne of þe v*er*rey place goþ bifore þe lyne of þe v*er*rey place & þ*er*fore þe space or þe equaciou*n* þat is bitwixe hem it bihoueþ be added // It is to knowe þ$^{t}$ eu*er*ech natural day .scilicet. day & nyȝt. is of .24. houres. Houres. soþly beþ tweyfolde to be vnderstonde*n*. þ$^{t}$ is houres equal & houres i*n*equal Soþly houres beþ seide equal whan þei beþ eu*er*more of oon gretenesse & eue*n*nesse. // Inequal soþly. whan þei occupieþ not eue*n* space of tyme. for þ$^{t}$ eu*er*more ech day & ech nyȝt of what eu*er* gretenesse or litelnesse hit be. is diuided i*n*to .12. p*ar*ties wherfore þe houres of hem beþ inequal // Also it is to know þ[t] þe firmament. þ[t] is þe heuen of fix sterres by rauyschinge of which. alle oþere neþer speres beþ meued. is meued by his propre meuynge from þe est into þe west. hit environeþ bitwixe þe day & ny3t euermore after al þe spere. & ouer .59. minutes 2[a] .8. of þe forseide spere. / [The text continues.]

*British Museum, MS. Royal 12. D. vi. folio 81 r.* (c. *1400*)

Fragment of English version of canon of Walter Anglus.

...of her radyes / but it is ful hard & tedious / þerfore me semed profitable & spedeful to reduce into oon table þe ascensiouns of þe signes by .6. houses / for þe latitude of .51. g[a] / & .50. mi[a] as is precisely ynow3 for Oxenford / by þe which table þe .12. houses mowe be made equate / & oþere þinges toforeseide li3tly be founden / To þe which table I haue sette two oþere schorte tables / by þe whiche þe same þinges mowe be founden in euery regioun / but by calculacioun more traueilous ¶ Þerfore þe firste table I deme nou3t vnworþily to be called þe table complete of .12. houses ¶ Alle men by kynde desiren to knowe & to haue kunnynge / but for kunnynge fewe coueyten to trauayle ¶ Also I hope forsoþe þ[t] no pluralite of conclusiouns schal dulle þe wittes / ne skarste of tables schal engendre heuynesse / Þerfore who so seeþ þis table I preye þ[t] he led wiþ enuye derke nou3t. þ[t] is to seyn / hyde nou3t or empeyre wiþ weyward spekinge or bitynges þ[t] is necessarie & vnknowen / & to me defyed wiþ grete trauayle & profite of studyeres in þe science of astronomie / But only what so euer he seeþ empeyred by defaute of þe writer / or happily þur3 negligence or hastynesse scornfully calculed. / turne he þe penne to reforme hit & amende....
[The text then continues with the Latin version.]

*Bodleian MS. Digby 67, ff. 6–12 b* (c. *1409*)

Copy of a translation (made *c.* 1385) of the *Exafrenon* of Richard of Wallingford. Another version of this translation is in MS. Digby Roll III.

Incipit Exafrenon Pronosticationum temporis.
(T)ho the perfecte knawlege of the domes of the crafte of astronomye. þe which be the rewlynge of kynde. ar browth forthe of the effectes of planetes. // First it is nedful for to knaw þe entrynge of the Sonne in to the first minute of Arietis. For þare begynnes principaly þe chaungynge of [*sic for* or] the turnynge of abowte of þe hole 3ere. // And after þat fro þe entrynge of the Sonne into Taurus be gynnes þe chaungynge of the moneth of Aprile. // And so of other signes of the zodiack. And other monethis of the 3ere. // And þat is þe tokynynge. Fore comonly in the myddis of the monethis þe tymes ben made dyuerse as to the effecte of the Sonne. // Also it is spedful for to knawe þe

entrynge of the Sonne in to euery gree of the zodiack fore particular domys þat folowes þe mone in coniunction & opposition & oþer planetis. // And þe knawleche of this: is ordaned for to knawe þe gree Ascendente opon þe orizont in þe same tyme. // Fore aftir the gree Ascendente thu moste mesure thyn howses. // § The secunde thou moste knawe þe crafte to seeche what gree of the zodiac rises vpon thyn orizont in the moment of the howre in þe which thu wele gyfe thyn dome fore dyuerse disposition of the meuynge of the firmament And planetis takis diuerse effectes in þin emisperys. // And þerfore the philosophirs founde a crafte to departe all þe firmament in .12. partes. The which partyes be called howses. // And þai began at the Ascendent. Fore it is principal. Fore aftir þe philosophirs in þe booke of þe heuen & of the warlde be the vertu of þe planetis it shewes moste in the Est. And so be the gree Ascendent: þu salt mesure all þat other .12. howses. // § The thride þu most knawe þe strenghtis & vertues substancial of all the planetis / Of the wich summe haue of þair awne kynde / wherfore ar clepid hote & colde / drye & moyste / as it shall show in þaire chapitres. // And oþer strenght is þ[t] thei haue of the kynde of the signe þat þay ben in. // And þai ar callid .5. dignytes. þe which ben. domus. exaltatio. triplicitas. terminus. & facies. &cetera what iche one of þees ben: þu salt see in þare chapitres. whan we come þerto. And this þe substance of domes. Fore vpon þe dignytes hanghtes moste fors & strenghtes of them. // Abowte the which philosofris travelis besily. //

The forthe is. þu moste knawe þe strenghtis þat be fallynge to þam be certayne movyngis in þaire differences or ellis in þaire epiciclis. As is station. progression. & retrogradation. And of þaire distance fro þe Sonne. As is risynge & goyng downe / & commynge toward or nygh the cenyth / & goynge awayewarde. Aftir þat it is in dyuerse parties of thayre differencis. // And all þose þu salt knawe. When þu hast made equations of the planetis be þe argumentis & þe centris / & oþer thyngis as it shewis be þayre canouns of þam : and be tables þat ar callid § Almanac. // The .5. fifth chapitre is þu moste knawe for to ordire in figurs of howsynge þe planetis in þaire signis. // And to knawe which is þe sextile aspecte w[t] coniuncion. // And also wich is þe vertue & þe effecte of ich on of þame. // And witt þu wele þat a planete is comforted by a good aspecte of an oþer planete. // And his vertue is feblid be an euil aspecte &c Also þu moste knawn to drawe owt the lorde of the ȝere / of the thynges of þe monethes / & of the howrs. // § The sexte & þe laste is. þu moste kunne deme & telle before thynges þat ar to come kyndely be auysament & gederynge togedir of all þees þat I haue sayde. // Neuerþeles so þat þu saye non thyng before certanely / but þu wytt wele þat þu haue lefte no þynge behynde þat I haue spokyn of. As says þe philosofre in Elenchis. Tho þat lookis euer to fewe

## *Table of Entry of Sun (as in MS.)*

Calculated corrections in bold figures below error

| | 1385 | | | | 1386 | | | | 1387 | | | | 1388 | | | | |
|---|---|---|---|---|---|---|---|---|---|---|---|---|---|---|---|---|---|
| Menses | d. | h. | m. | s. | d. | h. | m. | s. | d. | h. | m. | s. | d. | h. | m. | s. | Sign*a* |
| Januar*ius* | 10 | 3 | 14 | 21 | 10 | 19 | 30 | 37 | 11 | 1 | 19 | 13 | 9 | 9 | 11 | 7 | Aq*uarius* |
| | | **13** | | | | | **3** | | | **0** | **52** | **53** | **11** | **6** | **42** | **9** | |
| Februar*ius* | 9 | 4 | 46 | 16 | 9 | 10 | 35 | 32 | 9 | 16 | 24 | 44 | 9 | 22 | 13 | 4 | Pisc*es* |
| | | | | | | | | | | | | **48** | | | **14** | | |
| Marcius | 11 | 8 | 3 | 22 | 11 | 13 | 52 | 38 | 11 | 19 | 41 | 52 | 10 | 1 | 31 | 10 | Aries |
| | | | | | | | | | | | | **54** | **11** | | | | |
| Aprilis | 12 | 1 | 35 | 0 | 11 | 7 | 24 | 16 | 11 | 13 | 13 | 32 | 11 | 19 | 2 | 48 | Taur*us* |
| | **11** | | | | | | | | | | | | **10** | | | | |
| Maius | 12 | 7 | 32 | 0 | 12 | 13 | 21 | 16 | 12 | 19 | 10 | 32 | 12 | 0 | 59 | 48 | Ge*mini* |
| Juni*us* | 12 | 20 | 2 | 54 | 13 | 1 | 52 | 10 | 13 | 7 | 41 | 36 | 12 | 13 | 20 | 42 | Ca*ncer* |
| | | | | | | | | | | | | **26** | | | | | |
| Julius | 14 | 9 | 11 | 0 | 14 | 51 | 0 | 16 | 14 | 20 | 49 | 32 | 15 | 2 | 38 | 48 | Leo |
| | | | | | | **15** | | | | | | | **14** | | | | |
| August*us* | 14 | 15 | 7 | 0 | 14 | 20 | 56 | 16 | 15 | 2 | 45 | 32 | 15 | 8 | 34 | 48 | V*irgo* |
| | | | | | | | | | | | | | **14** | | | | |
| Septe*mber* | 14 | 8 | 18 | 51 | 14 | 14 | 18 | 7 | 14 | 20 | 7 | 33 | 15 | 1 | 56 | 39 | Lib*ra* |
| | | | **28** | | | | | | | | | **23** | **14** | | | | |
| October | 14 | 12 | 8 | 0 | 14 | 17 | 57 | 16 | 14 | 23 | 46 | 32 | 15 | 5 | 35 | 48 | Scor*pio* |
| | | | | | | | | | | | | | **14** | | | | |
| Nove*mber* | 13 | 3 | 13 | 0 | 13 | 9 | 2 | 16 | 13 | 14 | 51 | 32 | 12 | 20 | 40 | 48 | Sagi*ttarius* |
| Dece*mber* | 12 | 11 | 30 | 0 | 12 | 17 | 19 | 16 | 12 | 23 | 8 | 32 | 11 | 4 | 57 | 48 | Cap*ricornus* |
| | | | | | | | | | | | | | **12** | | | | |

| | h. | m. | s. | th. |
|---|---|---|---|---|
| 1409 | 4 | 17 | 42 | 31 |
| 1413 | 5 | 0 | 39 | 36 |
| 1417 | 5 | 43 | 36 | 42 |
| 1421 | 6 | 26 | 33 | 47 |
| 1425 | 7 | 9 | 30 | 52 |
| 1429 | 7 | 52 | 27 | 57 |
| 1433 | 8 | 35 | 25 | 3 |
| 1437 | 9 | 18 | 22 | 8 |
| 1441 | 10 | 1 | 19 | 13 |
| 1445 | 10 | 44 | 16 | 18 |
| 1449 | 11 | 27 | 13 | 24 |
| 1453 | 12 | 10 | 10 | 29 |
| 1457 | 12 | 10 | 7 | ** |
| | | **53** | **7** | **34** |
| 1461 | 13 | 53 | 4 | ** |
| | | **36** | **4** | **39** |
| 1465 | 14 | 36 | 1 | ** |
| | | **19** | **1** | **45** |

| | h. | m. | s. | th. |
|---|---|---|---|---|
| 1469 | 15 | 19 | ** | ** |
| | | **1** | **58** | **50** |
| 1473 | 15 | 1 | 5* | ** |
| | | **44** | **55** | **55** |
| 1477 | 16 | 44 | 5* | * |
| | | **27** | **53** | **0** |
| 1481 | 17 | 27 | 50 | * |
| | | **10** | **50** | **6** |
| 1485 | 17 | 10 | 47 | 11 |
| | | **53** | | |
| 1489 | 18 | 53 | 44 | ** |
| | | **36** | **44** | **16** |
| 1493 | 19 | 36 | 41 | ** |
| | | **19** | **41** | **21** |
| 1497 | 20 | 19 | 28 | ** |
| | | **2** | **28** | **27** |
| 1501 | 20 | ** | ** | ** |
| | | **45** | **25** | **32** |

* Entry illegible in manuscript.

thynges: ar lightly begiled. // Wher[f. 6v.]fore certanly thei are right wisly scorned. // And kan not þe Science þat þai trowe þai kan / & kan it not. // But certanly of vnku*n*nynge men þai are not dispised. but the science is haldyn vanyne. & but a iaperye / As sayes Albumazare & tellis. //

(T)he firste Chapitre is for to knawe þe verray entre of þe Sonne in to þe firste minute of Arietis. & i*n* to othere signes. // For as Albumazar says in his book De floribu*s*. Greet arrours: may fall abowte þe takyng*e* of þe lorde of the ȝere / & of the moneth. yf þat itt be reklesly loked / or takyn. & reson whi. For þe werkyn in þe tables of þe Almanac is not v*er*rey. And þat is. Fore þe eq*ui*noct*is* & the solsticijs are mevable as it shewis. Fore þe solsticis of wyntir was su*m*mes tyme in þe feste of the natiuite of Criste Nowe it is gone bakward vnto þe feste of sancte lucie. And therfore I haue made a table of all þe enterynges of the Sonne in to dyu*er*s signes withowtyn erroure: boþe fore tyme passid / & fore to come / lightly & withowtyn trauayle. And cause of goynge bakward of þe solsticys / & þe eq*ui*noct*is* And the q*ua*ntite of þe goynge bakwarde iche ȝere is founden be wise countoures i*n* þis tyme þat the seson goyes not agayne ento þe same poynte of þe zodiac p*re*cise. in .365. dayes & .6. howris as þe fownder of the kalender supposid þat it shuld doo. but it passid owre þe .5. parte of an howre almoste. And þerfore þe solsticy & the eq*ui*noctij / & þe festis þat are sett fixe i*n* þe kalendre as wise men of grece / of Arabie / & of Rome haue p*ro*vidid be certane exp*e*rience. // The firste table þatt (^ I) sett here. is þe table of the monethis of þe kalend*er* be .4. ȝere sewande. And þai ar made on þis manere. Firste I sowthe þe entryng*e* of þe Sonne in til eu*ery* siene be .4. yers. Of the which þe firste is þe bysexte. And the tables are of myn awne makyng. For the tables of the Abbot of sancte Albones made full of errorues. And þat I trewe bee throwgh the vices of writers / & nott of the vowshipfull clerke & p*re*late þat þe trectes made. // The .2$^{e}$. table is made to drawe owt of þe firste table. And on þis man*ere* salt þu wirk w$^{t}$ þis table. Yf þu wile haue þe entre of þe Sonne i*n* til any signe: entre in w$^{t}$ þe ȝere wher it be bissexte or þe firste / or þe secu*n*de / or þe thirde. Anente the moneth þat þe signe is in. And write owt þe day. þe howre. & þe mynute. And þose ar þe trewe entres of þe Sonne i*n* to þe signes fore þe ȝere þat þai ar titelid fore. / þat is to say. þe firste ȝere was fore þe first ȝere after þe bissexte. // The secu*n*de: for þe .2$^{e}$. // The thirde. fore þe .3$^{e}$. // And þe fowrte iv as þe bissexte. þat to say .1388. ȝere of c*ri*ste. / And fore þes fowre .4. ȝere þe entres was as þai stonde in þe table. // And fore .1389. ȝere in þe .2$^{e}$. table is 42. mynutes .57. 2$^{a}$ & .5. 3$^{a}$. / And þat sal be drawne owt of ich on of þe .4. ȝers sewande &c*etera* bi iiij$^{a}$. [The text then continues on folio 7r. with the second chapter.]

# GENERAL INDEX

# INDEX OF MANUSCRIPTS CITED

Milton Keynes UK
Ingram Content Group UK Ltd.
UKHW051814181024
449640UK00021B/286

9 781107 404274